Robust Control Design

RSP SERIES IN CONTROL THEORY AND APPLICATIONS

Series Editor: **Professor Witold Pedrycz**
Department of Electrical and Computer Engineering, University of Alberta

Fuzzy Control and Fuzzy Systems, Second, Extended Edition
Witold Pedrycz

Robust Control Design

RSP SERIES IN CONTROL THEORY AND APPLICATIONS

Series Editor: **Professor Witold Pedrycz**
Department of Electrical and Computer Engineering, University of Alberta

Fuzzy Control and Fuzzy Systems, Second, Extended Edition
Witold Pedrycz

Robust Control Design

An Optimal Control Approach

Feng Lin
Tongji University, China
and Wayne State University, USA

BICENTENNIAL 1807 WILEY 2007 BICENTENNIAL

John Wiley & Sons, Ltd

Published by John Wiley & Sons Ltd, The Atrium, Southern Gate, Chichester, West Sussex PO19 8SQ, England
Telephone (+44) 1243 779777

Email (for orders and customer service enquiries): cs-books@wiley.co.uk
Visit our Home Page on www.wiley.com

This work is a co-publication between Research Studies Press Limited and John Wiley & Sons, Ltd.

Other Wiley Editorial Offices

John Wiley & Sons Inc., 111 River Street, Hoboken, NJ 07030, USA

Jossey-Bass, 989 Market Street, San Francisco, CA 94103-1741, USA

Wiley-VCH Verlag GmbH, Boschstr. 12, D-69469 Weinheim, Germany

John Wiley & Sons Australia Ltd, 42 McDougall Street, Milton, Queensland 4064, Australia

John Wiley & Sons (Asia) Pte Ltd, 2 Clementi Loop #02-01, Jin Xing Distripark, Singapore 129809

John Wiley & Sons Canada Ltd, 6045 Freemont Blvd, Mississauga, ONT, Canada L5R 4J3

Wiley also publishes its books in a variety of electronic formats. Some content that appears in print may not be available in electronic books.

Anniversary Logo Design: Richard J. Pacifico

Library of Congress Cataloging in Publication Data

Lin, Feng.
Robust control design : an optimal control approach / Feng Lin.
p. cm.
Includes index.
ISBN 978-0-470-03191-9 (cloth)
1. Automatic control. I. Title.
TJ213.L455 2007
629.8—dc22
2007020129

British Library Cataloguing in Publication Data

A catalogue record for this book is available from the British Library

ISBN 978-0-470-03191-9 (HB)

Typeset in 10/12pt Sabon by Integra Software Services Pvt. Ltd, Pondicherry, India

This book is printed on acid-free paper responsibly manufactured from sustainable forestry in which at least two trees are planted for each one used for paper production.

To Sherry, Robert, and Nicholas

Contents

Contents

Preface

The reasons for writing this book are twofold: (1) I want to write a book that summarizes the different approaches to robust control design; and (2) I need a textbook that covers the topics of modern control theory, suitable for a second control course for senior undergraduate students and first year graduate students.

There are several books published on robust control. Most focus on a particular approach. These books are most suitable for researchers specialized in the particular area and techniques. I often have trouble understanding the theoretical underpinnings of these books. From my contacts with control engineers in the automotive industry who try to solve very practical problems, I learned that they too have trouble understanding these books. Furthermore, engineers will not be able to determine which approach could be the best for the problems at hand before they fully understand the numerous methods available to them. That means that they need to read several highly theoretical books, which is a daunting task, even for people like me who have worked in the control field for more than twenty years. Assuming that an engineer indeed reads books on different approaches to robust control, it is still not easy to compare different perspectives, especially from a practical point of view. Therefore, I feel that a book that describes the major approaches to robust control in the simplest terms possible, spells out pros and cons of different approaches, and illustrates their applications to practical problems will be an excellent book for control engineers to read. The main body of this book, starting at Chapter 5, is devoted to this task.

This book offers three main approaches to robust control. The first one concerns an optimal control approach. It translates a robust control problem into an optimal control problem and then solves this problem to obtain a solution to the robust control problem. The second approach is that of

Kharitonov. It checks the robust stability of a linear time invariant system by considering its characteristic polynomial. The uncertainty of the system is parameterized in the form of the characteristic equation. The third approach is referred to as the H_∞/H_2. Its task is to find a controller that minimizes the H_∞/H_2 norm of the controlled system so that the range of tolerable uncertainty can be maximized. We show that the optimal control approach is inherently suitable for control synthesis while the Kharitonov approach is inherently suitable for control analysis.

Chapters 2–4 deal with basic modern control theory using the state space model. This part of the book is motivated by another observation: I have been searching for a suitable textbook for a second course on control for many years. Such a course should cover most important topics in modern control theory and is offered to senior undergraduate students and first year graduate students. I have tried different textbooks (and there are many of them), but none fit my needs. These textbooks are usually bulky and expensive. They cover too many topics and are impossible to finish in one semester. On the other hand, many books do not cover topics that I feel are most essential, such as the Kalman filter. Therefore, I wrote the first five chapters of this book with the intent of addressing the needs of such courses.

This book is aimed at students and readers who want to get sufficient background on control theory, especially robust control theory, in order to use it in practical applications. The presentation on modern control theory in Chapters 2–4 is short, but covers all the important results needed for applications of the theory. Proofs are provided for most of the results presented in this book. For example, the results on the Kalman filter are proved without requiring the knowledge on stochastic processes. Some details are omitted if they are tedious and not insightful. Many examples are given in this book to illustrate results and applications.

This book emphasizes control design and its applications. We want to develop control theory that is not only elegant, but also useful. In fact, usefulness shall be the sole criterion for judging an approach. To illustrate the usefulness of the optimal control approach to robust control design, we provide three detailed applications to vibration systems, robot manipulators, and V/STOL aircraft. They are presented in Chapters 9, 10, and 11, respectively.

I would like to thank the editor and the publisher for their constant encouragement. I would also like to thank my co-workers, colleagues, and students for their invaluable contributions to the content of this book.

A special thank you goes to my PhD advisor, Professor W. Murray Wonham of University of Toronto. Professor Wonham not only taught me control theory, but he taught me how to do research and how to become a good researcher.

Feng Lin
Troy
December 2006

Notation

$A \wedge B$	A and B
$A \vee B$	A or B
$\neg A$	not A
$A \Rightarrow B$	A implies B ($A \Rightarrow B$ means $\neg A \vee B$),
$A \Leftrightarrow B$	A if and only if B ($A \Leftrightarrow B$ means $A \Rightarrow B$ and $B \Rightarrow A$),
$(\forall x)P(x)$	for all x, $P(x)$ is true
$(\exists x)P(x)$	there exists an x, $P(x)$ is true
$x \in X$	x belongs to X
$x \notin X$	x does not belong to X
$X \cup Y$	union of X and Y
$X \cap Y$	intersection of X and Y
$X \subseteq Y$	X is a subset of Y
$X \subset Y$	X is a proper subset of Y
$x \in R^n$	x are the n-dimensional state variables of a system
$\|A\|$	determinant of A
$adj(A)$	adjoint of A
$trace(A)$	trace of A
$rank(A)$	rank of A
A^{T}	transpose of A
$\bar{A}$	conjugate of A
A^{-1}	inverse of A
$\lambda(A)$	set of eigenvalues of A
$\rho(A)$	spectrum radius of A, $\rho(A) = \max \|\lambda(A)\|$
$\overline{\sigma}(A)$	largest singular value of A, $\overline{\sigma}(A) = \max \sqrt{\lambda(A^{\mathrm{T}}A)}$
$\underline{\sigma}(A)$	smallest singular value of A, $\underline{\sigma}(A) = \min \sqrt{\lambda(A^{\mathrm{T}}A)}$
B^{+}	pseudo-inverse of B

$\langle x, y\rangle$	inner product of x and y
$\|x\|$	norm of x
$\|x\|_p$	p-norm for vector x, $\|x\|_p = (\sum_{i=1}^{n} \|x_i\|^p)^{1/p}$
$\|A\|_p$	induced p-norm for matrix A, $\|A\|_p = \sup_{x\neq 0} \frac{\|Ax\|_p}{\|x\|_p}$
$\|A\|_F$	Frobenius norm for matrix A, $\|A\|_F = \sqrt{trace(A^T A)}$
$E[\psi]$	mean or expectation of ψ

1

Introduction

This book is about robust control design. To truly understand robust control design, we first need to understand basic concepts of systems and control. We will introduce systems and control theory in this chapter. We will also give an overview of the book.

1.1 SYSTEMS AND CONTROL

A Google search in August 2006 found more than 5 billion entries for the word 'system'. So what is a system? There are many definitions, depending on the areas of application or interest. For example, according to The Free Dictionary by Farlax (http://www.thefreedictionary.com/system), a system is:

1. A group of interacting, interrelated, or interdependent elements forming a complex whole.
2. A functionally related group of elements, especially:
   (a) the human body regarded as a functional physiological unit.
   (b) an organism as a whole, especially with regard to its vital processes or functions.
   (c) a group of physiologically or anatomically complementary organs or parts: the nervous system; the skeletal system.

Robust Control Design: An Optimal Control Approach F. Lin

(d) a group of interacting mechanical or electrical components.
(e) a network of structures and channels, as for communication, travel, or distribution.
(f) a network of related computer software, hardware, and data transmission devices.

3. An organized set of interrelated ideas or principles.
4. A social, economic, or political organizational form.
5. A naturally occurring group of objects or phenomena: the solar system.
6. A set of objects or phenomena grouped together for classification or analysis.
7. A condition of harmonious, orderly interaction.
8. An organized and coordinated method; a procedure.
9. The prevailing social order; the establishment.

All the above definitions are appropriate for some applications. However, in this book, we define a system as an assemblage of objects, real or abstract, that has some inputs and some outputs (Figure 1.1).

There are many examples of systems: an automobile whose input is the position of the gas pedal and whose output is the speed, a bank account whose input is the fund deposited and whose output is the interest generated, a traffic light whose input is the command indicated by green, yellow, or red lights and whose output is the traffic flow, and a dryer whose input is different dry circles and whose output is dry cloth.

To better understand systems, we shall classify them into different types. We will not classify systems according to their physical appearance, but rather according to their mathematical properties. Mathematically, we can view a system as a mapping $S: U \rightarrow Y$ from its input u to its output $y = S(u)$.

The first classification is whether a system is linear or nonlinear. A system is linear if its input–output relation is linear; that is, for all inputs u_1 and u_2

$$y_1 = S(u_1) \wedge y_2 = S(u_2) \Rightarrow \alpha_1 y_1 + \alpha_2 y_2 = S(\alpha_1 u_1 + \alpha_2 u_2) \tag{1.1}$$

where α_1 and α_2 are any constants, $\wedge$ means 'and', and $\Rightarrow$ means 'implies'. Equation (1.1) says that if y_1 is the output when the input is u_1 and y_2 is the output when the input is u_2, then $\alpha_1 y_1 + \alpha_2 y_2$ is the output when the input is $\alpha_1 u_1 + \alpha_2 u_2$. If there exist some inputs u_1 and u_2 such that Equation (1.1) is not satisfied, then the system is nonlinear.

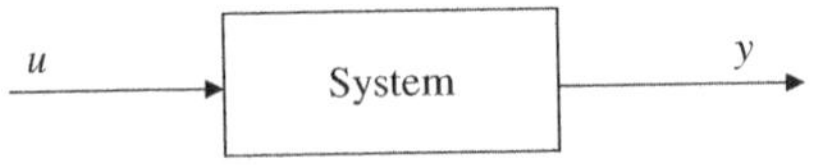

Figure 1.1 A system with input u and output y.

Let us consider the system of a bank account. If the interest rate is fixed at 3%, then the system is linear because the interest generated by the account is proportional to the balance of the account: \$100 will generate \$3, \$1000 000 will generate \$30 000, etc. However, in order to attract large deposits, a bank may use progressive interest rates. For example, the first \$10 000 of the balance earns an interest rate of 2% and the rest earns an interest rate of 4%. The account of this type is nonlinear because the interest generated by the account is not proportional to the balance of the account: \$100 will generate \$2 and \$1000 000 will generate $\$10\,000 \times 0.02 + 990\,000 \times 0.04 = \$39\,800$.

The second classification of systems is whether a system is time-invariant or time-varying. A system is time-invariant if its input–output relation does not change over time; that is, for any input u applied at different times,

$$y(t) = S(u(t)) \Rightarrow y(t+T) = S(u(t+T)) \tag{1.2}$$

where T is any constant time delay. If there exist some input u and some constant T such that Equation (1.2) is not satisfied, then the system is time-varying.

Consider again the system of a bank account. The system is time-invariant if the interest rate does not change over time. It is time-varying if the interest rate changes over time, which is most common in our daily experience.

The third classification of systems is whether a system has single input and single output (SISO) or multiple inputs and multiple outputs (MIMO). This classification requires no further explanation.

The last classification of systems is whether a system is a continuous-time or a discrete-time system. A system is a continuous-time system if its input and output are functions of a continuous time variable. All physical systems are continuous-time systems. However, nowadays, many physical systems are controlled by computers rather than by analogue devices. For computer control, input and output signals must be sampled. After a continuous-time signal $x(t)$ is sampled, it becomes a discrete-time signal $x(t_k)$, where t_k is the kth sampling time. In this book, we will study only continuous-time systems.

Our goal is to control a system to achieve some objectives. Generally speaking, the control objectives can be classified to ensure either stability or optimality, or both of a system. Stability means that the system will not 'blow up'; that is, the output of the system will not become unbounded as long as its input is bounded. This is a basic requirement of most systems that we encounter. Optimality means that the system performance will be optimal in some sense. For example, we may want an automobile to consume the least fuel; or we may want a bank account to generate most interest. In this book, we will discuss stability in Chapter 3 and optimality in Chapter 4.

To achieve stability or optimality, some control needs to be used. Generally speaking, two types of control can be used: (1) feedback or closed-loop control; and (2) open-loop control.

In feedback control, the controller knows the output of the system and uses this information in its control. A feedback control system is shown in Figure 1.2. Most control systems we see in our daily life are feedback control systems. For example, most control systems in an automobile, such as engine control, throttle control, cruise control, and power train control are feedback controls. So are temperature controls in modern houses or controls in ovens and refrigerators.

In open-loop control, the controller does not know the output of the system, as shown in Figure 1.3. Open-loop control is used if it is hard or meaningless to measure the output. There are not many, but some examples of open-loop control in existence. Most traffic controls are open-loop control because the controllers do not know the traffic flow that is being controlled. In most cases, washers and dryers are open-loop controlled, because it is hard to measure the cleanness or dryness of cloth.

Needless to say, feedback control has many advantages over open-loop control. Many unstable systems can be stabilized by feedback controls but cannot be stabilized by open-loop control. Feedback can often handle disturbance much better than open-loop control. Optimization can also be achieved using feedback. Since open-loop control is relatively easy to design and less frequently used in practice, almost all controls addressed in the control theory are feedback control. Most methods developed in control theory are for feedback control. This is also true in this book. We will investigate feedback control systems in this book.

To control a system, we first need to obtain a mathematical model of the system. In the development of the control theory, two main modelling frameworks have been proposed. One uses transfer functions and the other

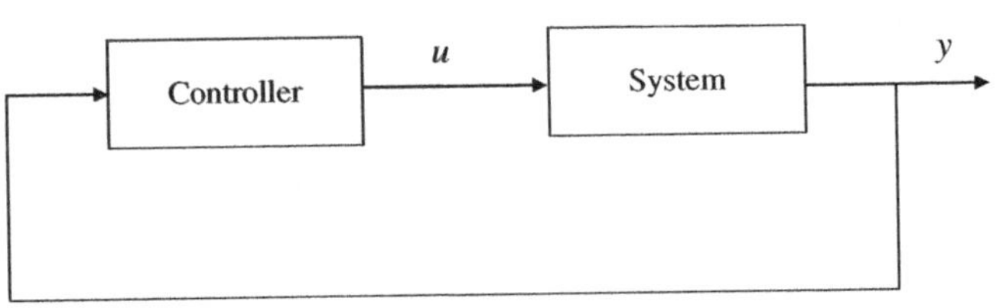

Figure 1.2 A feedback control system.

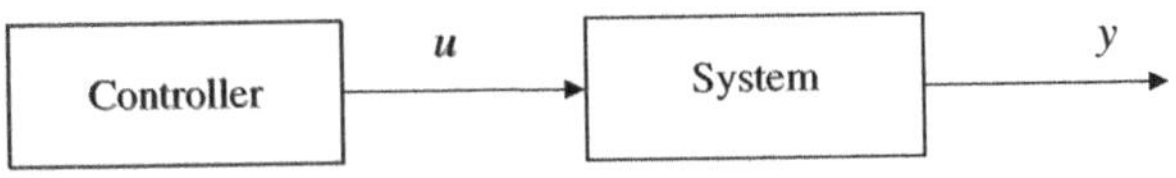

Figure 1.3 An open-loop control system.

uses state space representations. The methods developed using transfer functions are sometimes called 'classical control'. The methods developed using state space representations are sometimes called 'modern control'. In this book, we will mainly use state space representations to model systems.

Our focus is on robust control design. Robust control is related to modelling and model uncertainties. No matter how hard we try, no model is completely accurate. Every model has errors or uncertainties. If a control will work under uncertainties, we say that the control is robust. Robust control design tries to design a control that has good tolerance to modelling errors. There are several approaches available for robust control and robust control design. In this book, we will present two popular approaches: the parametric approach and the H_∞/H_2 approach. More importantly, we will present a new approach to robust control design. This new approach is 'indirect' in the following sense: it translates a robust control problem into an optimal control problem. Since we know how to solve a large class of optimal control problems, this optimal control approach allows us to solve some robust control problems that cannot be easily solved otherwise. Furthermore, this approach is easy to understand and easy to apply to practical problems.

To build the foundation for the optimal control approach, we will first present the fundamentals of control theory, stability theory, and optimal control.

1.2 MODERN CONTROL THEORY

We will start this book with a comprehensive review of modern control theory in Chapter 2. We will use general state space models to describe systems:

$$\dot{x} = f(x, u, t)$$
$$y = g(x, u, t)$$

where $f: R^n \times R^m \times R \to R^n$ and $g: R^n \times R^m \times R \to R^p$ are nonlinear functions. $\dot{x} = f(x, u, t)$ are state equations and $y = g(x, u, t)$ are output equations. Derivation of these equations is illustrated in Appendix A, where we model various electrical, mechanical and other practical systems.

Chapter 2 will focus on a linear time-invariant system of the form

$$\dot{x} = Ax + Bu$$
$$y = Cx + Du$$

We will study its responses and their properties. We will also study its transfer function. We note that for a given system, its state space representation or realization is not unique. Different representations are related by some similarity transformations. Some representations are more useful in control, including the Jordan canonical form, controllable canonical form, and observable canonical form.

Controllable and observable canonical forms are related to two important properties of systems: controllability and observability. Intuitively, a system is called controllable if all its states can be controlled in the sense that they can be driven to anywhere using some input, and a system is called observable if all its states can be observed in the sense that their values can be determined from the output. For linear time-invariant systems, these two properties can be easily checked by checking the ranks of some controllability or observability matrices.

The importance of controllability is due to the fact that if a system is controllable, then we can move or place its poles or eigenvalues in arbitrary places in the complex plane by using state feedback. We show how this can be done in three steps. First, we show how to design a state feedback for a system in controllable canonical form. We then show how to do this for general single-input systems. Finally, we show how to design a state feedback for a multi-input system.

Using state feedback requires that all state variables are available for control. This in turn requires that there are sensors to measure all state variables. This requirement is sometimes impractical and most times too expensive to satisfy. Furthermore, this requirement is also unnecessary because even if the state variables are not directly measurable, they can be estimated from the output of the system, if the system is observable. Such estimation is achieved by an observer. An observer is a linear time-invariant system whose inputs are the input and output of the system to be observed, and whose output is the estimate of the state variables. The performance of the observer is determined by its poles, which can be placed arbitrarily if the system is observable. The nice thing about feedback control is that the use of the observer does not change the poles determined by the state feedback. This separation principle allows us to design state feedback and an observer separately.

1.3 STABILITY

In Chapter 3, we will review the basic theory of stability. Intuitively, stability means that, without inputs, a system's response will converge to some equilibrium. Consider a general nonlinear system

$$\dot{x} = A(x)$$

where $x \in R^n$ are the state variables and $A : R^n \to R^n$ is a (nonlinear) function. Assume $A(0) = 0$, the equilibrium point $x_0 = 0$ is asymptotically stable if there exists a neighbourhood of $x_0 = 0$ such that if the system starts in the neighbourhood then its trajectory converges to the equilibrium point $x_0 = 0$ as $t \to \infty$.

Determining stability of a system is not easy if the system is nonlinear. One approach often used is the Lyapunov approach, which can be explained as follows: given a system, let us define some suitable 'energy' function of the system. This function must have the property that it is zero at the origin and positive elsewhere. Assume further that the system dynamics are such that the energy of the system is monotonically decreasing with time and hence eventually reduces to zero. Then the trajectories of the system have no other place to go but the origin. Therefore, the system is asymptotically stable. This generalized energy function is called a Lyapunov function. The Lyapunov approach will be used in deriving the results on our optimal control approach to robust control design.

On the other hand, for a linear time-invariant system

$$\dot{x} = Ax$$

its stability is determined by its characteristic polynomial

$$\varphi(s) = |sI - A| = a_n s^n + a_{n-1} s^{n-1} + \cdots + a_1 s + a_0$$

and its corresponding roots, which are eigenvalues or poles. A linear time-invariant system is asymptotically stable if and only if all the roots of its characteristic polynomial are in the open left half of the s-plane.

If the numerical values of matrix A are known, then we can always find the numerical values of its eigenvalues and hence determine the stability of the system. However, if symbolic values are used or, for any other reasons, we do not want to calculate the eigenvalues explicitly, then two other criterions can be used to determine the stability of a system.

The Routh–Hurwitz criterion is a method to determine the locations of roots of a polynomial $\varphi(s) = a_n s^n + a_{n-1} s^{n-1} + \cdots + a_1 s + a_0$ with constant real coefficients with respect to the left half of the s-plane without actually solving for the roots. It involves first constructing a Routh table and then checking the number of sign changes of the elements of the first column of the table, which is equal to the number of roots outside the open left half of the complex plane.

The second criterion is the Nyquist criterion. Unlike the Routh–Hurwitz criterion, the Nyquist criterion is a frequency domain method based on the frequency response of a linear time-invariant system. To use the Nyquist criterion to check the stability of a system with the characteristic equation given by $1 + G(s)H(s) = 0$, we first construct a Nyquist plot of $G(s)H(s)$.

For the system to be stable, the Nyquist plot of $G(s)H(s)$ must encircle the $(-1, j0)$ point as many times as the number of poles of $G(s)H(s)$ that are in the right half of the s-plane.

Chapter 3 also discusses two other properties of a linear time-invariant system: stabilizability and detectability. A system is stabilizable if all unstable eigenvalues are controllable. Obviously, stabilizability is weaker than controllability. It is the weakest condition that allows us to stabilize a system using feedback. Dually, a system is detectable if all unstable eigenvalues are observable.

1.4 OPTIMAL CONTROL

After we stabilize a system, the next thing we want to do is to optimize the system performance. Optimal control will be discussed in Chapter 3. This topic is not only important in its own right, but also serves as the basis of our optimal control approach to robust control design.

We formulate an optimal control problem for a general nonlinear system

$$\dot{x} = f(x, u)$$

so as to minimize the following cost functional

$$J(x, t) = \int_t^{t_f} L(x, u)d\tau$$

where t is the current time, t_f is the terminating time, $x = x(t)$ is the current state, and $L(x, u)$ characterizes the cost objective.

We will derive the solution to the optimal control problem from the principle of optimality, which states that if a control is optimal from some initial state, then it must satisfy the following property: after any initial period, the control for the remaining period must also be optimal with regard to the state resulting from the control of the initial period. Applying the principle of optimality to the optimal control problem, we can derive the Hamilton–Jacobi–Bellman equation that must be satisfied by any solution to the optimal control problem.

It is not always easy to solve the Hamilton–Jacobi–Bellman equation, especially for nonlinear systems. However, if the system is linear and the cost function is quadratic with infinite horizon; that is

$$\dot{x} = Ax + Bu$$

$$J(x, t) = \int_t^{\infty} (x^T Qx + u^T Ru)d\tau$$

then the Hamilton–Jacobi–Bellman equation is reduced to the following algebraic Riccati equation

$$SA + A^{T}S + Q - SBR^{-1}B^{T}S = 0$$

Solving the above equation for S, we can obtain the solution to the optimal control problem as

$$u^{*} = -R^{-1}B^{T}Sx$$

The above optimal control problem is also called a linear quadratic regulator (LQR) problem.

The problem dual to the optimal control problem is to design an optimal observer, more commonly known as the Kalman or Kalman–Bucy filter. Deriving results on the Kalman filter often requires knowledge and background on stochastic processes. However, we will provide a new method to derive the Kalman filter in Chapter 4 without using results on stochastic processes.

1.5 OPTIMAL CONTROL APPROACH

The main focus of this book is of course on the optimal control approach to robust control design. We will discuss this approach starting in Chapter 5, where we present the optimal control approach for linear systems. The system to be controlled is described by

$$\dot{x} = A(p)x + Bu$$

where p represents uncertainty. The goal is to design a state feedback to stabilize the system for all possible p within given bounds. The solution to this robust problem depends on whether the uncertainty satisfies a matching condition, which requires that the uncertainty is within the range of B.

If the uncertainty satisfies the matching condition, then the solution to the robust control problem always exists and can be obtained easily by solving an LQR problem. The LQR problem is obtained by including the bounds on the uncertainty in the cost functional. The proof that the solution to the LQR problem is a solution to the robust control problem is based on the properties of the optimal control, as described by the Hamilton–Jacobi–Bellman equation. Furthermore, if the matching condition is satisfied, we can also solve a robust pole placement problem by placing the poles of the controlled system to the left of $-\gamma$, where γ is some arbitrary positive real number, as long as the uncertainty is within the bounds.

If the uncertainty does not satisfy the matching condition, then the problem is much more complex. We first need to decompose the uncertainty into the matched part and the unmatched part. We will use an augmented control to deal with the unmatched uncertainty. Robust control may or may not be possible, depending on whether a sufficient condition is satisfied. This conclusion is in sync with the results obtained by other researchers in the field.

Chapter 5 also discusses how to handle uncertainty in the input matrix; that is, the uncertain system has the form

$$\dot{x} = A(p)x + BD(p)u$$

Method for this case is similar but the derivation is more complex.

The optimal control approach to nonlinear systems will be presented in Chapter 6. The idea is similar to the idea for linear systems: we will translate a robust control problem into an optimal control problem. However, because the system is nonlinear, it is more difficult to solve the optimal control problem. Hence, some numerical solutions or other methods may need to be used, although this is outside the scope of this book.

As in the case for linear systems, the procedure for systems satisfying the matching condition is quite different from the procedure for systems not satisfying the matching condition. For systems satisfying the matching condition, the solution to the optimal control problem is guaranteed to be a solution to the robust control problem. Therefore, as long as we can find an analytic or numerical solution to the optimal control problem, we have a solution to the robust control problem. For systems not satisfying the matching condition, the solution to the optimal control problem is a solution to the robust control problem only if a certain sufficient condition is satisfied. If the unmatched part of the uncertainty is too large, the sufficient condition is unlikely to be satisfied. Again, this is not surprising in view of results obtained by other researchers.

1.6 KHARITONOV APPROACH

A book on robust control design would not be complete without presenting the parametric approach, sometimes called the Kharitonov approach. The Kharitonov approach is an excellent method for robust analysis of control systems. To some degree, it can also be used for robust control design. We will discuss the Kharitonov approach in Chapter 7.

The Kharitonov approach considers a system with the following characteristic polynomial

$$\varphi(s, p) = p_0 + p_1 s + \cdots + p_{n-1} s^{n-1} + p_n s^n$$

where $p_i \in [p_i^-, p_i^+]$, $i = 0, 1, \ldots, n$ are coefficients whose values are uncertain, but we know their lower and upper bounds. The Kharitonov theorem states that the stability of the following four polynomials is necessary and sufficient for the stability of all polynomials with the uncertainty within the bounds:

$$K_1(s) = p_0^- + p_1^- s + p_2^+ s^2 + p_3^+ s^3 + p_4^- s^4 + p_5^- s^5 + \cdots$$
$$K_2(s) = p_0^- + p_1^+ s + p_2^+ s^2 + p_3^- s^3 + p_4^- s^4 + p_5^+ s^5 + \cdots$$
$$K_3(s) = p_0^+ + p_1^- s + p_2^- s^2 + p_3^+ s^3 + p_4^+ s^4 + p_5^- s^5 + \cdots$$
$$K_4(s) = p_0^+ + p_1^+ s + p_2^- s^2 + p_3^- s^3 + p_4^+ s^4 + p_5^+ s^5 + \cdots$$

To prove the Kharitonov theorem, we need a few preliminary results. These preliminary results will also be proven in Chapter 7.

To compare the optimal control approach with the Kharitonov approach, we note that the optimal control approach is inherently a design tool, in the sense that it will design a controller that can robustly stabilize the system; while the Kharitonov approach is inherently an analysis tool, in the sense that, given a (closed-loop) system, it will analyse and verify if the system is robustly stable.

1.7 H_∞ AND H_2 CONTROL

It is not easy to summarize the H_∞/H_2 control in one chapter, but that is what we will do in Chapter 8. We will start with the introduction of function spaces. In particular, H_∞ denotes the Banach space of all complex valued functions $F: C \to C$ that are analytic and bounded in the open right half of the complex plane and are bounded on the imaginary axis jR with its H_∞ norm defined as

$$\|F\|_\infty = \sup_{\omega \in R} |F(j\omega)|$$

H_2 denotes the Hilbert space of all complex valued functions $F: C \to C$ that are analytic and bounded in the open right half of the complex plane and the following integral is bounded

$$\int_{-\infty}^{\infty} \overline{F(j\omega)} F(j\omega) d\omega < \infty$$

The H_2 norm can then be defined as

$$\|F\|_2 = \sqrt{\frac{1}{2\pi} \int_{-\infty}^{\infty} \overline{F(j\omega)} F(j\omega) d\omega}$$

We will show how to calculate the H_∞ and H_2 norms.

To discuss robustness under uncertainty, we will separate the uncertainty from the nominal system and put the uncertainty in the feedback loop. We will prove a small-gain theorem which states intuitively that the perturbed closed-loop system is stable if the H_∞ norm of the loop is less than one. From the small-gain theorem, we can determine the bounds on the uncertainty that guarantee the stability of the perturbed system.

We will then show that H_2/H_∞ control synthesis boils down to designing a controller for the nominal system such that its H_∞/H_2 norm is minimized. Note that the H_2/H_∞ approach is very different from the optimal control approach. In the optimal control approach, we start with the bounds on uncertainties. We then design a controller based on these bounds. As the result, if the controller exists, then it is guaranteed to robustly stabilize the perturbed system. On the other hand, in the H_2/H_∞ approach, the bounds on uncertainties are not given in advance. The synthesis will try to achieve the largest tolerance range for the uncertainty. However, there is no guarantee that the range is large enough to cover all possible uncertainties. In other words, the H_2/H_∞approach cannot guarantee the robustness of the resulting controller. The approach will do its best to make the resulting controller robust. Whether this best is good enough depends on the nature of the uncertainty.

1.8 APPLICATIONS

We will present three practical applications of the optimal control approach to robust control design. These applications will be presented in Chapters 9, 10, and 11.

The first application is robust active damping for stability enhancement of vibration systems. Many practical systems such as buildings, flexible structures, and vehicles, exhibit vibration. How to reduce (damp) vibration is an important control problem. We will be interested in active damping that uses external force to actively control the system to reduce the vibration. The system will be modelled as

$$M_0\ddot{x} + A_0x = B_0u + C_0f_0(x, \dot{x})$$

where M_0 is the mass matrix, A_0 is the stiffness matrix, and $f_0(x, \dot{x})$ is the uncertainty. We will introduce a special inner product and the associated energy norm. The solution to the robust damping problem will be obtained by translating it into an optimal control problem. The control law will be obtained by solving an LQR problem.

The second application is robust control of robot manipulators. The dynamics of a robot manipulator is modelled as

$$M(q)\ddot{q} + V(q, \dot{q}) + U(\dot{q}) + W(q) = \tau$$

where q is the generalized coordinate vector, τ is the generalized force vector, $M(q)$ is the inertia matrix, $V(q, \dot{q})$ is the Coriolis/centripetal vector, $W(q)$ is the gravity vector, and $U(\dot{q})$ is the friction vector. Based on this model, we will formulate the robust control problem when the load and other parameters are uncertain. The resulting robust control problem satisfies the matching condition. However, there is also uncertainty in the input matrix. We will use the method in Chapter 5 to solve the robust control problem. We will apply the control law obtained to a two-joint SCARA-type robot and simulate the controlled system.

The third and last application is the hovering control of a vertical/short takeoff and landing (V/STOL) aircraft. The aircraft state is simply the positions, $\tilde{x}, \tilde{y}$ of the aircraft centre of mass, the roll angle θ of the aircraft, and the corresponding velocities $\dot{\tilde{x}}, \dot{\tilde{y}}, \dot{\theta}$. The control inputs U_t, U_m are, respectively, the thrust (directed out the bottom of the aircraft) and the rolling moment about the aircraft centre of mass. The dynamics of the aircraft can be written as

$$\begin{aligned}
m\ddot{\tilde{x}} &= -U_t \sin\theta + \varepsilon_0 U_m \cos\theta \\
m\ddot{\tilde{y}} &= U_t \cos\theta + \varepsilon_0 U_m \sin\theta - mg \\
J\ddot{\theta} &= U_m
\end{aligned}$$

where $\varepsilon_0 > 0$ is a coefficient describing the coupling between the rolling moment and the lateral force on the aircraft. We will design a robust control to take care of the coupling between the rolling moment and the lateral force on the aircraft. We will solve a nonlinear optimal control problem analytically to obtain a nonlinear robust control law.

1.9 USE OF THIS BOOK

By selecting different chapters, this book can be used in the following three courses.

Chapters 1–5 and Appendix A can be used for an undergraduate/graduate course on modern control theory. These parts cover the following topics:

1. Modelling and responses of systems (Appendix A and Chapter 2).

2. Properties of linear time-invariant systems (Chapters 2 and 3), including controllability, observability, stability, stabilizability, and detectability.
3. Control synthesis for linear time-invariant systems (Chapter 2): pole placement and observer design.
4. Introduction to optimal control and the Kalman filter (Chapter 4).
5. Introduction to robust control design (Chapter 5).

Chapters 5–8 can be used for a graduate level course on robust control design. Such a course will cover the following topics:

1. Optimal control approach to robust control design for linear systems (Chapter 5).
2. Optimal control approach to robust control design for nonlinear systems (Chapter 6).
3. Robust control of parametric systems using the Kharitonov theorem (Chapter 7)
4. H_∞ and H_2 robust control design (Chapter 8).

Finally, Chapters 5–6 and Chapters 9–11 can be used for an application-orientated course on robust control design using the optimal control approach, which covers the following topics:

1. Optimal control approach to robust control design for linear systems (Chapter 5).
2. Optimal control approach to robust control design for nonlinear systems (Chapter 6).
3. Robust active damping for vibration systems (Chapter 9).
4. Robust control of robot manipulators (Chapter 10).
5. Hovering control of (V/STOL) aircraft (Chapter 11).

2

Fundamentals of Control Theory

In this chapter we discuss the fundamentals of control theory. We investigate control for both linear and nonlinear systems, however we focus on linear systems. Systems will be modelled in state space representation with state variables, input variables, and output variables. Given inputs and initial conditions of a linear system, its responses in terms of states and outputs can be determined and investigated. We also discuss similarity transformations and show how to convert a system into its Jordan canonical form by means of similarity transformations. We then study controllability and observability of linear systems. If a system is controllable, then we can assign its poles to arbitrary locations using state feedback control. When state variables are not available directly for control, an observer must be built to estimate the state variables from the input and output variables of the system. Such an observer may not always exist. A necessary and sufficient condition for the existence of the observer is observability. Assuming a system is observable, we can build either a full-order observer or a reduced-order observer. A full-order observer has the same order as the system while a reduced-order observer has an order less than the order of the system. Either a full-order observer or a reduced-order observer can be used in a feedback loop to form a closed-loop control system.

Robust Control Design: An Optimal Control Approach F. Lin

2.1 STATE SPACE MODEL

In classical control theory, a transfer function is used to describe the input and output relation of a system and hence serves as a model of the system. Such a transfer function model is most suitable for linear time-invariant systems with a single input and a single output. If the system to be controlled is nonlinear, or time-varying, or has multiple inputs or outputs, then it will be difficult, if not impossible, to model it by a transfer function. Therefore, for nonlinear, time-varying, or multi-input–multi-output systems, we often need to use state space representation to model the systems.

The state variables of a system are defined as a minimum set of variables such that the knowledge of these variables at any time t_0, plus the information on the input subsequently applied, is sufficient to determine the state variables of the system at any time $t > t_0$. For example, in mechanical systems, state variables are usually the positions and velocities of objects. If a system has n state variables, we say that the order of the system is n. We often use an n-dimensional vector x to denote the state variables: $x \in R^n$. We use $u \in R^m$ to denote the m-dimensional input variables and $y \in R^p$ to denote p-dimensional output variables. A state space model of a system can then be written as

$$\dot{x} = f(x, u, t)$$

$$y = g(x, u, t)$$

where $f : R^n \times R^m \times R \rightarrow R^n$ and $g : R^n \times R^m \times R \rightarrow R^p$ are nonlinear functions. $\dot{x} = f(x, u, t)$ is a set of n first-order differential equations. We call $\dot{x} = f(x, u, t)$ the state equations and $y = g(x, u, t)$ the output equation.

If a system is time-invariant, then the time t will not appear explicitly in functions f and g. In other words, the state space representation can be written as

$$\dot{x} = f(x, u)$$

$$y = g(x, u)$$

If a system is also linear, then the functions f and g are linear functions. Hence the state space representation can be written as

$$\dot{x} = Ax + Bu$$

$$y = Cx + Du$$

where A, B, C, D are matrices of appropriate dimensions. Since A, B, C, D are constants, the system is time-invariant. We sometimes denote this linear time-invariant system by

$$\begin{pmatrix} A & B \\ C & D \end{pmatrix}$$

Most practical systems are nonlinear. However, many nonlinear systems can be approximated by linear systems using linearization methods. For obvious reasons, theory of control of linear systems is much better developed than that of control of nonlinear systems. Chapters 6 and 11 of this book deal with nonlinear systems, while the rest of the book deals with linear systems.

2.2 RESPONSES OF LINEAR SYSTEMS

To determine the response or output of a linear system for given initial conditions and/or inputs, we need to solve the state equation. Let us first consider the response due to the initial condition: given a system

$$\dot{x} = Ax \tag{2.1}$$

with the initial condition at t_0 as $x(t_0) = x_0$, what is the response $x(t)$ at t? To derive this response, we need to use the matrix exponential defined as

$$e^{At} = I + (At) + \frac{1}{2!}(At)^2 + \frac{1}{3!}(At)^3 + \cdots + \frac{1}{n!}(At)^n + \cdots \tag{2.2}$$

The matrix exponential has the following properties.

Properties of the matrix exponential

1. $\frac{\mathrm{d}}{dt}e^{At} = Ae^{At} = e^{At}A$
2. $e^{A0} = I$
3. $(e^{At})^{-1} = e^{-At}$

Proof

1. $$\begin{aligned}\frac{\mathrm{d}}{dt}e^{At} &= \frac{\mathrm{d}}{dt}\left(I + At + \frac{1}{2!}A^2t^2 + \cdots + \frac{1}{(n-1)!}A^{n-1}t^{n-1} + \cdots\right)\\ &= A + A^2t + \frac{1}{2!}A^3t^2 + \cdots + \frac{1}{(n-1)!}A^nt^{n-1} + \cdots\\ &= A\left(I + At + \frac{1}{2!}A^2t^2 + \cdots + \frac{1}{(n-1)!}A^{n-1}t^{n-1} + \cdots\right) = Ae^{At}\\ &= \left(I + At + \frac{1}{2!}A^2t^2 + \cdots + \frac{1}{(n-1)!}A^{n-1}t^{n-1} + \cdots\right)A = e^{At}A\end{aligned}$$

2. $e^{A0} = I + (A0) + \frac{1}{2!}(A0)^2 + \frac{1}{3!}(A0)^3 + \cdots + \frac{1}{n!}(A0)^n + \cdots = I$
3. $I = e^{A0} = e^{At-At} = e^{At}e^{-At}$

Q.E.D.

Based on the matrix exponential, the response of system (2.1) can be written as

$$x(t) = e^{A(t-t_0)}x_0 \tag{2.3}$$

This is because $x(t) = e^{A(t-t_0)}x_0$ satisfies the state equation:

$$\dot{x}(t) = \frac{\mathrm{d}}{\mathrm{d}t}e^{A(t-t_0)}x_0 = Ae^{A(t-t_0)}x_0 = Ax(t)$$

It also satisfies the initial condition:

$$x(t_0) = e^{A(t_0-t_0)}x_0 = e^0x_0 = x_0$$

Let us now consider the response due to both the initial condition and the input. For the system given by

$$\dot{x} = Ax + Bu$$

with the initial condition $x(t_0) = x_0$ and the input $u(t), t \geq t_0$, the response is

$$x(t) = e^{A(t-t_0)}x_0 + \int_{t_0}^{t} e^{A(t-\tau)}Bu(\tau)\mathrm{d}\tau \tag{2.4}$$

This can be shown as follows. First $x(t) = e^{A(t-t_0)}x_0 + \int_{t_0}^{t} e^{A(t-\tau)}Bu(\tau)\mathrm{d}\tau$ satisfies the state equation.

$$\begin{aligned}\dot{x}(t) &= Ae^{A(t-t_0)}x_0 + e^{A(t-t)}Bu(t) + \int_{t_0}^{t} Ae^{A(t-\tau)}Bu(\tau)\mathrm{d}\tau \\ &= A(e^{A(t-t_0)}x_0 + \int_{t_0}^{t} e^{A(t-\tau)}Bu(\tau)\mathrm{d}\tau) + Bu(t) \\ &= Ax(t) + Bu(t)\end{aligned}$$

Second

$$x(t) = e^{A(t-t_0)}x_0 + \int_{t_0}^{t} e^{A(t-\tau)}Bu(\tau)\mathrm{d}\tau$$

satisfies the initial condition.

$$x(t_0) = e^{A(t_0-t_0)}x_0 + \int_{t_0}^{t_0} e^{A(t_0-\tau)}Bu(\tau)\mathrm{d}\tau = x_0$$

To calculate the response of Equations (2.3) or (2.4), we need to compute the matrix exponential e^{At}. Doing this by using the definition (2.2) will not be practical. We need to find some effective ways to compute e^{At}. One such way is to use the Laplace transform. It is not difficult to see that the Laplace transform of e^{At}, denoted by $\mathbf{L}[e^{At}]$, is

$$\mathbf{L}[e^{At}] = (sI - A)^{-1}$$

Hence, e^{At} is the inverse Laplace transform of $(Is - A)^{-1}$.

$$e^{At} = \mathbf{L}^{-1}[(sI - A)^{-1}]$$

Let us show the computation in the following example.

Example 2.1

Let

$$A = \begin{bmatrix} 1 & 0 \\ 0 & 2 \end{bmatrix}$$

then

$$sI - A = \begin{bmatrix} s & 0 \\ 0 & s \end{bmatrix} - \begin{bmatrix} 1 & 0 \\ 0 & 2 \end{bmatrix} = \begin{bmatrix} s-1 & 0 \\ 0 & s-2 \end{bmatrix}$$

$$(sI - A)^{-1} = \begin{bmatrix} s-1 & 0 \\ 0 & s-2 \end{bmatrix}^{-1} = \begin{bmatrix} \dfrac{1}{s-1} & 0 \\ 0 & \dfrac{1}{s-2} \end{bmatrix}$$

The inverse Laplace Transform can be calculated as

$$e^{At} = \mathbf{L}^{-1}[(sI - A)^{-1}] = \begin{bmatrix} e^{t} & 0 \\ 0 & e^{2t} \end{bmatrix}$$

Let us consider another example.

Example 2.2

Let

$$A = \begin{bmatrix} 0 & 1 \\ -2 & -3 \end{bmatrix}$$

then

$$sI-A=\begin{bmatrix} s & 0 \\ 0 & s \end{bmatrix}-\begin{bmatrix} 0 & 1 \\ -2 & -3 \end{bmatrix}=\begin{bmatrix} s & -1 \\ 2 & s+3 \end{bmatrix}$$

Since

$$\begin{vmatrix} s & -1 \\ 2 & s+3 \end{vmatrix}=s(s+3)+2=s^2+3s+2=(s+1)(s+2)$$

$$(sI-A)^{-1}=\begin{bmatrix} s & -1 \\ 2 & s+3 \end{bmatrix}^{-1}=\frac{1}{(s+1)(s+2)}\begin{bmatrix} s+3 & 1 \\ -2 & s \end{bmatrix}$$

$$=\begin{bmatrix} \dfrac{2}{s+1}+\dfrac{-1}{s+2} & \dfrac{1}{s+1}+\dfrac{-1}{s+2} \\ \dfrac{-2}{s+1}+\dfrac{2}{s+2} & \dfrac{-1}{s+1}+\dfrac{2}{s+2} \end{bmatrix}.$$

Taking the inverse Laplace transform, we have

$$e^{At}=\mathrm{L}^{-1}[(sI-A)^{-1}]=\begin{bmatrix} 2e^{-t}-e^{-2t} & e^{-t}-e^{-2t} \\ -2e^{-t}+2e^{-2t} & -e^{-t}+2e^{-2t} \end{bmatrix}$$

From the solution to the state equation

$$x(t)=e^{A(t-t_0)}x_0+\int_{t_0}^{t} e^{A(t-\tau)}Bu(\tau)\mathrm{d}\tau$$

we can calculate the output response of the system as follows.

$$y(t)=Cx(t)+Du(t)=Ce^{A(t-t_0)}x_0+\int_{t_0}^{t} e^{A(t-\tau)}Bu(\tau)\mathrm{d}\tau+Du(t)$$

In the response $y(t)$, the part $Ce^{A(t-t_0)}x_0$ is due to the initial condition and is called the zero-input response (the response when the input is zero) , while the part $\int_{t_0}^{t} e^{A(t-\tau)}Bu(\tau)\mathrm{d}\tau+Du(t)$ is due to the input and is called the zero-state response (the response when the initial state is zero).

Example 2.3

Let us calculate the response of the following system with the initial condition $x(0)=\begin{bmatrix} 1 \\ -1 \end{bmatrix}$ and the input $u(t)=1$.

$$\dot{x}=\begin{bmatrix} 0 & 1 \\ -2 & -3 \end{bmatrix}x+\begin{bmatrix} 0 \\ 1 \end{bmatrix}u$$

$$y=[1 \;\; 1]x$$

From Example 2.2

$$e^{At} = \begin{bmatrix} 2e^{-t} - e^{-2t} & e^{-t} - e^{-2t} \\ -2e^{-t} + 2e^{-2t} & -e^{-t} + 2e^{-2t} \end{bmatrix}$$

Hence

$$\begin{aligned}
x(t) &= e^{A(t)}x_0 + \int_0^t e^{A(t-\tau)}Bu(\tau)\mathrm{d}\tau \\
&= e^{A(t)}x_0 + \int_0^t e^{A\lambda}B\mathrm{d}\lambda \\
&= \begin{bmatrix} 2e^{-t} - e^{-2t} & e^{-t} - e^{-2t} \\ -2e^{-t} + 2e^{-2t} & -e^{-t} + 2e^{-2t} \end{bmatrix} \begin{bmatrix} 1 \\ -1 \end{bmatrix} \\
&\quad + \int_0^t \begin{bmatrix} 2e^{-\lambda} - e^{-2\lambda} & e^{-\lambda} - e^{-2\lambda} \\ -2e^{-\lambda} + 2e^{-2\lambda} & -e^{-\lambda} + 2e^{-2\lambda} \end{bmatrix} \begin{bmatrix} 0 \\ 1 \end{bmatrix} \mathrm{d}\lambda \\
&= \begin{bmatrix} e^{-t} \\ -e^{-t} \end{bmatrix} + \int_0^t \begin{bmatrix} e^{-\lambda} - e^{-2\lambda} \\ -e^{-\lambda} + 2e^{-2\lambda} \end{bmatrix} \mathrm{d}\lambda \\
&= \begin{bmatrix} e^{-t} \\ -e^{-t} \end{bmatrix} + \left. \begin{bmatrix} -e^{-\lambda} + \frac{1}{2}e^{-2\lambda} \\ e^{-\lambda} - e^{-2\lambda} \end{bmatrix} \right|_0^t \\
&= \begin{bmatrix} e^{-t} \\ -e^{-t} \end{bmatrix} + \begin{bmatrix} -e^{-t} + \frac{1}{2}e^{-2t} \\ e^{-t} - e^{-2t} \end{bmatrix} - \begin{bmatrix} -\frac{1}{2} \\ 0 \end{bmatrix} \\
&= \begin{bmatrix} \frac{1}{2} + \frac{1}{2}e^{-2t} \\ -e^{-2t} \end{bmatrix}
\end{aligned}$$

The output response is

$$y = [1 \quad 2] \begin{bmatrix} \frac{1}{2} + \frac{1}{2}e^{-2t} \\ -e^{-2t} \end{bmatrix} = \frac{1}{2} - \frac{1}{2}e^{-2t}$$

Another way to find the response of a linear time-invariant system is to first find its transfer function (matrix). To do this, we take the Laplace transform of

$$\dot{x} = Ax + Bu$$
$$y = Cx + Du$$

assuming zero initial conditions.

$$sX(s) = AX(s) + BU(s)$$

$$\Rightarrow (sI - A)X(s) = BU(s)$$
$$\Rightarrow X(s) = (sI - A)^{-1}BU(s)$$

and

$$\begin{aligned} Y(s) &= CX(s) + DU(s) \\ &= C(sI - A)^{-1}BU(s) + DU(s) \\ &= (C(sI - A)^{-1}B + D)U(s) \end{aligned}$$

Therefore, the transfer function is given by

$$\begin{aligned} G(s) &= C(sI - A)^{-1}B + D \\ &= C\frac{Adj(sI - A)}{|sI - A|}B + D \end{aligned}$$

where $Adj(sI - A)$ is the adjoint of $sI - A$. From the above expression, it is clear that all poles of $G(s)$ are eigenvalues of A. However, an eigenvalue of A may not be a pole of $G(s)$ because cancellation with the numerate may occur.

Example 2.4

For

$$\begin{aligned} \dot{x} &= \begin{bmatrix} 0 & 1 \\ -2 & -3 \end{bmatrix} x + \begin{bmatrix} 1 \\ 1 \end{bmatrix} u \\ y &= \begin{bmatrix} 1 & 0 \end{bmatrix} x \end{aligned}$$

its transfer function is

$$\begin{aligned} G(s) &= C(sI - A)^{-1}B + D \\ &= \begin{bmatrix} 1 & 0 \end{bmatrix} \begin{bmatrix} s & -1 \\ 2 & s+3 \end{bmatrix}^{-1} \begin{bmatrix} 1 \\ 1 \end{bmatrix} \\ &= \frac{1}{(s+1)(s+2)} \begin{bmatrix} 1 & 0 \end{bmatrix} \begin{bmatrix} s+3 & 1 \\ -2 & s \end{bmatrix} \begin{bmatrix} 1 \\ 1 \end{bmatrix} \\ &= \frac{1}{(s+1)(s+2)} \begin{bmatrix} s+3 & 1 \end{bmatrix} \begin{bmatrix} 1 \\ 1 \end{bmatrix} \\ &= \frac{s+4}{(s+1)(s+2)}. \end{aligned}$$

For linear time-invariant systems, $e^{A(t-t')}$ plays the role of a state transition matrix, denoted by $\Phi(t, t')$. By this we mean that if at time t' the state of the system is $x(t')$, then without inputs, the state at time t is $x(t) = e^{A(t-t')}x(t')$. In general $x(t) = \Phi(t, t')x(t')$.

If a system is linear, but time-varying; that is,

$$\dot{x} = A(t)x$$

where $A(t)$ is a function of time, then it is more difficult to find its state transition matrix as shown in the following example.

Example 2.5

Consider a linear time-varying system

$$\begin{bmatrix} \dot{x}_1 \\ \dot{x}_2 \end{bmatrix} = \begin{bmatrix} t & 0 \\ 0 & 1 \end{bmatrix} \begin{bmatrix} x_1 \\ x_2 \end{bmatrix}$$

We can rewrite the state equation as $\dot{x}_1 = tx_1$ and $\dot{x}_2 = x_2$. We know the solution to $\dot{x}_2 = x_2$: $x_2(t) = e^{t-t'}x_2(t')$. To solve $\dot{x}_1 = tx_1$, we separate the variables:

$$\begin{aligned}
&\dot{x}_1 = tx_1 \\
\Rightarrow\ & \frac{dx_1}{dt} = tx_1 \\
\Rightarrow\ & \frac{dx_1}{x_1} = tdt \\
\Rightarrow\ & \ln x_1 \Big|_{x_1(t')}^{x_1(t)} = \frac{1}{2}t^2\Big|_{t'}^{t} \\
\Rightarrow\ & \ln x_1(t) - \ln x_1(t') = \frac{1}{2}t^2 - \frac{1}{2}t'^2 \\
\Rightarrow\ & \ln \frac{x_1(t)}{x_1(t')} = \frac{1}{2}t^2 - \frac{1}{2}t'^2 \\
\Rightarrow\ & \frac{x_1(t)}{x_1(t')} = e^{\frac{1}{2}t^2 - \frac{1}{2}t'^2} \\
\Rightarrow\ & x_1(t) = e^{\frac{1}{2}t^2 - \frac{1}{2}t'^2}x_1(t')
\end{aligned}$$

Combine two equations:

$$\begin{bmatrix} x_1(t) \\ x_1(t) \end{bmatrix} = \begin{bmatrix} e^{\frac{1}{2}t^2 - \frac{1}{2}t'^2} & 0 \\ 0 & e^{t-t'} \end{bmatrix} \begin{bmatrix} x_1(t') \\ x_1(t') \end{bmatrix}$$

In other words, the state transition matrix is given by

$$\Phi(t, t') = \begin{bmatrix} e^{\frac{1}{2}t^2 - \frac{1}{2}t'^2} & 0 \\ 0 & e^{t-t'} \end{bmatrix}$$

We may not always be able to find the analytical expression of the state transition matrix of a system. If we do, then we can write the response of a linear time-varying system

$$\dot{x} = A(t)x + B(t)u$$

as follows:

$$x(t) = \Phi(t, t_0)x_0 + \int_{t_0}^{t} \Phi(t, \tau)B(\tau)u(\tau)\mathrm{d}\tau$$

Example 2.6

For the following linear time-varying system

$$\begin{bmatrix} \dot{x}_1 \\ \dot{x}_2 \end{bmatrix} = \begin{bmatrix} t & 0 \\ 0 & 1 \end{bmatrix} \begin{bmatrix} x_1 \\ x_2 \end{bmatrix} + \begin{bmatrix} 1 \\ 1 \end{bmatrix} u$$

the response when the initial condition is

$$x(0) = \begin{bmatrix} 1 \\ -1 \end{bmatrix}$$

and the input is $u(t) = 1$ can be computed as

$$\begin{aligned} x(t) &= \Phi(t, t_0)x_0 + \int_0^t \Phi(t, \tau)B(\tau)u(\tau)\mathrm{d}\tau \\ &= \Phi(t, 0)x(0) + \int_0^t \Phi(t, \tau)B(\tau)u(\tau)\mathrm{d}\tau \\ &= \begin{bmatrix} e^{\frac{1}{2}t^2} & 0 \\ 0 & e^t \end{bmatrix} \begin{bmatrix} 1 \\ -1 \end{bmatrix} + \int_0^t \begin{bmatrix} e^{\frac{1}{2}t^2 - \frac{1}{2}\tau^2} & 0 \\ 0 & e^{t-\tau'} \end{bmatrix} \begin{bmatrix} 1 \\ 1 \end{bmatrix} \mathrm{d}\tau \\ &= \begin{bmatrix} e^{\frac{1}{2}t^2} \\ -e^t \end{bmatrix} + \int_0^t \begin{bmatrix} e^{\frac{1}{2}t^2 - \frac{1}{2}\tau^2} \\ e^{t-\tau'} \end{bmatrix} \mathrm{d}\tau \end{aligned}$$

Since the above integral does not have an analytic solution, it can only be solved numerically. We can use various computer programs, such as MATLAB to 'simulate' the response of such a system.

2.3 SIMILARITY TRANSFORMATION

In a state space model of a system, the inputs and outputs are given and cannot be changed. However, the states are intermediate variables and they are not unique, but can be changed without affecting the input–output relation of the system. In other words, a transfer function may have many state space representations or realizations using different state variables. We will develop a systematic way to change state variables (or sometimes referred to as coordinate change). This is achieved by similarity transformation. Consider a system given by

$$\dot{x} = Ax + Bu$$
$$y = Cx + Du$$

We would like to change the state variables from x to z. We assume z has the same dimension as x; and x and z are linearly related: $x = Tz$, where T is an $n \times n$ transformation matrix. In order for this transformation to be one-to-one, T needs to be invertible, that is, $z = T^{-1}x$. Let us derive the state equation and output equation with state variables z.

$$\begin{aligned}\dot{z} &= T^{-1}\dot{x}\\ &= T^{-1}(Ax + Bu)\\ &= T^{-1}(ATz + Bu)\\ &= T^{-1}ATz + T^{-1}Bu\\ y &= Cx + Du\\ &= CTz + Du\end{aligned}$$

If we denote $\widetilde{A} = T^{-1}AT$, $\widetilde{B} = T^{-1}B$, $\widetilde{C} = CT$, $\widetilde{D} = D$, then the state equation and output equation can be written as

$$\dot{z} = \widetilde{A}z + \widetilde{B}u$$
$$y = \widetilde{C}z + \widetilde{D}u$$

Obviously, the old model

$$\begin{pmatrix} A & B \\ C & D \end{pmatrix}$$

and the new model

$$\begin{pmatrix} \widetilde{A} & \widetilde{B} \\ \widetilde{C} & \widetilde{D} \end{pmatrix}$$

are related. First, they have the same characteristic polynomial and hence the same eigenvalues:

$$
\begin{aligned}
|sI-\widetilde{A}| &= |sI-T^{-1}AT| \\
&= |sT^{-1}T-T^{-1}AT| \\
&= |T^{-1}(sI-A)T| \\
&= |T^{-1}||(sI-A)||T| \\
&= |T^{-1}||T||(sI-A)| \\
&= |T^{-1}T||(sI-A)| \\
&= |(sI-A)|
\end{aligned}
$$

Second, both systems have the same transfer function and hence the same input–output relation:

$$
\begin{aligned}
\widetilde{G}(s) &= \widetilde{C}(sI-\widetilde{A})^{-1}\widetilde{B}+\widetilde{D} \\
&= CT(sI-T^{-1}AT)^{-1}T^{-1}B+D \\
&= CT(T^{-1}(sI-A)T)^{-1}T^{-1}B+D \\
&= CTT^{-1}(sI-A)^{-1}TT^{-1}B+D \\
&= C(sI-A)^{-1}B+D \\
&= G(s)
\end{aligned}
$$

The above result also shows that state space representation (also called realization) of a system is not unique. Among these representations, some canonical forms are of particular interest. They are the Jordan canonical form, controllable canonical form, and observable canonical form. We discuss the Jordan canonical form first. The controllable canonical form and observable canonical form will be discussed when we discuss controllability and observability.

There are three types of Jordan canonical form: one for systems with distinct and real eigenvalues, one for systems with repeated and real eigenvalues, and one for systems with complex eigenvalues.

If matrix A has distinct and real eigenvalues $\lambda_1\ \lambda_2\ \ldots\ \lambda_n$, then we can find a transformation matrix T such that

$$
\widetilde{A}=T^{-1}AT=\begin{bmatrix} \lambda_1 & 0 & \ldots & 0 \\ 0 & \lambda_2 & \ldots & 0 \\ \ldots & & & \\ 0 & 0 & \ldots & \lambda_n \end{bmatrix}
$$

In fact, T is the matrix consisting of n eigenvectors

$$T = [v_1 \ v_2 \ \ldots \ v_n]$$

where the column vector v_i, $i = 1, 2, \ldots, n$ is the eigenvector corresponding to the eigenvalue λ_i satisfying

$$Av_i = \lambda_i v_i \tag{2.5}$$

It is proved in linear algebra that v_i are independent of each other if λ_i are distinct. Hence, the inverse of T exists. Denote

$$T^{-1} = \begin{bmatrix} w_1 \\ w_2 \\ \ldots \\ w_n \end{bmatrix}$$

where w_i, $i = 1, 2, \ldots, n$ is a row vector. Since

$$\begin{aligned} T^{-1}T &= \begin{bmatrix} w_1 \\ w_2 \\ \ldots \\ w_n \end{bmatrix} [v_1 \ v_2 \ldots \ v_n] \\ &= \begin{bmatrix} w_1 v_1 & w_1 v_2 & \ldots & w_1 v_n \\ w_2 v_1 & w_2 v_2 & \ldots & w_2 v_n \\ \ldots & & & \\ w_n v_1 & w_n v_2 & \ldots & w_n v_n \end{bmatrix} \\ &= \begin{bmatrix} 1 & 0 & \ldots & 0 \\ 0 & 1 & \ldots & 0 \\ \ldots & & & \\ 0 & 0 & \ldots & 1 \end{bmatrix} \end{aligned}$$

we have

$$w_i v_j = \begin{cases} 1 & \text{if } i = j \\ 0 & \text{otherwise} \end{cases} \tag{2.6}$$

Therefore, by Equations (2.5) and (2.6),

$$\begin{aligned} \widetilde{A} &= T^{-1}AT \\ &= \begin{bmatrix} w_1 \\ w_2 \\ \ldots \\ w_n \end{bmatrix} A [v_1 \ v_2 \ \ldots \ v_n] \end{aligned}$$

$$= \begin{bmatrix} w_1 \\ w_2 \\ \cdots \\ w_n \end{bmatrix} \begin{bmatrix} Av_1 & Av_2 & \ldots & Av_n \end{bmatrix}$$

$$= \begin{bmatrix} w_1 \\ w_2 \\ \cdots \\ w_n \end{bmatrix} \begin{bmatrix} \lambda_1 v_1 & \lambda_2 v_2 & \ldots & \lambda_n v_n \end{bmatrix}$$

$$= \begin{bmatrix} w_1\lambda_1 v_1 & w_1\lambda_2 v_2 & \ldots & w_1\lambda_n v_n \\ w_2\lambda_1 v_1 & w_2\lambda_2 v_2 & \ldots & w_2\lambda_n v_n \\ \cdots & & & \\ w_n\lambda_1 v_1 & w_n\lambda_2 v_2 & \ldots & w_n\lambda_n v_n \end{bmatrix}$$

$$= \begin{bmatrix} \lambda_1 & 0 & \ldots & 0 \\ 0 & \lambda_2 & \ldots & 0 \\ \cdots & & & \\ 0 & 0 & \ldots & \lambda_n \end{bmatrix}$$

The reasons that we are interested in the Jordan canonical form are: (1) all state variables are 'decoupled' in the sense that one does not depend on another; (2) it is straightforward to find the eigenvalues of the system, and hence the stability of the system (see Chapter 3).

Example 2.7

Consider the following system

$$\begin{bmatrix} \dot{\theta} \\ \dot{\omega} \\ \dot{i} \end{bmatrix} = \begin{bmatrix} 0 & 1 & 0 \\ 0 & 0 & 4.438 \\ 0 & -12 & -24 \end{bmatrix} \begin{bmatrix} \theta \\ \omega \\ i \end{bmatrix} + \begin{bmatrix} 0 & 0 \\ 0 & -7.396 \\ 20 & 0 \end{bmatrix} \begin{bmatrix} v \\ T \end{bmatrix}$$

$$\theta = \begin{bmatrix} 1 & 0 & 0 \end{bmatrix} \begin{bmatrix} \theta \\ \omega \\ i \end{bmatrix}$$

This is a model of a DC motor, where the states θ, ω, i are the angle position, angle velocity, and current, respectively; the inputs v and T are the voltage and torque, respectively; and the output is the angle position.

Let us use similarity transformation to transform the system into Jordan canonical form. We can use MATLAB to find the eigenvalues and eigenvectors of A, and the transformation matrix T. The results are shown in Figure 2.1, which shows that the MATLAB command to find eigenvalues and eigenvectors is 'eig(.)'. In the figure, T is the transition matrix, *invT* is the inverse of T, and J is the Jordan canonical form.

```
Using Toolbox Path Cache. type "help toolbox_path_cache" for mode info.
To get started, select "MATLAB HELP" from the help menu.
 >>A=[0 1 0; 0 0 4.438;  0 -12  -24]
 A=
                        0      1.0000          0
                        0           0     4.4380
                        0    -12.0000   -24.0000
 >> [T, J]= eig(A)
 T=
                   1.0000     -0.3329     0.0094
                        0      0.8236    -0.2019
                        0     -0.4591     0.9794
 J=
                        0           0          0
                        0     -2.4740          0
                                    0   -21.5260
 >>invT=inv(T)
 invT=
                   1.0000      0.4507     0.0833
                        0      1.3718     0.2828
                        0      0.6431     1.1537
```

Figure 2.1 MATLAB results of Example 2.1

The new state space representation in Jordan canonical form is given by

$$\widetilde{A} = T^{-1}AT = \begin{bmatrix} 0 & 0 & 0 \\ 0 & -2.4740 & 0 \\ 0 & 0 & -21.5260 \end{bmatrix}$$

$$\widetilde{B} = T^{-1}B = \begin{bmatrix} 1.6667 & -3.3330 \\ 5.6565 & -10.1459 \\ 23.0734 & -4.7566 \end{bmatrix}$$

$$\widetilde{C} = CT = \begin{bmatrix} 1 & -0.3329 & 0.0094 \end{bmatrix}$$

$$\widetilde{D} = D = 0$$

Next, consider the case that matrix A has repeated and real eigenvalues. Without loss of generality, assume that λ_1 is the eigenvalues repeated k times. It is known in linear algebra that if $rank(\lambda_1 I - A) = l$, then there are $n - l$ independent eigenvectors corresponding to λ_1. If $n - l = k$, then there are k independent eigenvectors corresponding to λ_1 and the procedure described above can be applied; that is,

$$T = \begin{bmatrix} v_1 & v_2 & \dots & v_n \end{bmatrix}$$

where $v_1 \ldots v_k$ are independent eigenvectors corresponding to λ_1. The Jordan canonical form is given by

$$\widetilde{A} = T^{-1}AT = \begin{bmatrix} \lambda_1 & 0 & \ldots & 0 \\ 0 & \lambda_2 & \ldots & 0 \\ \ldots & & & \\ 0 & 0 & \ldots & \lambda_n \end{bmatrix}$$

where $\lambda_1 = \lambda_2 = \ldots = \lambda_k$.

It is quite possible that $n - l < k$. If so, then we cannot find k independent eigenvectors corresponding to λ_1. What we need to do is to find some generalized eigenvectors to construct the Jordan canonical form. Let v^0 be an eigenvector of λ_1. Its generalized eigenvectors can be obtained by solving

$$(A - \lambda_1 I)v^1 = v^0$$

$$(A - \lambda_1 I)v^2 = v^1$$

$$\ldots$$

$$(A - \lambda_1 I)v^{i+1} = v^i$$

Solutions can be found by continuing the chain as long as necessary. Using the generalized eigenvectors to form the transform matrix

$$T = \begin{bmatrix} v^0 & v^1 & \ldots & v_n \end{bmatrix}$$

we can get the Jordan canonical form. It is no longer diagonal; it has some terms of value 1 above the diagonal.

$$\widetilde{A} = T^{-1}AT = \begin{bmatrix} \lambda_1 & 1 & \ldots & 0 \\ 0 & \lambda_1 & \ldots & 0 \\ \ldots & & & \\ 0 & 0 & \ldots & \lambda_n \end{bmatrix}$$

Example 2.8

Consider the following system (which is the linearized model of an inverted pendulum)

$$\dot{x} = \begin{bmatrix} 0 & 1 & 0 & 0 \\ 0 & 0 & -9.8 & 0 \\ 0 & 0 & 0 & 1 \\ 0 & 0 & 19.6 & 1 \end{bmatrix} x + \begin{bmatrix} 0 \\ 1 \\ 0 \\ -1 \end{bmatrix} u$$

$$y = \begin{bmatrix} 1 & 0 & 0 & 0 \\ 0 & 0 & 1 & 0 \end{bmatrix} x$$

The matrix A has four eigenvalues: $\lambda_1 = 0, \lambda_2 = 0, \lambda_3 = 4.4272, \lambda_4 = -4.4272$. $\lambda_1 = \lambda_2 = 0$ are repeated eigenvalues. Since $rank(\lambda_1 I - A) = 3$, there is only one independent eigenvector corresponding to $\lambda_1 (= \lambda_2)$. Using MATLAB, we can calculate three independent eigenvectors as

$$v^0 = v_1 = \begin{bmatrix} 1 \\ 0 \\ 0 \\ 0 \end{bmatrix} \quad v_3 = \begin{bmatrix} -0.0985 \\ -0.4362 \\ 0.1971 \\ 0.8724 \end{bmatrix} \quad v_4 = \begin{bmatrix} 0.0985 \\ -0.4362 \\ -0.1971 \\ 0.8724 \end{bmatrix}$$

We need to find one generalized eigenvector by solving $(A - \lambda_1 I)v^1 = v^0$.

$$\begin{bmatrix} 0 & 1 & 0 & 0 \\ 0 & 0 & -9.8 & 0 \\ 0 & 0 & 0 & 1 \\ 0 & 0 & 19.6 & 1 \end{bmatrix} v^1 = \begin{bmatrix} 1 \\ 0 \\ 0 \\ 0 \end{bmatrix} \quad \Rightarrow \quad v^1 = \begin{bmatrix} 0 \\ 1 \\ 0 \\ 0 \end{bmatrix}$$

The resulting transform matrix is

$$T = \left[v^0 \; v^1 \; v_3 \; v_4 \right] = \begin{bmatrix} 1 & 0 & -0.0985 & 0.0985 \\ 0 & 1 & -0.4362 & -0.4362 \\ 0 & 0 & 0.1971 & -0.1971 \\ 0 & 0 & 0.8724 & 0.8724 \end{bmatrix}$$

The new state space representation in Jordan canonical form is given by

$$\widetilde{A} = T^{-1}AT = \begin{bmatrix} 0 & 1 & 0 & 0 \\ 0 & 0 & 0 & 0 \\ 0 & 0 & 4.4272 & 0 \\ 0 & 0 & 0 & -4.4272 \end{bmatrix}$$

$$\widetilde{B} = T^{-1}B = \begin{bmatrix} 0 \\ 0.5 \\ -0.5731 \\ -0.5731 \end{bmatrix}$$

$$\widetilde{C} = CT = \begin{bmatrix} 1 & 0 & -0.0985 & 0.0985 \\ 0 & 0 & 0.1971 & -0.1971 \end{bmatrix}$$

$$\widetilde{D} = D = \begin{bmatrix} 0 \\ 0 \end{bmatrix}$$

Finally, consider the case that matrix A has complex eigenvalues. Obviously, we cannot transform a system into a system with complex numbers because they are not 'real' and cannot be implemented practically. So, we need to modify the method. Without loss of generality, assume that

$\lambda_1 = \sigma + j\omega$ is the complex eigenvalue and $\lambda_2 = \sigma - j\omega$ is its complex conjugate. Since A is a real matrix, its eigenvalues must appear as a pair of complex conjugates. The corresponding eigenvectors must also appear as complex conjugates: $v_1 = p + jq$, $v_2 = p - jq$. Using the transform matrix

$$T = \begin{bmatrix} p & q & v_3 & \dots & v_n \end{bmatrix}$$

we have the following Jordan canonical form

$$\widetilde{A} = T^{-1}AT = \begin{bmatrix} \sigma & \omega & 0 & \dots & 0 \\ -\omega & \sigma & 0 & \dots & 0 \\ 0 & 0 & \lambda_3 & \dots & 0 \\ \dots & & & \dots & \\ 0 & 0 & 0 & \dots & \lambda_n \end{bmatrix}$$

Example 2.9

Consider the following system

$$\dot{x} = \begin{bmatrix} 1 & -1 & 2 \\ -2 & 5 & 4 \\ -4 & -5 & 8 \end{bmatrix} x + \begin{bmatrix} 1 \\ -1 \\ 2 \end{bmatrix} u$$

$$y = \begin{bmatrix} 1 & 0 & 1 \end{bmatrix} x$$

The eigenvalues of A are $\lambda_1 = 5.8468 + j4.2243$, $\lambda_2 = 5.8468 - j4.2243$, and $\lambda_3 = 2.3063$. The corresponding eigenvectors are

$$v_1 = \begin{bmatrix} -0.2207 + j0.0714 \\ -0.1513 + j0.5861 \\ -0.7614 \end{bmatrix} \qquad v_2 = \begin{bmatrix} -0.2207 - j0.0714 \\ -0.1513 - j0.5861 \\ -0.7614 \end{bmatrix}$$

$$v_3 = \begin{bmatrix} -0.8567 \\ 0.1118 \\ -0.5036 \end{bmatrix}$$

Hence, the transform matrix is

$$T = \begin{bmatrix} -0.2207 & 0.0714 & -0.8567 \\ -0.1513 & 0.5861 & 0.1118 \\ -0.7614 & 0 & -0.5036 \end{bmatrix}$$

The Jordan canonical form is given by

$$\widetilde{A} = T^{-1}AT = \begin{bmatrix} 5.8468 & 4.2243 & 0 \\ -4.2243 & 5.8468 & 0 \\ 0 & 0 & 2.306 \end{bmatrix}$$

$$\widetilde{B} = T^{-1}B = \begin{bmatrix} -2.0963 \\ -2.0943 \\ -0.8019 \end{bmatrix}$$

$$\widetilde{C} = CT = \begin{bmatrix} -0.9821 & 0.0714 & -1.3603 \end{bmatrix}$$

$$\widetilde{D} = D = 0$$

2.4 CONTROLLABILITY AND OBSERVABILITY

Before we discuss controllability and observability of linear systems, let us first illustrate the idea using the following example.

Example 2.10

Suppose we have the following system

$$\dot{x} = \begin{bmatrix} 2 & 3 & 2 & 1 \\ -2 & -3 & 0 & 0 \\ -2 & -2 & -4 & 0 \\ -2 & -2 & -2 & -5 \end{bmatrix} x + \begin{bmatrix} 1 \\ -2 \\ 2 \\ -1 \end{bmatrix} u$$

$$y = [7\,6\,4\,2]x$$

Its transfer function is given by

$$H(s) = C(sI - A)^{-1}B + D = \frac{s^3 + 9s^2 + 26s + 24}{s^4 + 10s^3 + 35s^2 + 50s + 24} = \frac{1}{s+1}$$

Clearly, the state equation of the system is of fourth order, but the transfer function is first order because of pole-zero cancellation. To understand why this happens, let us perform the following similarity transformation to transform the system into its Jordan canonical form: $z = T^{-1}x$ with

$$T = \begin{bmatrix} 1 & -1 & 0 & 0 \\ -1 & 2 & -1 & 0 \\ 0 & -1 & 2 & -1 \\ 0 & 0 & -1 & 2 \end{bmatrix}$$

For the new state variable z, the state and output equations are

$$\dot{z} = \begin{bmatrix} -1 & 0 & 0 & 0 \\ 0 & -2 & 0 & 0 \\ 0 & 0 & -3 & 0 \\ 0 & 0 & 0 & -4 \end{bmatrix} z + \begin{bmatrix} 1 \\ 0 \\ 1 \\ 0 \end{bmatrix} u$$

$$y = \begin{bmatrix} 1 & 1 & 0 & 0 \end{bmatrix} z$$

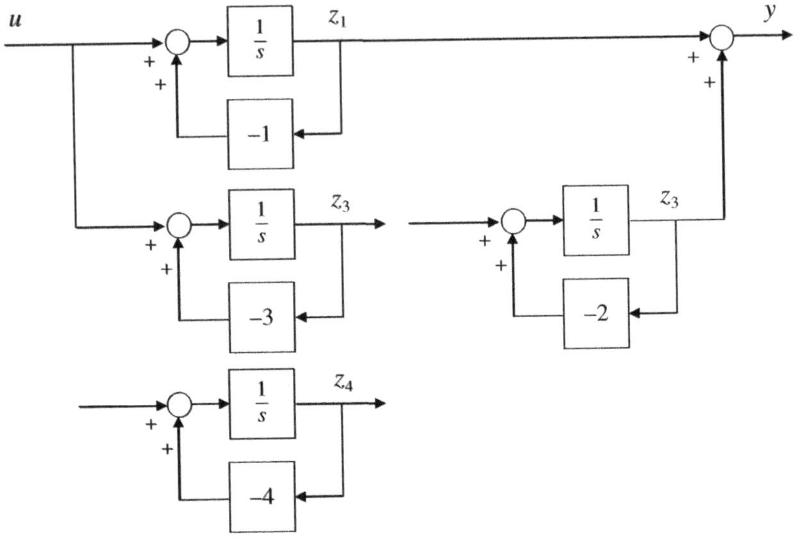

Figure 2.2 State diagram of the system in Example 2.10.

The diagram of this system is shown in Figure 2.2. Only states z_1 and z_3 can be controlled by the input u and only states z_1 and z_2 can be observed from output y.

Intuitively, a system is called controllable if all its states can be controlled in the sense that they can be driven to anywhere using some input, and a system is called observable if all its states can be observed in the sense that their values can be determined from the output. As we will see, controllability and observability are very important in control design. Controllability ensures that we can move the eigenvalues or poles of a system to any desirable locations to achieve stability or optimality by state feedback. Observability ensures that we can estimate or reconstruct the state variables from the output and hence make state feedback feasible. For the convenience of presentation, we first discuss observability.

Observability

A linear time-invariant system

$$\dot{x} = Ax + Bu$$

$$y = Cx + Du$$

is observable if the initial state $x(0) = x_0$ can be uniquely deduced from the knowledge of the input $u(t)$ and output $y(t)$ over the interval $t \in [0, \Delta]$ for some $\Delta > 0$.

Remarks

1. In the above definition of observability, the initial state $x(0) = x_0$ is arbitrary: we can find x_0 no matter where the system starts.
2. If $x(0) = x_0$ can be deduced, then we can calculate the state response at any time by

$$x(t) = e^{At}x_0 + \int_0^t e^{A(t-\tau)}Bu(\tau)\, d\tau$$

3. Since in the output response

$$y(t) = Ce^{At}x_0 + \int_0^t e^{A(t-\tau)}Bu(\tau)\, d\tau + Du(t)$$

the zero-state response $\int_0^t e^{A(t-\tau)}Bu(\tau)\, d\tau + Du(t)$ is known for a given input, and is independent of the initial condition x_0; observability is determined by the zero-input response $y(t) = Ce^{At}x_0$ and hence determined by the matrix pair (A, C).

From now on, in view of Remark 3 above, when we discuss observability, we mean the observability of (A, C). We say that a state $x_0 \neq 0$ is unobservable if the zero-input response with $x(0) = x_0$ is zero for all $t \geq 0$: $y(t) = Ce^{At}x_0 = 0$.

Theorem 2.1

A linear time-invariant system (A, C) is observable if and only if it has no unobservable state.

Proof

(ONLY IF) If there exists $x_0 \neq 0$ such that $y(t) = Ce^{At}x_0 = 0$ for all $t \geq 0$, then we can find two states x_1 and $x_2 = x_1 + x_0$ such that $x_2 \neq x_1$, but their zero-input responses are the same: let $y_1(t) = Ce^{At}x_1$ be the response to x_1 and $y_2(t) = Ce^{At}x_2$ be the response to x_2. Clearly

$$y_2(t) = Ce^{At}x_2 = Ce^{At}(x_1 + x_0) = Ce^{At}x_1 + Ce^{At}x_0 = Ce^{At}x_1 = y_1(t)$$

Therefore, no matter what we do, we cannot distinguish x_1 from x_2. In other words, the system (A, C) is not observable.

(IF) If there exists no $x_0 \neq 0$ such that $y(t) = Ce^{At}x_0 = 0$ for all $t \geq 0$, then

$$
\begin{aligned}
&(\forall x_0 \neq 0)(\exists t \geq 0) y(t) = Ce^{At}x_0 \neq 0 \\
&\Rightarrow (\forall x_0 \neq 0)(\exists t \geq 0)(Ce^{At}x_0)^{\mathrm{T}}(Ce^{At}x_0) > 0 \\
&\Rightarrow (\forall x_0 \neq 0)(\exists \Delta > 0) \int_0^{\Delta} (Ce^{At}x_0)^{\mathrm{T}}(Ce^{At}x_0)\mathrm{d}t > 0 \\
&\Rightarrow (\forall x_0 \neq 0)(\exists \Delta > 0) x_0^{\mathrm{T}} \left(\int_0^{\Delta} (Ce^{At})^{\mathrm{T}}(Ce^{At})\mathrm{d}t \right) x_0 > 0
\end{aligned}
$$

In other words, there exists $\Delta > 0$ such that the matrix $M(\Delta) = \int_0^{\Delta} (Ce^{At})^{\mathrm{T}}(Ce^{At})\mathrm{d}t$ is positive definite. In particular, $M(\Delta)^{-1}$ exists. Using this result, we can deduce x_0 from $y(t) = Ce^{At}x_0$ as follows.

$$
\begin{aligned}
&y(t) = Ce^{At}x_0 \\
&\Rightarrow (Ce^{At})^{\mathrm{T}}y(t) = (Ce^{At})^{\mathrm{T}}Ce^{At}x_0 \\
&\Rightarrow \int_0^{\Delta} (Ce^{At})^{\mathrm{T}}y(t)\mathrm{d}t = \int_0^{\Delta} (Ce^{At})^{\mathrm{T}}Ce^{At}x_0\mathrm{d}t \\
&\Rightarrow \int_0^{\Delta} (Ce^{At})^{\mathrm{T}}y(t)\mathrm{d}t = \left(\int_0^{\Delta} (Ce^{At})^{\mathrm{T}}Ce^{At}\mathrm{d}t \right) x_0 \\
&\Rightarrow \int_0^{\Delta} (Ce^{At})^{\mathrm{T}}y(t)\mathrm{d}t = M(\Delta)x_0 \\
&\Rightarrow x_0 = M(\Delta)^{-1} \int_0^{\Delta} (Ce^{At})^{\mathrm{T}}y(t)\mathrm{d}t
\end{aligned}
$$

Hence, the system (A, C) is observable.

Q.E.D

By Theorem 2.1, checking observability is equivalent to checking if there exists an unobservable state $x_0 \neq 0$. However checking this condition is still not trivial by its definition. Fortunately, the problem can be further reduced as stated in the following theorem.

Theorem 2.2

$x_0 \neq 0$ is an unobservable state if and only if

$$\begin{bmatrix} C \\ CA \\ CA^2 \\ \cdots \\ CA^{n-1} \end{bmatrix} x_0 = 0$$

Proof

(ONLY IF) If $x_0 \neq 0$ is an unobservable state, then for all $t \geq 0$, $y(t) = Ce^{At}x_0$ is zero and all the derivatives of $y(t)$ are also zero; that is, $y(t) = 0$, $y^1(t) = 0, \ldots, y^{n-1}(t) = 0$.

$$y(t) = Ce^{At}x_0 = 0 \Rightarrow y(0) = Cx_0 = 0$$
$$y^1(t) = CAe^{At}x_0 = 0 \Rightarrow y^1(0) = CAx_0 = 0$$
$$y^2(t) = CA^2e^{At}x_0 = 0 \Rightarrow y^2(0) = CA^2x_0 = 0$$
$$\cdots$$
$$y^{n-1}(t) = CA^{n-1}e^{At}x_0 = 0 \Rightarrow y^{n-1}(0) = CA^{n-1}x_0 = 0$$

Therefore,

$$\begin{bmatrix} C \\ CA \\ CA^2 \\ \cdots \\ CA^{n-1} \end{bmatrix} x_0 = 0$$

(IF) Let us first recall the Caylay–Hamilton theorem: for any matrix A, let $\varphi(s)$ be its characteristic polynomial

$$\varphi(s) = |sI - A| = s^n + a_{n-1}s^{n-1} + \cdots + a_1 s + a_0$$

Then

$$\varphi(A) = A^n + a_{n-1}A^{n-1} + \cdots + a_1 A + a_0 I = 0$$

This implies that for all $k \geq 0$, A^k can be expressed as a linear combination of $I, A, A^2, \ldots, A^{n-1}$. For example

$$A^n = -(a_{n-1}A^{n-1} + \cdots + a_1 A + a_0 I)$$

$$
\begin{aligned}
A^{n+1} &= -A(a_{n-1}A^{n-1} + \cdots + a_1 A + a_0 I) \\
&= -a_{n-1}A^n - (a_{n-2}A^{n-1} + \cdots + a_1 A^2 + a_0 A) \\
&= a_{n-1}(a_{n-1}A^{n-1} + \cdots + a_1 A + a_0 I) - (a_{n-2}A^{n-1} + \cdots + a_1 A^2 + a_0 A) \\
&= (a_{n-1}^2 - a_{n-2})A^{n-1} + \cdots + (a_1 - a_0)A + a_0 I
\end{aligned}
$$

Since

$$e^{At} = I + (At) + \frac{1}{2!}(At)^2 + \frac{1}{3!}(At)^3 + \cdots + \frac{1}{n!}(At)^n + \cdots$$

is a linear combination of A^k, $k \geq 0$, it can also be expressed as a linear combination of I, A, $A^2, \ldots, A^{n-1}$

$$e^{At} = \rho_0(t)I + \rho_1(t)A + \rho_2(t)A^2 + \cdots + \rho_{n-1}(t)A^{n-1}$$

Now, if

$$\begin{bmatrix} C \\ CA \\ CA^2 \\ \cdots \\ CA^{n-1} \end{bmatrix} x_0 = 0$$

then for all $t \geq 0$

$$
\begin{aligned}
y(t) &= Ce^{At}x_0 \\
&= C(\alpha_0(t)I + \alpha_1(t)A + \alpha_2(t)A^2 + \cdots + \alpha_{n-1}(t)A^{n-1})x_0 \\
&= \alpha_0(t)Cx_0 + \alpha_1(t)CAx_0 + \alpha_2(t)CA^2x_0 + \cdots + \alpha_{n-1}(t)x_0 A^{n-1}x_0 \\
&= 0
\end{aligned}
$$

Hence, $x_0 \neq 0$ is an unobservable state.

Q.E.D.

By Theorems 2.1 and 2.2, we conclude

$$
\begin{aligned}
& (A, C) \text{is observable} \\
& \Leftrightarrow (\forall x_0 \neq 0) \begin{bmatrix} C \\ CA \\ \cdots \\ CA^{n-1} \end{bmatrix} x_0 \neq 0 \\
& \Leftrightarrow rank \begin{bmatrix} C \\ CA \\ \cdots \\ CA^{n-1} \end{bmatrix} = n
\end{aligned}
$$

Let us define the observability matrix as

$$\mathcal{O} = \begin{bmatrix} C \\ CA \\ \cdots \\ CA^{n-1} \end{bmatrix}$$

then we conclude that a system is observable if and only if its observability matrix is of full rank.

$$(A, C) \text{ is observable} \quad \Leftrightarrow \quad rank(\mathcal{O}) = n$$

Example 2.11

Let us consider the system in Example 2.10:

$$\dot{x} = \begin{bmatrix} 2 & 3 & 2 & 1 \\ -2 & -3 & 0 & 0 \\ -2 & -2 & -4 & 0 \\ -2 & -2 & -2 & -5 \end{bmatrix} x + \begin{bmatrix} 1 \\ -2 \\ 2 \\ -1 \end{bmatrix} u$$

$$y = \begin{bmatrix} 7 \, 6 \, 4 \, 2 \end{bmatrix} x$$

Its observability matrix is

$$\mathcal{O} = \begin{bmatrix} C \\ CA \\ CA^2 \\ CA^3 \end{bmatrix} = \begin{bmatrix} 7 & 6 & 4 & 2 \\ -10 & -9 & -6 & -3 \\ 16 & 15 & 10 & 5 \\ -28 & -27 & -18 & -9 \end{bmatrix}$$

The rank of the observability matrix is 2. In Example 2.10, we know that two states are observable. Since not all states are observable, the system is not observable.

Example 2.12

Consider the DC motor of Example 2.7:

$$\begin{bmatrix} \dot{\theta} \\ \dot{\omega} \\ \dot{i} \end{bmatrix} = \begin{bmatrix} 0 & 1 & 0 \\ 0 & 0 & 4.438 \\ 0 & -12 & -24 \end{bmatrix} \begin{bmatrix} \theta \\ \omega \\ i \end{bmatrix} + \begin{bmatrix} 0 & 0 \\ 0 & -7.396 \\ 20 & 0 \end{bmatrix} \begin{bmatrix} v \\ T \end{bmatrix}$$

$$y = \begin{bmatrix} 1 & 0 & 0 \end{bmatrix} \begin{bmatrix} \theta \\ \omega \\ i \end{bmatrix}$$

The observability matrix is

$$\text{O} = \begin{bmatrix} C \\ CA \\ CA^2 \end{bmatrix} = \begin{bmatrix} 1 & 0 & 0 \\ 0 & 1 & 0 \\ 0 & 0 & 4.4380 \end{bmatrix}$$

Clearly, $rank(\text{O}) = 3$ and the system is observable. Here we assume that θ is the output. If instead of θ, i is the output, then

$$C = [0 \quad 0 \quad 1]$$

and

$$\text{O} = \begin{bmatrix} C \\ CA \\ CA^2 \end{bmatrix} = \begin{bmatrix} 0 & 0 & 1 \\ 0 & -12 & -24 \\ 0 & 288 & 522.7 \end{bmatrix}$$

Since $rank(\text{O}) = 2$, the system is not observable.

Example 2.13

Consider the inverted pendulum in Example 2.8:

$$\dot{x} = \begin{bmatrix} 0 & 1 & 0 & 0 \\ 0 & 0 & -9.8 & 0 \\ 0 & 0 & 0 & 1 \\ 0 & 0 & 19.6 & 1 \end{bmatrix} x + \begin{bmatrix} 0 \\ 1 \\ 0 \\ -1 \end{bmatrix} u$$

$$y = [1 \quad 0 \quad 0 \quad 0]x$$

The observability matrix is

$$\text{O} = \begin{bmatrix} C \\ CA \\ CA^2 \\ CA^3 \end{bmatrix} = \begin{bmatrix} 1 & 0 & 0 & 0 \\ 0 & 1 & 0 & 0 \\ 0 & 0 & -9.8 & 0 \\ 0 & 0 & 0 & -9.8 \end{bmatrix}$$

Clearly, $rank(\text{O}) = 4$ and the system is observable. If we change the output to $C = [\ 0 \quad 0 \quad 1 \quad 0\ ]$, then

$$\text{O} = \begin{bmatrix} C \\ CA \\ CA^2 \\ CA^3 \end{bmatrix} = \begin{bmatrix} 0 & 0 & 1 & 0 \\ 0 & 0 & 0 & 1 \\ 0 & 0 & 19.6 & 1 \\ 0 & 0 & 19.6 & 20.6 \end{bmatrix}$$

Since $rank(\text{O}) = 2$, the system is not observable.

Let us now investigate controllability. As we will see, for linear time-invariant systems, controllability is ‘dual’ to observability.

Controllability

A linear time-invariant system

$$\dot{x} = Ax + Bu$$
$$y = Cx + Du$$

is controllable if for every state x_1, and for every $\Delta > 0$, there exists an input function $u(t)$, $t \in [0, \Delta]$ such that under this input, the state of the system moves from 0 at $t = 0$ to x_1 at $t = \Delta$.

Remarks

If a system is controllable, then for every $\Delta > 0$, there exists an input function $u(t)$, $t \in [0, \Delta]$ such that under this input, the state of the system can move from any state x_2 at $t = 0$ to any state x_1 at $t = \Delta$. Such $u(t)$, $t \in [0, \Delta]$ can be found in the following way. Let $u_1(t)$, $t \in [0, \Delta]$ be the input that moves the state of the system from 0 at $t = 0$ to x_1 at $t = \Delta$ and $u_2(t)$, $t \in [0, \Delta]$ be the input that moves the state of the system from 0 at $t = 0$ to x_2 at $t = \Delta$ (their existence is guaranteed by controllability). Then by the linearity of the system, $u(t) = u_1(t) - u_2(t)$, $t \in [0, \Delta]$ moves the state of the system from x_2 at $t = 0$ to x_1 at $t = \Delta$.

To check controllability, let the system start at state $x(0) = 0$ and we investigate its zero-state response at $t = \Delta$

$$x(\Delta) = \int_0^\Delta e^{A(\Delta-\tau)} Bu(\tau) d\tau = \int_0^\Delta e^{A\tau} Bu(\Delta - \tau) d\tau$$

This response depends on matrices A and B, not on matrices C and D. Therefore, when we discuss controllability, we discuss the controllability of (A, B). We say that a state $x_0 \neq 0$ is uncontrollable if the zero-state response $x(\Delta)$ is orthogonal to x_0, that is, $x_0^T x(\Delta) = 0$, for all $\Delta \geq 0$ and for all $u(t)$, $t \in [0, \Delta]$.

Theorem 2.3

A linear time-invariant system (A, B) is controllable if and only if it has no uncontrollable state.

Proof

(ONLY IF) If there exists $x_0 \neq 0$ such that for all $\Delta \geq 0$ and for all $u(t)$, $t \in [0, \Delta]$, the zero-state response $x(\Delta)$ is orthogonal to x_0, then obviously no control or input can move the state of the system from 0 at $t = 0$ to x_0 at $t = \Delta$. Hence, the system (A, B) is not controllable.

(IF) If there exists no $x_0 \neq 0$ such that $x_0^T x(\Delta) = 0$, for all $\Delta \geq 0$ and for all $u(t)$, $t \in [0, \Delta]$, then

$$N(\Delta) = \int_0^\Delta e^{A\tau} B (e^{A\tau} B)^T \, d\tau > 0$$

Otherwise

$$\begin{aligned}
&(\exists x_0 \neq 0) x_0^T \left(\int_0^\Delta e^{A\tau} B (e^{A\tau} B)^T d\tau\right) x_0 = 0 \\
&\Rightarrow (\exists x_0 \neq 0) \int_0^\Delta x_0^T e^{A\tau} B (x_0^T e^{A\tau} B)^T d\tau = 0 \\
&\Rightarrow (\exists x_0 \neq 0)(\forall \tau \in [0, \Delta]) x_0^T e^{A\tau} B = 0 \\
&\Rightarrow (\exists x_0 \neq 0) \int_0^\Delta x_0^T e^{A\tau} B u(\Delta - \tau) d\tau = 0 \\
&\Rightarrow (\exists x_0 \neq 0) x_0^T \int_0^\Delta e^{A\tau} B u(\Delta - \tau) d\tau = 0 \\
&\Rightarrow (\exists x_0 \neq 0) x_0^T x(\Delta) = 0
\end{aligned}$$

this is a contradiction.

Therefore, $N(\Delta)^{-1}$ exists. Now, for every state x_1, and for every $\Delta > 0$, let the input function $u(t)$, $t \in [0, \Delta]$ to be such that

$$u(\Delta - \tau) = (e^{A\tau} B)^T N(\Delta)^{-1} x_1$$

Under this input, the state of the system moves from 0 at $t = 0$ to x_1 at $t = \Delta$ because

$$\begin{aligned}
x(\Delta) &= \int_0^\Delta e^{A\tau} B u(\Delta - \tau) d\tau \\
&= \int_0^\Delta e^{A\tau} B (e^{A\tau} B)^T N(\Delta)^{-1} x_1 d\tau \\
&= \left(\int_0^\Delta e^{A\tau} B (e^{A\tau} B)^T \, d\tau\right) N(\Delta)^{-1} x_1 \\
&= N(\Delta) N(\Delta)^{-1} x_1 \\
&= x_1
\end{aligned}$$

Hence, the system (A, B) is controllable.

Q.E.D

By Theorem 2.3, checking controllability is equivalent to checking if there exists an uncontrollable state $x_0 \neq 0$, which can be done by applying the following theorem.

Theorem 2.4

$x_0 \neq 0$ is an uncontrollable state if and only if

$$x_0^{\mathrm{T}}\begin{bmatrix} B & AB & A^2B & \dots & A^{n-1}B \end{bmatrix} = 0$$

Proof

By the definition of uncontrollable state, $x_0 \neq 0$ is an uncontrollable state if and only if $x_0^{\mathrm{T}}x(\Delta) = 0$, for all $\Delta \geq 0$ and for all $u(t), t \in [0, \Delta]$.

Clearly, $x_0^{\mathrm{T}}x(\Delta) = x_0^{\mathrm{T}}\int_0^{\Delta} e^{A\tau}Bu(\Delta - \tau)\mathrm{d}\tau = \int_0^{\Delta} x_0^{\mathrm{T}}e^{A\tau}Bu(\Delta - \tau)\mathrm{d}\tau = 0$, for all $\Delta \geq 0$ and for all $u(t), t \in [0, \Delta]$ if and only if $x_0^{\mathrm{T}}e^{A\tau}B = 0$, for all $\Delta \geq 0$. By Theorem 2.2

$$x_0^{\mathrm{T}}e^{A\tau}B = 0$$

$$\Leftrightarrow B^{\mathrm{T}}e^{A^{\mathrm{T}}\tau}x_0 = 0$$

$$\Leftrightarrow \begin{bmatrix} B^{\mathrm{T}} \\ B^{\mathrm{T}}A^{\mathrm{T}} \\ B^{\mathrm{T}}(A^{\mathrm{T}})^2 \\ \dots \\ B^{\mathrm{T}}(A^{\mathrm{T}})^{n-1} \end{bmatrix} x_0 = 0$$

$$\Leftrightarrow x_0^{\mathrm{T}}\begin{bmatrix} B & AB & A^2B & \dots & A^{n-1}B \end{bmatrix} = 0$$

Q.E.D.

By Theorems 2.3 and 2.4, we conclude

$$(A, B) \text{ is controllable}$$
$$\Leftrightarrow (\forall x_0 \neq 0) x_0^{\mathrm{T}}\begin{bmatrix} B & AB & A^2B & \dots & A^{n-1}B \end{bmatrix} \neq 0$$
$$\Leftrightarrow rank\begin{bmatrix} B & AB & A^2B & \dots & A^{n-1}B \end{bmatrix} = n$$

Let us define the controllability matrix as

$$\mathrm{C} = \begin{bmatrix} B & AB & A^2B & \dots & A^{n-1}B \end{bmatrix}$$

then we conclude that a system is controllable if and only if its controllability matrix is of full rank.

$$(A, B) \text{ is controllable} \quad \Leftrightarrow \quad rank(\mathsf{C}) = n$$

We can see that the controllability and observability problems are dual. In particular, we have the following duality

$$(A, B) \text{ is controllable} \quad \Leftrightarrow \quad (A^T, B^T) \text{is observable}$$

Example 2.14

Consider the system in Example 2.10:

$$\dot{x} = \begin{bmatrix} 2 & 3 & 2 & 1 \\ -2 & -3 & 0 & 0 \\ -2 & -2 & -4 & 0 \\ -2 & -2 & -2 & -5 \end{bmatrix} x + \begin{bmatrix} 1 \\ -2 \\ 2 \\ -1 \end{bmatrix} u$$
$$y = \begin{bmatrix} 7 & 6 & 4 & 2 \end{bmatrix} x$$

Its controllability matrix is

$$\mathsf{C} = \begin{bmatrix} B & AB & A^2B & A^3B \end{bmatrix} = \begin{bmatrix} 1 & -1 & 1 & -1 \\ -2 & 4 & -10 & 28 \\ 2 & -6 & 18 & -54 \\ -1 & 3 & -9 & 27 \end{bmatrix}$$

Since $rank(\mathsf{C}) = 2 < 4$, the system is not controllable. In fact, two states are controllable and two states are not.

2.5 POLE PLACEMENT BY STATE FEEDBACK

As we will show in the next two chapters, stability and optimality of a system are closely related to the location of poles or eigenvalues of the system. If the poles are not in the desired locations, can we move and place them in the right places? This is the problem to be solved by pole placement. Pole placement can be achieved by feedback control. In this section, we assume that all states are available for feedback control. Hence, we consider only the state equation

$$\dot{x} = Ax + Bu$$

The poles of this system are eigenvalues of A, denoted by $\lambda(A)$. We use state feedback control $u = Kx + v$, where Kx is linear state feedback and v is some external input. Under this feedback control, the controlled system is given by

$$\dot{x} = (A + BK)x + Bv$$

The poles of the controlled system are $\lambda(A+BK)$. The question is whether or not we can relocate the poles to arbitrary locations in the complex plane as we desire. The answer is that this can be done if and only if (A, B) is controllable. To see this, let us first consider single-input–single-output systems with transfer function of the form.

$$G(s) = \frac{b_0 + b_1 s + \cdots + b_{n-1} s^{n-1}}{a_0 + a_1 s + \cdots + a_{n-1} s^{n-1} + s^n}$$

We can realize this system in state space representation in the following form:

$$\dot{x} = \begin{bmatrix} 0 & 1 & \dots & 0 \\ \dots & & & \\ 0 & 0 & \dots & 1 \\ -a_0 & -a_1 & \dots & -a_{n-1} \end{bmatrix} x + \begin{bmatrix} 0 \\ \dots \\ 0 \\ 1 \end{bmatrix} u \tag{2.7}$$
$$y = \begin{bmatrix} b_0 & b_1 & \cdots & b_{n-1} \end{bmatrix} x$$

The above state space representation is called the controllable canonical form. The characteristic equation of the system is

$$\varphi(s) = s^n + a_{n-1} s^{n-1} + \cdots + a_1 s + a_0$$

Our goal is to find a state feedback

$$u = Kx + v = \begin{bmatrix} k_0 & k_1 & \dots & k_{n-1} \end{bmatrix} x + v$$

so that the poles of the controlled system is in the desired locations represented by the desired characteristic equation

$$\overline{\varphi}(s) = s^n + \overline{a}_{n-1} s^{n-1} + \cdots + \overline{a}_1 s + \overline{a}_0$$

This can be achieved by letting the feedback matrix be

$$K = \begin{bmatrix} k_0 & k_1 & \dots & k_{n-1} \end{bmatrix} = \begin{bmatrix} a_0 - \overline{a}_0 & a_1 - \overline{a}_1 & \dots & a_{n-1} - \overline{a}_{n-1} \end{bmatrix}$$

To prove this, let us substitute $u = Kx + v$ into the state equation in (2.7),

$$\dot{x} = \begin{bmatrix} 0 & 1 & \dots & 0 \\ \dots & & & \\ 0 & 0 & \dots & 1 \\ -a_0 & -a_1 & \dots & -a_{n-1} \end{bmatrix} x + \begin{bmatrix} 0 \\ \dots \\ 0 \\ 1 \end{bmatrix} \left(\begin{bmatrix} k_0 & k_1 & \dots & k_{n-1} \end{bmatrix} x + v \right)$$

$$= \begin{bmatrix} 0 & 1 & \dots & 0 \\ \dots & & & \\ 0 & 0 & \dots & 1 \\ -a_0 + k_0 & -a_1 + k_1 & \dots & -a_{n-1} + k_{n-1} \end{bmatrix} x + \begin{bmatrix} 0 \\ \dots \\ 0 \\ 1 \end{bmatrix} v$$

The characteristic equation of the above controlled system is

$$s^n + (a_{n-1} - k_{n-1})s^{n-1} + \dots + (a_1 - k_1)s + (a_0 - k_0)$$
$$= s^n + \overline{a}_{n-1}s^{n-1} + \dots + \overline{a}_1 s + \overline{a}_0$$

Therefore, if the system is represented in the controllable canonical form, it is straightforward to design a state feedback to place its poles in arbitrary locations in the complex plane to achieve stability or optimality. The next question is whether a system can be represented in the controllable canonical form. To answer this question, let us first verify that the controllable canonical form is always controllable. For

$$A = \begin{bmatrix} 0 & 1 & \dots & 0 \\ \dots & & & \\ 0 & 0 & \dots & 1 \\ -a_0 & -a_1 & \dots & -a_{n-1} \end{bmatrix} \qquad B = \begin{bmatrix} 0 \\ \dots \\ 0 \\ 1 \end{bmatrix}$$

its controllability matrix is

$$\mathsf{C} = \begin{bmatrix} 0 & 0 & \dots & 1 \\ \dots & & & \\ 0 & 1 & \dots & * \\ 1 & -a_{n-1} & \dots & * \end{bmatrix}$$

Since the elements on the diagonal are 1 and all elements above the diagonal are 0, the determinant of C is -1. It is independent of the elements below the diagonal, which is denoted by *.

Secondly, we can show that any similarity transformation does not change the controllability of a system. For $\widetilde{A} = T^{-1}AT$, $\widetilde{B} = T^{-1}B$, its controllability matrix is

$$\widetilde{\mathsf{C}} = \begin{bmatrix} \widetilde{B} & \widetilde{A}\widetilde{B} & \widetilde{A}^2\widetilde{B} & \dots & \widetilde{A}^{n-1}\widetilde{B} \end{bmatrix}$$

$$
\begin{aligned}
&= [T^{-1}B \quad T^{-1}ATT^{-1}B \quad (T^{-1}AT)^2T^{-1}B \quad \dots \quad (T^{-1}AT)^{n-1}T^{-1}B] \\
&= [T^{-1}B \quad T^{-1}AB \quad T^{-1}A^2B \quad \dots \quad T^{-1}A^{n-1}B] \\
&= T^{-1}[B \quad AB \quad A^2B \quad \dots \quad A^{n-1}B] \\
&= T^{-1}\mathsf{C}
\end{aligned}
$$

Hence, $rank(\widetilde{\mathsf{C}}) = rank(\mathsf{C})$. In other words, (A, B) is controllable if and only if $(\widetilde{A}, \widetilde{B})$ is controllable.

The above two results show that a system can be transformed into the controllable canonical form to place its poles arbitrarily if and only if the system is controllable.

If system (A, B) is controllable, then the matrix T_c that transforms (A, B) into its controllable canonical form, denoted by (A_c, B_c), can be found as follows. $\widetilde{\mathsf{C}} = T_c^{-1}\mathsf{C}$ implies $T_c = \widetilde{\mathsf{C}}\mathsf{C}^{-1}$. Hence

$$
T_c = [B \quad AB \quad A^2B \quad \cdots \quad A^{n-1}B][B_c \quad A_cB_c \quad A_c^2B_c \quad \cdots \quad A_c^{n-1}B_c]^{-1}
$$

After the transformation, the system in the controllable canonical form is given by

$$
\dot{z} = A_c z + B_c u
$$

We can design the state feedback $u = K_c z + v$ as discussed above. Since $x = T_c z$, the state feedback for x is given by $u = K_c T_c^{-1} x + v = Kx + v$.

Based on the above discussion, we derive the following procedure for pole placement.

Procedure 1 (Pole placement of single-input systems)

Given: a controllable system (A, B) and a desired characteristic polynomial

$$
\overline{\varphi}(s) = s^n + \overline{a}_{n-1}s^{n-1} + \cdots + \overline{a}_1 s + \overline{a}_0
$$

1. Find the characteristic polynomial of A

$$
\varphi(s) = s^n + a_{n-1}s^{n-1} + \cdots + a_1 s + a_0
$$

2. Write the controllable canonical form (A_c, B_c)
3. Find the transform matrix

$$
T_c = [B \quad AB \quad A^2B \quad \cdots \quad A^{n-1}B][B_c \quad A_cB_c \quad A_c^2B_c \quad \cdots \quad A_c^{n-1}B_c]^{-1}
$$

4. Determine the feedback matrix for (A_c, B_c)

$$K_c = \begin{bmatrix} a_0 - \overline{a}_0 & a_1 - \overline{a}_1 & \dots & a_{n-1} - \overline{a}_{n-1} \end{bmatrix}$$

5. Determine the feedback matrix for (A, B)

$$K = K_c T_c^{-1}$$

6. The state feedback is given by $u = Kx + v$.

Example 2.15

Consider the inverted pendulum in Example 2.8:

$$\dot{x} = \begin{bmatrix} 0 & 1 & 0 & 0 \\ 0 & 0 & -9.8 & 0 \\ 0 & 0 & 0 & 1 \\ 0 & 0 & 19.6 & 1 \end{bmatrix} x + \begin{bmatrix} 0 \\ 1 \\ 0 \\ -1 \end{bmatrix} u$$

We want to find a state feedback to place the poles at -5, -10, $-2+j2$, $-2-j2$. Let us first check if the system is controllable.

$$\mathsf{C} = \begin{bmatrix} 0 & 1 & 0 & 9.8 \\ 1 & 0 & 9.8 & 0 \\ 0 & -1 & 0 & -19.6 \\ -1 & 0 & -19.6 & 0 \end{bmatrix}$$

Since $rank(\mathsf{C}) = 4$, the system is controllable. The characteristic polynomial of A is

$$\varphi(s) = s^4 - 19.6s^2$$

The controllable canonical form is

$$A_c = \begin{bmatrix} 0 & 1 & 0 & 0 \\ 0 & 0 & 1 & 0 \\ 0 & 0 & 0 & 1 \\ 0 & 0 & 19.6 & 0 \end{bmatrix} \qquad B_c = \begin{bmatrix} 0 \\ \dots \\ 0 \\ 1 \end{bmatrix}$$

The controllability matrix of (A_c, B_c) is

$$\mathsf{C}_c = \begin{bmatrix} 0 & 0 & 0 & 1 \\ 0 & 0 & 1 & 0 \\ 0 & 1 & 0 & 19.6 \\ 1 & 0 & 19.6 & 0 \end{bmatrix}$$

The transform matrix is

$$T_c = CC_c^{-1} = \begin{bmatrix} -9.8 & 0 & 1 & 0 \\ 0 & -9.8 & 0 & 1 \\ 0 & 0 & -1 & 0 \\ 0 & 0 & 0 & -1 \end{bmatrix}$$

The desirable characteristic equation is

$$\begin{aligned}\overline{\varphi}(s) &= (s+5)(s+10)(s+2+j2)(s+2-j2) \\ &= s^4 + 19s^3 + 118s^2 + 320s + 400\end{aligned}$$

The feedback matrix for (A_c, B_c) is

$$\begin{aligned}K_c &= [\ 0-400 \quad 0-320 \quad -19.6-118 \quad 0-19\ ] \\ &= [\ -400 \quad -320 \quad -137.6 \quad -19\ ]\end{aligned}$$

The feedback matrix for (A, B) is

$$K = K_c T_c^{-1} = [\ 40.8163 \quad 32.6531 \quad 178.4163 \quad 51.6531\ ]$$

Now we know how to place poles for single-input systems, let us consider multi-input systems.

Before we discuss pole placement for multi-input systems, we first note that for single-input systems, the feedback matrix K is unique, because K is a $1 \times n$ row vector with n elements to be determined for n poles to be placed. However, for multi-input systems, this is no longer the case. If $u \in R^m$ has m dimensions, then K is a $m \times n$ matrix with $m \times n$ elements. There are more elements than n poles so the choice of elements is not unique.

To overcome this difficulty, we need to restrict the solution and 'convert' a multi-input system into a single-input system. Write

$$BK = \begin{bmatrix} b_{11} & \dots & b_{1m} \\ \dots & & \\ b_{n1} & \dots & b_{nm} \end{bmatrix} \begin{bmatrix} k_{11} & \dots & k_{1n} \\ \dots & & \\ k_{m1} & \dots & k_{mn} \end{bmatrix} = \begin{bmatrix} b_{11} & \dots & b_{1m} \\ \dots & & \\ b_{n1} & \dots & b_{nm} \end{bmatrix} \begin{bmatrix} r_1 \\ \dots \\ r_m \end{bmatrix}$$
$$[k_1 \dots k_n]$$

Define a new B matrix as

$$\overline{B} = \begin{bmatrix} b_{11} & \dots & b_{1m} \\ \dots & & \\ b_{n1} & \dots & b_{nm} \end{bmatrix} \begin{bmatrix} r_1 \\ \dots \\ r_m \end{bmatrix}$$

Then $\overline{B}$ is a $n \times 1$ column vector. $(A, \overline{B})$ is a single-input system. If we can find a feedback matrix $[k_1 \ \ldots \ k_n]$ to place the poles in the desired locations, then

$$K = \begin{bmatrix} r_1 \\ \ldots \\ r_m \end{bmatrix} [k_1 \ldots k_n]$$

is the feedback matrix for (A, B). This is because

$$A + BK = A + \begin{bmatrix} b_{11} & \ldots & b_{1m} \\ \ldots & & \\ b_{n1} & \ldots & b_{nm} \end{bmatrix} \begin{bmatrix} r_1 \\ \ldots \\ r_m \end{bmatrix} [\, k_1 \ \ldots \ k_n \,] = A + \overline{B}[k_1 \ldots k_n]$$

Hence, if we pick some suitable

$$\begin{bmatrix} r_1 \\ \ldots \\ r_m \end{bmatrix}$$

then we can convert a multi-input system into a single-input system. So what

$$\begin{bmatrix} r_1 \\ \ldots \\ r_m \end{bmatrix}$$

shall we pick? The criterion for picking

$$\begin{bmatrix} r_1 \\ \ldots \\ r_m \end{bmatrix}$$

is such that $(A, \overline{B})$ is controllable as long as (A, B) is controllable.

In summary, we have the following procedure for pole placement of multi-input systems.

Procedure 2 (Pole placement of multi-input systems)

Given: a controllable system (A, B) and a desired characteristic polynomial

$$\overline{\varphi}(s) = s^n + \overline{a}_{n-1}s^{n-1} + \ldots + \overline{a}_1 s + \overline{a}_0$$

1. Randomly pick

$$\begin{bmatrix} r_1 \\ \ldots \\ r_m \end{bmatrix}$$

2. Check if $(A, \overline{B})$ is controllable. If yes, continue, otherwise, go back to Step 1.
3. Determine the feedback matrix $[k_1 \ldots k_n]$ for $(A, \overline{B})$.
4. Determine the feedback matrix for (A, B)

$$K = \begin{bmatrix} r_1 \\ \cdots \\ r_m \end{bmatrix} [k_1 \ldots k_n]$$

Example 2.16

Consider the following multi-input system

$$\dot{x} = \begin{bmatrix} 0 & 1 & 0 \\ 0 & 0 & 0 \\ 0 & 0 & 2 \end{bmatrix} x + \begin{bmatrix} 0 & 0 \\ 1 & 0 \\ 0 & 1 \end{bmatrix} u$$

We want to find a state feedback to place the poles at -5, -10, -20. Let us first check if the system is controllable.

$$\mathsf{C} = \begin{bmatrix} 0 & 0 & 1 & 0 & 0 & 0 \\ 1 & 0 & 0 & 0 & 0 & 0 \\ 0 & 1 & 0 & 2 & 0 & 4 \end{bmatrix}$$

Since $rank(\mathsf{C}) = 3$, the system is controllable. Let us pick

$$\begin{bmatrix} r_1 \\ r_2 \end{bmatrix} = \begin{bmatrix} -2 \\ 3 \end{bmatrix}$$

The converted system is

$$A = \begin{bmatrix} 0 & 1 & 0 \\ 0 & 0 & 0 \\ 0 & 0 & 2 \end{bmatrix} \qquad \overline{B} = \begin{bmatrix} 0 \\ -2 \\ 3 \end{bmatrix}$$

which is controllable. The characteristic polynomial of A is

$$\varphi(s) = s^3 - 2s^2$$

The controllable canonical form is

$$A_c = \begin{bmatrix} 0 & 1 & 0 \\ 0 & 0 & 1 \\ 0 & 0 & 2 \end{bmatrix} \qquad B_c = \begin{bmatrix} 0 \\ 0 \\ 1 \end{bmatrix}$$

The controllability matrix of (A_c, B_c) is

$$\mathsf{C}_c = \begin{bmatrix} 0 & 0 & 1 \\ 0 & 1 & 2 \\ 1 & 2 & 4 \end{bmatrix}$$

The transform matrix is

$$T_c = \begin{bmatrix} \overline{B} & A\overline{B} & A^2\overline{B} \end{bmatrix} \mathsf{C}_c^{-1} = \begin{bmatrix} 4 & -2 & 0 \\ 0 & 4 & 0 \\ 0 & 0 & 3 \end{bmatrix}$$

The desirable characteristic equation is

$$\overline{\varphi}(s) = (s+5)(s+10)(s+20) = s^3 + 35s^2 + 350s + 1000$$

The feedback matrix for (A_c, B_c) is

$$K_c = \begin{bmatrix} -1000 & -350 & -37 \end{bmatrix}$$

The feedback matrix for (A, B) is

$$K = \begin{bmatrix} r_1 \\ r_2 \end{bmatrix} K_c T_c^{-1} = \begin{bmatrix} 500 & 425 & 308 \\ -750 & -637.5 & -462 \end{bmatrix}$$

2.6 POLE PLACEMENT USING OBSERVER

If a system is controllable, then state feedback can be used to place poles to arbitrary locations in the complex plane to achieve stability and optimality. However, in practice, not all states can be directly measured. Therefore, in most applications, it is not possible to use state feedback directly. What we can do is to estimate the state variables and to use feedback based on estimates of states. In this section, we present the following results. (1) We can estimate the state of a system if and only if the system is observable. (2) If the system is observable, we can construct an observer to estimate the state. (3) Pole placement for state feedback and observer design can be done separately.

Let us first consider a naïve approach to state estimation. Suppose that a system (A, B) is driven by an input u. We cannot measure its state x. To estimate x, we can build a duplicate system (say, in a computer) and let it be driven by the same input, as shown in Figure 2.3.

In the figure, x indicates the actual states and $\hat{x}$ the estimates. To see how good the estimates are, let us investigate the estimation error $\tilde{x} = x - \hat{x}$, which satisfies the following differential equation.

$$\dot{\tilde{x}} = \dot{x} - \dot{\hat{x}} = Ax + Bu - (A\hat{x} + Bu) = Ax - A\hat{x} = A\tilde{x}$$

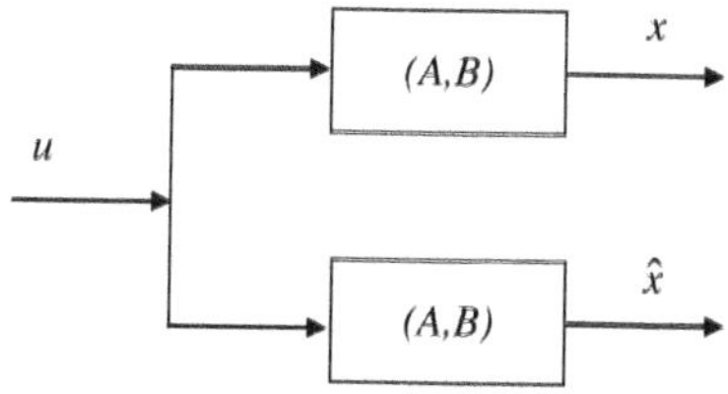

Figure 2.3 A naïve approach to state estimation.

Clearly, the dynamics of $\tilde{x}$ is determined by the eigenvalues of A, which may not be satisfactory. One way to get a better estimation is to find the error of estimation and feed it back to improve the estimation. However, we cannot measure $\tilde{x}$ directly, but we can measure $C\tilde{x} = Cx - C\hat{x} = y - C\hat{x}$. So, let us use this feedback and construct an 'observer', as shown in Figure 2.4.

The dynamics of the observer in Figure 2.4 is given by

$$\begin{aligned} \dot{\hat{x}} &= A\hat{x} + Bu - G(y - \hat{y}) \\ &= A\hat{x} + Bu - G(y - C\hat{x}) \\ &= (A + GC)\hat{x} + Bu - Gy \end{aligned}$$

where G is the observer matrix to feedback the error. Now, the estimation error $\tilde{x}$ satisfies the following differential equation.

$$\begin{aligned} \dot{\tilde{x}} &= \dot{x} - \dot{\hat{x}} \\ &= Ax + Bu - ((A + GC)\hat{x} + Bu - Gy) \\ &= Ax - (A + GC)\hat{x} + Gy \\ &= Ax - (A + GC)\hat{x} + GCx \end{aligned}$$

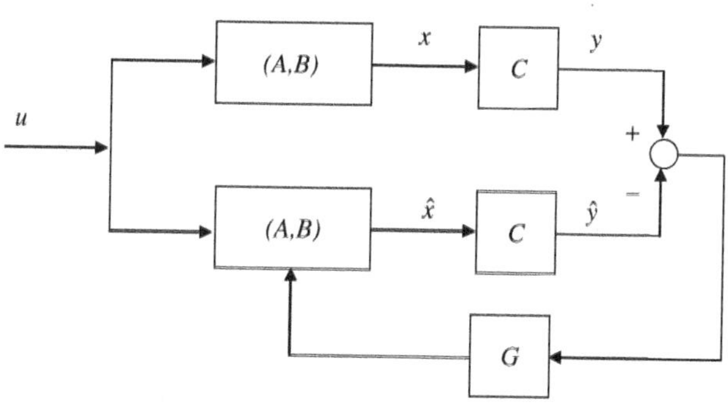

Figure 2.4 State estimation using an observer.

$$= (A+GC)x - (A+GC)\hat{x}$$
$$= (A+GC)\tilde{x}$$

The dynamics of $\tilde{x}$ is determined by the eigenvalues of $A+GC$. Since the observer matrix G can be selected in the design process, we can place the poles of the observer in the right locations to meet the desired performance requirements. The pole placement problem for the observer can be solved as a 'dual' of the pole placement problem for the state feedback. Note that a matrix and its transpose have the same eigenvalues

$$\lambda(A+GC) = \lambda(A^{\mathrm{T}} + C^{\mathrm{T}}G^{\mathrm{T}})$$

Hence, if we view $(A^{\mathrm{T}}, C^{\mathrm{T}})$ as (A, B) and G^{T} as K, then the pole placement problem for the observer is transferred to the pole placement problem for the state feedback. By duality and our previous results

$$\begin{aligned} &\text{The poles } A\text{–}GC \text{ of can be arbitrarily assigned} \\ &\Leftrightarrow (A^T, C^T) \text{ is controllable} \\ &\Leftrightarrow (A, C) \text{is observable.} \end{aligned}$$

Procedure 3 (Full-order observer design)

Given: an observable system (A, C) and a desired characteristic polynomial of the observer $\overline{\varphi}(s) = s^n + \overline{a}_{n-1}s^{n-1} + \ldots + \overline{a}_1 s + \overline{a}_0$

1. Solve pole placement problem for state feedback of the dual system $(A^{\mathrm{T}}, C^{\mathrm{T}})$ to obtain K.
2. Let the observer matrix $G = K^{\mathrm{T}}$.
3. Construct the observer $\dot{\hat{x}} = (A+GC)\hat{x} + Bu - Gy$.

Example 2.17

Consider the following system

$$\dot{x} = \begin{bmatrix} -2 & 0 & 0 \\ 0 & 0 & -2 \\ 0 & 0 & -1 \end{bmatrix} x + \begin{bmatrix} 2 \\ 0 \\ -1 \end{bmatrix} u$$
$$y = \begin{bmatrix} 2 & 1 & 1 \end{bmatrix} x$$

We want to design an observer with poles at -20, $-10+j10$, $-10-j10$. Let us first check if the system is observable.

$$\mathsf{O} = \begin{bmatrix} 2 & 1 & 1 \\ -4 & 0 & -3 \\ 8 & 0 & 3 \end{bmatrix}$$

Since $rank(\mathsf{O}) = 3$, the system is observable. The dual system is controllable and is given by

$$A^{\mathrm{T}} = \begin{bmatrix} -2 & 0 & 0 \\ 0 & 0 & 0 \\ 0 & -2 & -1 \end{bmatrix} \qquad C^{\mathrm{T}} = \begin{bmatrix} 2 \\ 1 \\ 1 \end{bmatrix}$$

The characteristic polynomial of A^{T} is

$$\varphi(s) = s^3 + 3s^2 + 2s$$

The controllable canonical form is

$$A_c = \begin{bmatrix} 0 & 1 & 0 \\ 0 & 0 & 1 \\ 0 & -2 & -3 \end{bmatrix} \qquad B_c = \begin{bmatrix} 0 \\ 0 \\ 1 \end{bmatrix}$$

The controllability matrix of (A_c, B_c) is

$$\mathsf{C}_c = \begin{bmatrix} 0 & 0 & 1 \\ 0 & 1 & -3 \\ 1 & -3 & 7 \end{bmatrix}$$

The transform matrix is

$$T_c = \mathsf{O}^{\mathrm{T}}\mathsf{C}_c^{-1} = \begin{bmatrix} 0 & 2 & 2 \\ 2 & 3 & 1 \\ -4 & 0 & 1 \end{bmatrix}$$

The desirable characteristic equation is

$$\overline{\varphi}(s) = (s+20)(s+10+j10)(s+10-j10) = s^3 + 40s^2 + 600s + 4000$$

The feedback matrix for (A_c, B_c) is

$$K_c = [-4000 \quad -598 \quad -37]$$

The feedback matrix for $(A^{\mathrm{T}}, C^{\mathrm{T}})$ is

$$K = K_c T_c^{-1} = [-738 \quad 292.7 \quad 1146.3]$$

The observer matrix is

$$G = K^{\mathrm{T}} = \begin{bmatrix} -738 \\ 292.7 \\ 1146.3 \end{bmatrix}$$

The observer is

$$\dot{\hat{x}} = (A + GC)\hat{x} + Bu - Gy$$
$$= \begin{bmatrix} -1478 & -738 & -738 \\ 585.4 & 292.7 & 290.7 \\ 2292.6 & 1146.3 & 1145.3 \end{bmatrix} \hat{x} + \begin{bmatrix} 2 \\ 0 \\ -1 \end{bmatrix} u - \begin{bmatrix} -738 \\ 292.7 \\ 1146.3 \end{bmatrix} y$$

Example 2.18

Consider the following multi-output system

$$\dot{x} = \begin{bmatrix} 0 & 0 & 0 \\ 1 & 0 & 0 \\ 0 & 0 & 2 \end{bmatrix} x + \begin{bmatrix} 0 \\ 1 \\ 1 \end{bmatrix} u$$

$$y = \begin{bmatrix} 0 & 1 & 0 \\ 0 & 0 & 1 \end{bmatrix} x$$

We want to design an observer with poles at -5, -10, -20. The pole placement for state feedback of the dual system

$$A^{\mathrm{T}} = \begin{bmatrix} 0 & 1 & 0 \\ 0 & 0 & 0 \\ 0 & 0 & 2 \end{bmatrix} \qquad C^{\mathrm{T}} = \begin{bmatrix} 0 & 0 \\ 1 & 0 \\ 0 & 1 \end{bmatrix}$$

was solved in Example 2.16. The feedback matrix is

$$K = \begin{bmatrix} 500 & 425 & 308 \\ -750 & -637.5 & -462 \end{bmatrix}$$

By the duality, the observer matrix is

$$G = K^{\mathrm{T}} = \begin{bmatrix} 500 & -750 \\ 425 & -637.5 \\ 308 & -462 \end{bmatrix}$$

The observer is

$$\dot{\hat{x}} = (A + GC)\hat{x} + Bu - Gy$$
$$= \begin{bmatrix} 0 & 500 & -750 \\ 1 & 425 & -637.5 \\ 0 & 308 & -460 \end{bmatrix} \hat{x} + \begin{bmatrix} 0 \\ 1 \\ 1 \end{bmatrix} u - \begin{bmatrix} 500 & -750 \\ 425 & -637.5 \\ 308 & -462 \end{bmatrix} y$$

For an observer, its inputs are u and y, and its output is $\hat{x}$. The purpose of introducing the observer is to use its output to do state feedback. This is illustrated in Figure 2.5.

The overall closed-loop system is described by the following equations

$$\begin{aligned} \dot{x} &= Ax + Bu \\ y &= Cx \\ \dot{\hat{x}} &= (A + GC)\hat{x} + Bu - Gy \\ u &= K\hat{x} + v \end{aligned} \tag{2.8}$$

Since feedback control is based on the state estimates $\hat{x}$, not on the states themselves, we would like to know if the poles of the above system are the same as the poles of the system using direct state feedback:

$$\begin{aligned} \dot{x} &= Ax + Bu \\ u &= Kx + v \end{aligned} \tag{2.9}$$

To find poles of the system in (2.8), let us rewrite the state equations in terms of x and $\tilde{x} = x - \hat{x}$ as follows.

$$\begin{aligned} \dot{x} &= Ax + Bu \\ &= Ax + B(K\hat{x} + v) \\ &= Ax + BK(x - \tilde{x}) + Bv \\ &= (A + BK)x - BK\tilde{x} + Bv \\ \dot{\tilde{x}} &= \dot{x} - \dot{\hat{x}} \\ &= Ax + Bu - ((A + GC)\hat{x} + Bu - Gy) \\ &= Ax + Gy - (A + GC)\hat{x} \\ &= Ax + GCx - (A + GC)\hat{x} \end{aligned}$$

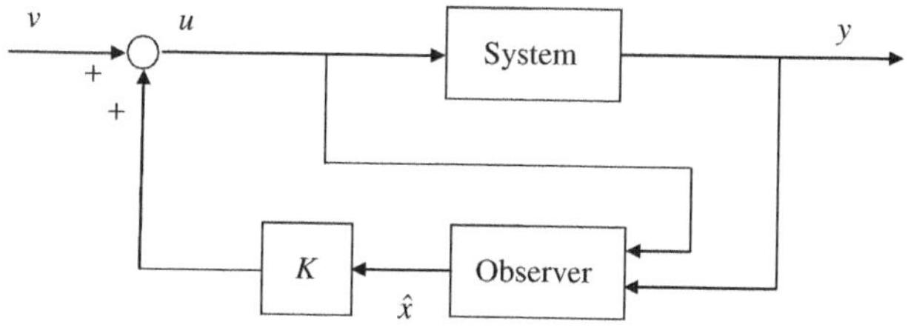

Figure 2.5 Feedback control using an observer.

$$= (A+GC)(x-\hat{x})$$
$$= (A+GC)\tilde{x}$$

Putting them in the matrix form, we have

$$\begin{bmatrix} \dot{x} \\ \dot{\tilde{x}} \end{bmatrix} = \begin{bmatrix} A+BK & -BK \\ 0 & A+GC \end{bmatrix} \begin{bmatrix} x \\ \tilde{x} \end{bmatrix} + \begin{bmatrix} B \\ 0 \end{bmatrix} v$$

Since

$$\lambda\left(\begin{bmatrix} A+BK & -BK \\ 0 & A+GC \end{bmatrix}\right) = \lambda(A+BK) \cup \lambda(A+GC)$$

the poles of the closed-loop system (2.8) consist of poles of system (2.9) and poles of the observer. In other words, the use of the observer does not change the poles determined by the state feedback. This separation principle allows us to design state feedback and observer separately.

Example 2.19

For the following system

$$\dot{x} = \begin{bmatrix} 0 & 1 & 0 \\ 0 & 0 & 1 \\ -2 & 3 & 5 \end{bmatrix} x + \begin{bmatrix} 0 \\ 0 \\ 1 \end{bmatrix} u$$
$$y = \begin{bmatrix} 1 & 0 & 0 \end{bmatrix} x$$

we would like to design an observer-based feedback control such that: (1) the closed-loop system has poles at -5, $-1+j$, $-1-j$; and (2) the observer has poles at -20, $-10+j10$, $-10-j10$.

Since the system is in the controllable canonical form, finding K is straightforward. The characteristic equation of A is

$$\varphi(s) = s^3 - 5s^2 - 3s + 2$$

The desired characteristic equation is

$$\overline{\varphi}(s) = (s+5)(s+1+j)(s+1-j) = s^3 + 7s^2 + 12s + 10$$

The feedback matrix is

$$K = \begin{bmatrix} 2-10 & -3-12 & -5-7 \end{bmatrix} = \begin{bmatrix} -8 & -15 & -12 \end{bmatrix}$$

Determining G is more involved. Since we have performed the procedure twice in Examples 2.17 and 2.18, we will leave it to the reader. We only provide the result here:

$$G = \begin{bmatrix} -45 \\ -828 \\ -8273 \end{bmatrix}$$

Finally, the overall closed-loop system is given by

$$\dot{x} = \begin{bmatrix} 0 & 1 & 0 \\ 0 & 0 & 1 \\ -2 & 3 & 5 \end{bmatrix} x + \begin{bmatrix} 0 \\ 0 \\ 1 \end{bmatrix} u$$

$$y = \begin{bmatrix} 1 & 0 & 0 \end{bmatrix} x,$$

$$\dot{\hat{x}} = \begin{bmatrix} -45 & 1 & 0 \\ -828 & 0 & 1 \\ -8275 & 3 & 3 \end{bmatrix} \hat{x} + \begin{bmatrix} 0 \\ 0 \\ 1 \end{bmatrix} u - \begin{bmatrix} -45 \\ -828 \\ -8273 \end{bmatrix} y$$

$$u = \begin{bmatrix} -8 & -15 & -12 \end{bmatrix} \hat{x} + v$$

In the above example, the observer estimates three states: x_1, x_2, x_3. However, x_1 can be measured directly because $y = x_1$. There is no reason to estimate x_1. So let us design a reduced-order observer which estimates only states that cannot be directly measured. Divide states x into those that can be directly measured, denoted by x_m and those that cannot, denoted by x_u. Assume

$$\begin{bmatrix} \dot{x}_m \\ \dot{x}_u \end{bmatrix} = \begin{bmatrix} A_{mm} & A_{mu} \\ A_{um} & A_{uu} \end{bmatrix} \begin{bmatrix} x_m \\ x_u \end{bmatrix} + \begin{bmatrix} B_m \\ B_u \end{bmatrix} u$$

$$y = \begin{bmatrix} I & 0 \end{bmatrix} \begin{bmatrix} x_m \\ x_u \end{bmatrix} = x_m$$

If the state-space representation is not in the above form, we can use a similarity transformation to transform it into the above form. Rewrite the state equation as

$$\begin{aligned} \dot{x}_u &= A_{uu}x_u + A_{um}x_m + B_u u \\ \dot{x}_m - A_{mm}x_m - B_m u &= A_{mu}x_u \end{aligned} \tag{2.10}$$

Define

$$\overline{Bu} = A_{um}x_m + B_u u$$

$$\overline{y} = \dot{x}_m - A_{mm}x_m - B_m u$$

Then Equation (2.10) becomes

$$\dot{x}_{\mathrm{u}} = A_{\mathrm{uu}} x_{u} + \overline{B}\,\overline{u}$$
$$\overline{y} = A_{\mathrm{mu}} x_{\mathrm{u}}$$

We can design an observer for the above system using the approach described earlier.

$$\dot{\hat{x}}_{\mathrm{u}} = (A_{\mathrm{uu}} + G_{\mathrm{u}} A_{\mathrm{mu}})\hat{x}_{\mathrm{u}} + \overline{B}\,\overline{u} - G_{\mathrm{u}}\overline{y}$$

Substitute $\overline{B}\,\overline{u}$ and $\overline{y}$:

$$\begin{aligned}\dot{\hat{x}}_{\mathrm{u}} &= (A_{\mathrm{uu}} + G_{\mathrm{u}} A_{\mathrm{mu}})\hat{x}_{\mathrm{u}} + A_{\mathrm{um}} x_{\mathrm{m}} + B_{\mathrm{u}} u - G_{\mathrm{u}}(\dot{x}_{\mathrm{m}} - A_{\mathrm{mm}} x_{\mathrm{m}} - B_{\mathrm{m}} u)\\ &= (A_{\mathrm{uu}} + G_{\mathrm{u}} A_{\mathrm{mu}})\hat{x}_{\mathrm{u}} + A_{\mathrm{um}} x_{\mathrm{m}} + B_{\mathrm{u}} u - G_{\mathrm{u}}\dot{x}_{\mathrm{m}} + G_{\mathrm{u}} A_{\mathrm{mm}} x_{\mathrm{m}} + G_{\mathrm{u}} B_{\mathrm{m}} u\end{aligned}$$

To remove the derivative in the right-hand side of the above equation, define

$$w = \hat{x}_{\mathrm{u}} + G_{\mathrm{u}} x_{\mathrm{m}}$$

In terms of state variable w, the reduced-order observer equation is

$$\begin{aligned}\dot{w} &= (A_{\mathrm{uu}} + G_{\mathrm{u}} A_{\mathrm{mu}})\hat{x}_{\mathrm{u}} + A_{\mathrm{um}} x_{\mathrm{m}} + B_{\mathrm{u}} u + G_{\mathrm{u}} A_{\mathrm{mm}} x_{\mathrm{m}} + G_{\mathrm{u}} B_{\mathrm{m}} u\\ &= (A_{\mathrm{uu}} + G_{\mathrm{u}} A_{\mathrm{mu}})(w - G_{\mathrm{u}} x_{\mathrm{m}}) + A_{\mathrm{um}} x_{\mathrm{m}} + B_{\mathrm{u}} u + G_{\mathrm{u}} A_{\mathrm{mm}} x_{\mathrm{m}} + G_{\mathrm{u}} B_{\mathrm{m}} u\\ &= (A_{\mathrm{uu}} + G_{\mathrm{u}} A_{\mathrm{mu}})w - (A_{\mathrm{uu}} + G_{\mathrm{u}} A_{\mathrm{mu}})G_{\mathrm{u}} x_{\mathrm{m}} + A_{\mathrm{um}} x_{\mathrm{m}} + B_{\mathrm{u}} u + G_{\mathrm{u}} A_{\mathrm{mm}} x_{\mathrm{m}} + G_{\mathrm{u}} B_{\mathrm{m}} u\\ &= (A_{\mathrm{uu}} + G_{\mathrm{u}} A_{\mathrm{mu}})w + (A_{\mathrm{um}} + G_{\mathrm{u}} A_{\mathrm{mm}} - A_{\mathrm{uu}} G_{\mathrm{u}} - G_{\mathrm{u}} A_{\mathrm{mu}} G_{\mathrm{u}})x_{\mathrm{m}} + (B_{\mathrm{u}} + G_{\mathrm{u}} B_{\mathrm{m}})u\\ &= (A_{\mathrm{uu}} + G_{\mathrm{u}} A_{\mathrm{mu}})w + (A_{\mathrm{um}} + G_{\mathrm{u}} A_{\mathrm{mm}} - A_{\mathrm{uu}} G_{\mathrm{u}} - G_{\mathrm{u}} A_{\mathrm{mu}} G_{\mathrm{u}})y + (B_{\mathrm{u}} + G_{\mathrm{u}} B_{\mathrm{m}})u\end{aligned}$$

The state feedback can be derived as

$$\begin{aligned}u &= \begin{bmatrix} K_{\mathrm{m}} & K_{\mathrm{u}} \end{bmatrix}\begin{bmatrix} x_{\mathrm{m}} \\ \hat{x}_{\mathrm{u}} \end{bmatrix} + Bv\\ &= K_{\mathrm{m}} x_{\mathrm{m}} + K_{\mathrm{u}}\hat{x}_{\mathrm{u}} + Bv\\ &= K_{\mathrm{m}} x_{\mathrm{m}} + K_{\mathrm{u}}(w - G_{\mathrm{u}} x_{\mathrm{m}}) + Bv\\ &= (K_{\mathrm{m}} - K_{\mathrm{u}} G_{\mathrm{u}})x_{\mathrm{m}} + K_{\mathrm{u}} w + Bv\\ &= (K_{\mathrm{m}} - K_{\mathrm{u}} G_{\mathrm{u}})y + K_{\mathrm{u}} w + Bv\end{aligned}$$

The procedure to design a reduced-order observer can be summarized as follows.

Procedure 4 (Reduced-order observer design)

Given: an observable system

$$\left(\begin{bmatrix} A_{mm} & A_{mu} \\ A_{um} & A_{uu} \end{bmatrix}, [I \quad 0]\right)$$

and a desired characteristic polynomial of the observer $\overline{\varphi}(s) = s^q + \overline{a}_{q-1}s^{q-1} + \cdots + \overline{a}_1 s + \overline{a}_0$, where q is the dimension of A_{uu}.

1. For (A_{uu}, A_{mu}), determine the observer matrix G_u using Procedure 3.
2. Construct the reduced-order observer

$$\begin{aligned}\dot{w} = &(A_{uu} + G_u A_{mu})w + (A_{um} + G_u A_{mm} - A_{uu}G_u - G_u A_{mu} G_u)y \\ &+ (B_u + G_u B_m)u\end{aligned}$$

3. Use the state feedback

$$u = (K_m - K_u G_u)y + K_u w + Bv$$

Example 2.20

For

$$\dot{x} = \begin{bmatrix} 1 & -1 & 0 \\ 1 & 0 & 1 \\ -2 & 2 & 0 \end{bmatrix} x + \begin{bmatrix} -1 \\ 0 \\ 1 \end{bmatrix} u$$
$$y = [1 \quad 0 \quad 0]x$$

we would like to design a reduced-order observer with poles at −1 and −2.

Clearly, x_1 can be directly measured but x_2 and x_3 cannot. Hence

$$A_{uu} = \begin{bmatrix} 0 & 1 \\ 2 & 0 \end{bmatrix}, \qquad A_{mu} = [-1 \quad 0]$$

We can design a G_u such that $A_{uu} + G_u A_{mu}$ has eigenvalues at −1 and −2. Such a G_u is given by

$$G_u = \begin{bmatrix} 3 \\ 4 \end{bmatrix}$$

The reduced-order observer is

$$\begin{aligned}\dot{w} &= (A_{uu} + G_u A_{mu})w + (A_{um} + G_u A_{mm} - A_{uu}G_u - G_u A_{mu} G_u)y + (B_u + G_u B_m)u \\ &= \begin{bmatrix} -3 & 1 \\ -2 & 0 \end{bmatrix} w + \begin{bmatrix} 9 \\ 8 \end{bmatrix} y_m + \begin{bmatrix} -3 \\ -3 \end{bmatrix} u\end{aligned}$$

Example 2.21

Let us consider the following system

$$\dot{x} = \begin{bmatrix} 2 & -1 & 1 \\ 1 & 1 & 2 \\ -2 & 0 & -2 \end{bmatrix} x + \begin{bmatrix} -2 \\ 0 \\ 1 \end{bmatrix} u$$
$$y = \begin{bmatrix} 1 & 1 & 1 \end{bmatrix} x$$

We would like to design: (1) a state feedback that places the poles at -7, $-1+j2$ and $-1-j2$; and (2) a reduced-order observer with poles at -20 and -30.

For the state feedback, the feedback matrix that places the poles at -7, $-1+j2$ and $-1-j2$ is given by (details will be left to the reader)

$$K = \begin{bmatrix} -10 & -55 & -30 \end{bmatrix}$$

For the observer, since $C \neq \begin{bmatrix} 1 & 0 & 0 \end{bmatrix}$, we first need to apply the following similarity transformation to the system.

$$T = \begin{bmatrix} 1 & -1 & -1 \\ 0 & 1 & 0 \\ 0 & 0 & 1 \end{bmatrix}$$

After the transformation, the new system is

$$\dot{z} = \begin{bmatrix} 1 & -1 & 0 \\ 1 & 0 & 1 \\ -2 & 2 & 0 \end{bmatrix} z + \begin{bmatrix} -1 \\ 0 \\ 1 \end{bmatrix} u$$
$$y = \begin{bmatrix} 1 & 0 & 0 \end{bmatrix} z$$

This is the same system as in Example 2.20. Using the same approach, we have

$$G_u = \begin{bmatrix} 50 \\ 602 \end{bmatrix}$$

The reduced-order observer is

$$\dot{w} = (A_{uu} + G_u A_{mu}) w + (A_{um} + G_u A_{mm} - A_{uu} G_u - G_u A_{mu} G_u) y + (B_u + G_u B_m) u$$
$$= \begin{bmatrix} -50 & 1 \\ -600 & 0 \end{bmatrix} w + \begin{bmatrix} 1949 \\ 30600 \end{bmatrix} y + \begin{bmatrix} -50 \\ -601 \end{bmatrix} u$$

The state feedback is

$$
\begin{aligned}
u &= Kx + v \\
&= KTz + v \\
&= KT\begin{bmatrix} z_m \\ \hat{z}_u \end{bmatrix} + v \\
&= KT\begin{bmatrix} z_m \\ w - G_u z_m \end{bmatrix} + v \\
&= KT\begin{bmatrix} I & 0 \\ -G_u & I \end{bmatrix}\begin{bmatrix} z_m \\ w \end{bmatrix} + v \\
&= KT\begin{bmatrix} I & 0 \\ -G_u & I \end{bmatrix}\begin{bmatrix} y \\ w \end{bmatrix} + v \\
&= \begin{bmatrix} 14280 & -45 & -20 \end{bmatrix}\begin{bmatrix} y \\ w_1 \\ w_2 \end{bmatrix} + v
\end{aligned}
$$

2.7 NOTES AND REFERENCES

In this chapter, we have summarized the fundamentals of modern control theory using state space models. The theory was developed in the 1960s and forms the foundation of many later developments, including robust control theory. We have provided the proofs of all major results and illustrated them as series of examples. Understanding these fundamental results will be sufficient to follow the rest of this book. However, if the reader would like to know more about the modern control theory, there are many reference books available. In particular, we recommend the books by Antsaklis and Michel [7], Belanger [26], Chui and Chen [44], and Rugh [140].

2.8 PROBLEMS

2.1 Calculate e^{At} for

(a)

$$A = \begin{bmatrix} 0 & 1 \\ -1 & -2 \end{bmatrix}$$

(b)

$$A = \begin{bmatrix} -5 & 0 & 0 \\ 0 & 0 & 1 \\ 0 & -1 & -2 \end{bmatrix}$$

(c)

$$A = \begin{bmatrix} -2 & 0 & 0 & 0 \\ 7 & 3 & 0 & 0 \\ 0 & 0 & -6 & 9 \\ 0 & 0 & 0 & 4 \end{bmatrix}$$

2.2 Find the response of the system

$$\dot{x} = \begin{bmatrix} -1 & 0 & 0 \\ 1 & -2 & 5 \\ 0 & 0 & -4 \end{bmatrix} x + \begin{bmatrix} 0 \\ 0 \\ 1 \end{bmatrix} u$$

$$y = \begin{bmatrix} 1 & 1 & 0 \end{bmatrix} x$$

when the initial condition is

$$x(0) = \begin{bmatrix} -10 \\ 15 \\ 30 \end{bmatrix}$$

and the input is the unit step function.

2.3 Use MATLAB SIMULINK to build a simulator for

$$\dot{x} = \begin{bmatrix} -8 & 7 & 0 & -3 & -5 \\ 0 & -4 & 0 & 6 & 8 \\ -3 & 7 & -3 & -6 & 5 \\ 0 & 0 & 0 & -5 & 0 \\ 0 & 0 & 0 & 10 & -1 \end{bmatrix} x + \begin{bmatrix} 1 \\ 0 \\ -3 \\ 6 \\ 2 \end{bmatrix} u$$

$$y = \begin{bmatrix} 2 & -4 & 0 & 0 & 9 \end{bmatrix} x$$

Run the system with different initial conditions and inputs (suggest using step, sinusoidal, and random inputs).

2.4 Consider the system given by

$$\dot{x} = \begin{bmatrix} -1 & 0 & 1 \\ 1 & -2 & 0 \\ 0 & 0 & -3 \end{bmatrix} x + \begin{bmatrix} 0 \\ 0 \\ 1 \end{bmatrix} u$$

$$y = \begin{bmatrix} 1 & 1 & 0 \end{bmatrix} x$$

Obtain the transfer function of the systems.

2.5 Find similarity transformations to transform the following systems into their Jordan canonical forms.

(a)

$$\dot{x} = \begin{bmatrix} -1 & 0 & 4 \\ -9 & -5 & -5 \\ 0 & 0 & -3 \end{bmatrix} x + \begin{bmatrix} -3 \\ 0 \\ 1 \end{bmatrix} u$$
$$y = \begin{bmatrix} 1 & 2 & 0 \end{bmatrix} x$$

(b)

$$\dot{x} = \begin{bmatrix} -9 & 4 & 7 & -5 \\ 6 & -2 & -7 & 9 \\ -5 & 0 & -9 & 5 \\ 0 & -7 & 9 & 4 \end{bmatrix} x + \begin{bmatrix} -4 \\ 0 \\ -7 \\ 3 \end{bmatrix} u$$
$$y = \begin{bmatrix} -8 & 6 & 2 & 0 \end{bmatrix} x$$

(c)

$$\dot{x} = \begin{bmatrix} 0 & -5 & -2 & -6 & 0 \\ 3 & 6 & -2 & 8 & 0 \\ 0 & 0 & -9 & 3 & 6 \\ -3 & -5 & 0 & 0 & 1 \\ 0 & 1 & -9 & 0 & 7 \end{bmatrix} x + \begin{bmatrix} -2 \\ 4 \\ 9 \\ 0 \\ -4 \end{bmatrix} u$$
$$y = \begin{bmatrix} 0 & 1 & 0 & 0 & 0 \end{bmatrix} x$$

2.6 Assume that matrix A has two complex eigenvalues $\lambda_1 = \sigma + j\omega$ and $\lambda_2 = \sigma - j\omega$. The other eigenvalues $\lambda_3, \ldots, \lambda_3$ are distinct and real. Let the corresponding eigenvectors be $v_1 = p + jq$, $v_2 = p - jq$, $v_3, \ldots, v_n$. Prove that using the transform matrix

$$T = \begin{bmatrix} p & q & v_3 & \ldots & v_n \end{bmatrix}$$

the corresponding Jordan canonical form is given by

$$\tilde{A} = T^{-1}AT = \begin{bmatrix} \sigma & \omega & 0 & \ldots & 0 \\ -\omega & \sigma & 0 & \ldots & 0 \\ 0 & 0 & \lambda_3 & \ldots & 0 \\ \ldots & & & \ldots & \\ 0 & 0 & 0 & \ldots & \lambda_n \end{bmatrix}$$

2.7 Prove that the transfer function of

$$\dot{x} = \begin{bmatrix} 0 & 1 & \dots & 0 \\ \dots & & & \\ 0 & 0 & \dots & 1 \\ -a_0 & -a_1 & \dots & -a_{n-1} \end{bmatrix} x + \begin{bmatrix} 0 \\ \dots \\ 0 \\ 1 \end{bmatrix} u$$

$$y = \begin{bmatrix} b_0 & b_1 & \dots & b_{n-1} \end{bmatrix} x$$

is given by

$$G(s) = \frac{b_0 + b_1 s + \dots + b_{n-1} s^{n-1}}{a_0 + a_1 s + \dots + a_{n-1} s^{n-1} + s^n}$$

2.8 Let us consider the following system

$$\dot{x} = \begin{bmatrix} 2 & -1 & 1 \\ 1 & 1 & 2 \\ -2 & 0 & -2 \end{bmatrix} x + \begin{bmatrix} -2 \\ 0 \\ 1 \end{bmatrix} u$$

$$y = \begin{bmatrix} 1 & 1 & 1 \end{bmatrix} x$$

Check controllability of the system and, if possible, design a state feedback that places the poles at -7, $-1+j2$ and $-1-j2$.

2.9 Design a state feedback that places the poles at -15, -10, $-2+j2$ and $-2-j2$ for the following system

$$\dot{x} = \begin{bmatrix} -9 & 4 & 4 & -5 \\ 6 & -2 & -7 & 9 \\ -5 & 0 & -9 & 5 \\ 0 & -7 & 0 & 4 \end{bmatrix} x + \begin{bmatrix} 1 \\ 0 \\ -2 \\ 3 \end{bmatrix} u$$

$$y = \begin{bmatrix} 1 & 0 & 2 & 0 \end{bmatrix} x$$

2.10 Using SIMULINK to simulate the closed-loop system obtained in Problem 2.9.

2.11 Consider the following system

$$\dot{x} = \begin{bmatrix} a_1 & a_2 \\ a_3 & a_4 \end{bmatrix} x + \begin{bmatrix} b_1 \\ b_2 \end{bmatrix} u$$

(a) Derive conditions (on a_i and b_j) such that the system is controllable.

(b) Assume that the system is controllable, design a state feedback that places the poles at $-p_1$ and $-p_2$.

2.12 Suppose an open-loop system is given by

$$\dot{x}(t) = \begin{bmatrix} 0 & 1 \\ 0 & 2 \end{bmatrix} x(t) + \begin{bmatrix} 1 \\ 1 \end{bmatrix} u(t)$$
$$y(t) = \begin{bmatrix} 0 & 1 \end{bmatrix} x(t)$$

(a) Is the above system controllable? Is it observable?
(b) Suppose the feedback $u(t) = -Kx(t) + v(t)$ is applied, where $v(t)$ is the external input for the closed-loop system with $K = [k_1\ k_2]$. For what values of k_1, k_2, is the closed-loop system controllable? For what values of k_1, k_2, is the closed-loop system observable?

2.13 For the following system, design a full-order observer with the desired poles at -8 and -9.

$$\dot{x} = \begin{bmatrix} 0 & 10 \\ 1 & 0 \end{bmatrix} x + \begin{bmatrix} 0 \\ 1 \end{bmatrix} u$$
$$y = \begin{bmatrix} 0 & 1 \end{bmatrix} x$$

2.14 For the following system

$$\dot{x} = \begin{bmatrix} 0 & 1 & 0 \\ 0 & 0 & 1 \\ -2 & 3 & 5 \end{bmatrix} x + \begin{bmatrix} 0 \\ 0 \\ 1 \end{bmatrix} u$$
$$y = \begin{bmatrix} 1 & 0 & 0 \end{bmatrix} x$$

design a full-order observer with poles at -5, $-1+j$, $-1-j$.

2.15 For the following system

$$\dot{x} = \begin{bmatrix} -1 & -5 & 4 \\ -9 & 6 & -5 \\ 0 & 0 & -3 \end{bmatrix} x + \begin{bmatrix} -3 \\ 0 \\ 1 \end{bmatrix} u$$
$$y = \begin{bmatrix} 1 & 2 & 0 \end{bmatrix} x$$

design a full-order observer that has poles at -20, $-10+j10$, $-10-j10$.

2.16 Consider the following system

$$\dot{x} = \begin{bmatrix} 0 & 1 & 0 \\ 0 & 0 & 1 \\ 1 & -1 & 2 \end{bmatrix} x + \begin{bmatrix} 0 \\ 0 \\ 1 \end{bmatrix} u$$
$$y = \begin{bmatrix} 1 & 0 & 0 \end{bmatrix} x$$

(a) Find the feedback to place poles at -4, $-1+j1$, and $-1-j1$.
(b) Design a full-order order observer with poles at -10, -15, and -20.
(c) Write equations for the closed-loop system with observer and feedback.

2.17 Use SIMULINK to simulate the closed-loop system obtained in Problem 2.16.

2.18 For the following system, design a reduced-order observer to place the poles at -15 and -10. Write the observer equation.

$$\dot{x} = \begin{bmatrix} -2 & 0 & 0 \\ 0 & 0 & -2 \\ 0 & 0 & -1 \end{bmatrix} x + \begin{bmatrix} 0 \\ 0 \\ 1 \end{bmatrix} u$$

$$y = \begin{bmatrix} 1 & 0 & 0 \end{bmatrix} x$$

2.19 For the following system, design a reduced-order observer to place the poles at -5 and -10. Write the observer equation.

$$\dot{x} = \begin{bmatrix} -2 & 4 & 0 \\ -1 & 0 & -2 \\ 0 & 0 & -1 \end{bmatrix} x + \begin{bmatrix} 0 \\ 1 \\ 1 \end{bmatrix} u$$

$$y = \begin{bmatrix} 1 & 1 & 0 \end{bmatrix} x$$

2.20 For the system in Problem 2.16, design a reduced-order order observer with poles at -10 and -15.

2.21 Repeat Problem 2.17 using the reduced-order observer of Problem 2.20. Compare the result with that of Problem 2.17.

3

Stability Theory

System performance to be achieved by control can be characterized either as stability or optimality. In this chapter, we discuss stability, while optimality will be covered in Chapter 4. Intuitively, stability means that, in the absence of inputs, a system's response will converge to some equilibrium. Stability is an essential requirement for any practical system. Unfortunately, some systems are unstable to begin with. A most notorious example is the economical system. Hence, the first objective of control is to bring a system into stability. We start with a general nonlinear system and define its stability. We present the Lyapunov stability theorem which will be used extensively in this book to prove stability. We then focus on linear systems, whose stability is determined by the locations of their poles. We discuss several stability criteria for checking stability. We also study stabilizability and detectability, which are weaker versions of controllability and observability discussed in Chapter 2.

3.1 STABILITY AND LYAPUNOV THEOREM

Consider a general nonlinear system

$$\dot{x} = A(x) \tag{3.1}$$

where $x \in R^n$ are the state variables and $A : R^n \rightarrow R^n$ is a (nonlinear) function. We assume that A is such that the system (3.1) has a unique

Robust Control Design: An Optimal Control Approach F. Lin

solution $x(t)$ over $[0, \infty)$ for all initial conditions $x(0)$ and that the solution depends continuously on $x(0)$. In general, this will be assumed for all systems discussed in this book.

A vector $x_0 \in R^n$ is an equilibrium point of the system (3.1) if

$$A(x_0) = 0$$

Without loss of generality, we can assume that $x_0 = 0$ is an equilibrium point of the system (3.1); that is, $A(0) = 0$. Otherwise we can perform a simple state transformation $z = x - x_0$ to obtain a new state equation

$$\dot{z} = \widetilde{A}(z) = A(z + x_0)$$

where $z_0 = 0$ is an equilibrium point ($\widetilde{A}(0) = A(x_0) = 0$). Clearly, the solution of the differential equation (3.1) shows that if $x(0) = 0$, then $x(t) = 0$, for all $t > 0$. However, this solution may or may not be stable.

Stability

The equilibrium point $x_0 = 0$ of the system (3.1) is stable if for all $\varepsilon > 0$, there exists a $\delta(\varepsilon) > 0$ such that

$$\|x(0)\| < \delta(\varepsilon) \Rightarrow \|x(t)\| < \varepsilon \quad \forall t \geq 0$$

In other words, the equilibrium point $x_0 = 0$ is stable if arbitrarily small perturbations of the initial state $x(0)$ from the equilibrium point result in arbitrarily small perturbation of the corresponding state trajectory $x(t)$.

Asymptotic Stability

The equilibrium point $x_0 = 0$ of the system (3.1) is asymptotically stable if it is stable and there exists some $c > 0$ such that if $\|x(0)\| < c$, then

$$x(t) \to 0 \quad \text{as} \quad t \to \infty$$

In other words, the equilibrium point $x_0 = 0$ is asymptotically stable if there exists a neighbourhood of $x_0 = 0$ such that if the system starts in the neighbourhood then its trajectory converges to the equilibrium point $x_0 = 0$ as $t \to \infty$.

The equilibrium point $x_0 = 0$ of the system (3.1) is globally asymptotically stable if $c > 0$ can be arbitrarily large; that is, all trajectories converge to the equilibrium point $x_0 = 0$:

$$x(t) \to 0 \quad \text{as} \quad t \to \infty$$

Determining stability of a system may not be an easy task if the system is nonlinear. One approach often used to determine stability is that of Lyapunov. Intuitively, the Lyapunov stability theorem can be explained as follows. Given a system with an equilibrium point $x_0 = 0$, let us define some suitable 'energy' function of the system. This function must have the property that it is zero at the origin (the equilibrium point $x_0 = 0$) and positive elsewhere. Assume further that the system dynamics are such that the energy of the system is monotonically decreasing with time and hence eventually reduces to zero. Then the trajectories of the system have no other places to go but the origin. Therefore, the system is asymptotically stable. This generalized energy function is called a Lyapunov function. If there exists a Lyapunov function, then we can prove the asymptotic stability using the following Lyapunov stability theorem.

Theorem 3.1

The equilibrium point $x_0 = 0$ of system (3.1) is asymptotically stable if there exists a Lyapunov function $V: R^n \to R$ such that

$$
\begin{aligned}
V(x) &> 0, & x &\neq 0\\
V(x) &= 0, & x &= 0\\
\dot{V}(x) &< 0, & x &\neq 0\\
\dot{V}(x) &= 0, & x &= 0
\end{aligned}
$$

is true in a neighbourhood of $x_0 = 0$, $N = \{x : \|x\| < c\}$ for some $c > 0$.

Proof

The precise mathematical proof is cumbersome and uninspiring. So we will provide the following intuitive proof by contradiction. If the equilibrium point $x_0 = 0$ of the system (3.1) is not asymptotically stable; that is, $x(t) \to 0$ as $t \to \infty$ is not true even if $\|x(0)\| < c$ for some $c > 0$, then $\dot{V}(x) < -\alpha$ for some $\alpha > 0$. Since

$$V(x(t)) = V(x(0)) + \int_0^t \dot{V}(x)\, d\tau = V(x(0)) + \int_0^t -\alpha d\tau = V(x(0)) - \alpha t$$

for a sufficiently large t, $V(x(t)) < 0$. This contradicts the assumption $V(x(t)) \geq 0$.

Q.E.D.

The key to proving stability of a system using the Lyapunov stability theorem is to construct a Lyapunov function. This construction must be done in a case-by-case basis. There is no general method for the construction. The following example illustrates the application of the Lyapunov stability theorem.

Example 3.1

Let us consider the following system:

$$\begin{aligned} \dot{x}_1 &= x_2 - 3x_1 \\ \dot{x}_2 &= -x_2^3 - 2x_1 \end{aligned}$$

To prove it is asymptotically stable, let us consider the following Lyapunov function:

$$V(x) = 2x_1^2 + x_2^2$$

Clearly

$$\begin{aligned} V(x) &> 0 \qquad x \neq 0 \\ V(x) &= 0 \qquad x = 0 \end{aligned}$$

On the other hand

$$\begin{aligned} \dot{V}(x) &= 4x_1\dot{x}_1 + 2x_2\dot{x}_2 \\ &= 4x_1(x_2 - 3x_1) + 2x_2(-x_2^3 - 2x_1) \\ &= 4x_1x_2 - 12x_1^2 - 2x_2^4 - 4x_1x_2 \\ &= -12x_1^2 - 2x_2^4 \end{aligned}$$

Therefore

$$\begin{aligned} \dot{V}(x) &< 0 \qquad x \neq 0 \\ \dot{V}(x) &= 0 \qquad x = 0 \end{aligned}$$

Hence, we conclude that the system is asymptotically stable.

Example 3.2

Consider the following system:

$$\dot{x}_1 = x_1(x_1^2 + x_2^2 - 1) - x_2$$

$$\dot{x}_2 = x_1 + x_2(x_1^2 + x_2^2 - 1)$$

Let us try the following Lyapunov function:

$$V(x) = x_1^2 + x_2^2$$

Clearly

$$V(x) > 0 \qquad x \neq 0$$
$$V(x) = 0 \qquad x = 0$$

Furthermore

$$\begin{aligned}\dot{V}(x) &= 2x_1\dot{x}_1 + 2x_2\dot{x}_2 \\ &= 2x_1(x_1(x_1^2 + x_2^2 - 1) - x_2) + 2x_2(x_1 + x_2(x_1^2 + x_2^2 - 1)) \\ &= 2x_1^2(x_1^2 + x_2^2 - 1) - 2x_1x_2 + 2x_2x_1 + 2x_2^2(x_1^2 + x_2^2 - 1) \\ &= 2(x_1^2 + x_2^2)(x_1^2 + x_2^2 - 1)\end{aligned}$$

In the neighbourhood of $N = \{x : \|x\| < 1\}$,

$$\dot{V}(x) < 0 \qquad x \neq 0$$
$$\dot{V}(x) = 0 \qquad x = 0$$

Hence, we conclude that the system is asymptotically stable. Note that the system is not globally asymptotically stable. To prove globally asymptotic stability, an additional condition must be satisfied, as to be discussed in Chapter 6.

3.2 LINEAR SYSTEMS

Although the Lyapunov approach is a nice way to check stability of a system, it is not always feasible, because it is often difficult, if not impossible, to construct a Lyapunov function. If no Lyapunov function can be found, then nothing can be said about the stability of a system. For nonlinear systems, this is essentially the case: there is no general criterion to check the stability of a nonlinear system.

However, for a linear time-invariant system

$$\dot{x} = Ax$$

we can do more. In fact there are several criteria available to check the stability of a linear time-invariant system. For a linear time-invariant system, its stability is determined by its characteristic polynomial

$$\varphi(s) = a_n s^n + a_{n-1}s^{n-1} + \cdots + a_1 s + a_0$$

and its corresponding roots, which are eigenvalues or poles. If a root is real, that is, $\lambda = \alpha$, then its corresponding zero-input response is

$$x(t) = ce^{\alpha t}$$

where c is some constant. Clearly, $x(t) = ce^{\alpha t} \to 0$ as $t \to \infty$ if and only if $\alpha < 0$. If a root is complex; that is, $\lambda = \sigma + j\omega$, then its corresponding zero-input response is

$$x(t) = ce^{\sigma t} \sin(\omega t + \theta)$$

where c and θ are constants. Clearly, $x(t) = ce^{\sigma t} \sin(\omega t + \theta) \to 0$ as $t \to \infty$ if and only if $\sigma < 0$. In both cases, the roots must be in the open left half of the s-plane.

Theorem 3.2

A linear time-invariant system

$$\dot{x} = Ax$$

is asymptotically stable if and only if all the roots of its characteristic polynomial are in the open left half of the s-plane.

Proof

Assume that a linear time-invariant system has $2m$ complex roots of the form $\sigma_i + j\omega_i$ and $n - 2m$ real roots of the form α_j. Then the zero-input response can be written as

$$x(t) = \sum_{i=1}^{m} c_i e^{\sigma_i t} \sin(\omega_i t + \theta_i) + \sum_{j=1}^{n-2m} c_j e^{\alpha_j t}$$

From the above discussion, it is clear that $x(t) \to 0$ as $t \to \infty$ if and only if for all $\sigma_i + j\omega_i$, $\sigma_i < 0$ and for all α_j, $\alpha_j < 0$.

Q.E.D.

Let us define the stable and unstable regions of the s-plane as shown in Figure 3.1, then for a system to be stable, all the roots of its characteristic polynomial (that is, its eigenvalues or poles) must be in the stable region.

For linear systems, asymptotical stability and globally asymptotic stability are equivalent. A system is asymptotically stable if and only if it is globally asymptotically stable.

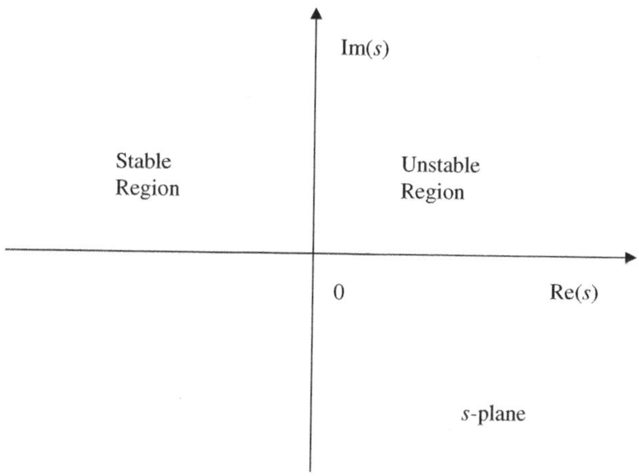

Figure 3.1 Stable and unstable regions in the s-plane.

3.3 ROUTH–HURWITZ CRITERION

The Routh–Hurwitz criterion is a method to determine the locations of roots of a polynomial with constant real coefficients with respect to the left half of the s-plane without actually solving for the roots. It is used to check the stability of linear time-invariant systems. Since it does not actually try to find the numerical solutions of the roots, it can handle symbolic polynomials. This is very useful in control synthesis.

Before we present the Routh–Hurwitz criterion, let us first prove a necessary condition for a polynomial

$$\varphi(s) = a_n s^n + a_{n-1} s^{n-1} + \cdots + a_1 s + a_0$$

to have all its roots in the open left half of the complex plane.

Lemma 3.1

If a polynomial $\varphi(s) = a_n s^n + a_{n-1} s^{n-1} + \cdots + a_1 s + a_0$ has all its roots in the open left half of the complex plane, then all its coefficients a_i, $i = 0, 1, 2, \ldots, n$ must have the same sign.

Proof

The result is obvious for first-order ($n = 1$) polynomials ($\varphi(s) = a_1 s + a_0$) and second-order ($n = 2$) polynomials ($\varphi(s) = a_2 s^2 + a_1 s + a_0$).

For high-order ($n > 2$) polynomials, $\varphi(s)$ can always be decomposed into a product of first-order polynomials and second-order polynomials. Hence, the result follows.

Q.E.D.

The condition of Lemma 3.1 is necessary but not sufficient. For example, the polynomial

$$\varphi(s) = 6s^3 + 2s^2 + 4s + 5$$

has all its coefficients positive, but its roots are -0.8006, $0.2337 + j0.9931$, and $0.2337 - j0.9931$.

To find a necessary and sufficient condition for a polynomial to have all its roots in the open left half of the complex plane, we construct the following Routh table. The first two rows of the Routh table are the coefficients of the polynomial

$$\begin{array}{ccccc} a_n & a_{n-2} & a_{n-4} & a_{n-6} & \cdots \\ a_{n-1} & a_{n-3} & a_{n-5} & a_{n-7} & \cdots \end{array}$$

The third row is calculated based on the first two rows as follows.

$$\begin{array}{ccccc} a_n & a_{n-2} & a_{n-4} & a_{n-6} & \cdots \\ a_{n-1} & a_{n-3} & a_{n-5} & a_{n-7} & \cdots \\ \dfrac{a_{n-1}a_{n-2} - a_n a_{n-3}}{a_{n-1}} & \dfrac{a_{n-1}a_{n-4} - a_n a_{n-5}}{a_{n-1}} & \dfrac{a_{n-1}a_{n-6} - a_n a_{n-7}}{a_{n-1}} & \dfrac{a_{n-1}a_{n-8} - a_n a_{n-9}}{a_{n-1}} & \cdots \end{array}$$

In general, the next row is calculated based on the previous two rows using the same rule. If there is no element left to be calculated, we fill the table with zeros. This process will continue until we have all $n+1$ rows. For example for a sixth-order polynomial

$$\varphi(s) = a_6 s^6 + a_5 s^5 + a_4 s^4 + a_3 s^3 + a_2 s^2 + a_1 s + a_0$$

its Routh table is given below.

$$\begin{array}{ccccc}
s^6 & a_6 & a_4 & a_2 & a_0 \\
s^5 & a_5 & a_3 & a_1 & 0 \\
s^4 & \dfrac{a_5 a_4 - a_6 a_3}{a_5} = \alpha & \dfrac{a_5 a_2 - a_6 a_1}{a_5} = \beta & \dfrac{a_5 a_0 - a_6 0}{a_5} = a_0 & 0 \\
s^3 & \dfrac{\alpha a_3 - a_5 \beta}{\alpha} = \chi & \dfrac{\alpha a_1 - a_5 a_0}{\alpha} = \delta & 0 & 0 \\
s^2 & \dfrac{\chi\beta - \alpha\delta}{\chi} = \phi & \dfrac{\chi a_0 - \alpha 0}{\chi} = a_0 & 0 & 0 \\
s^1 & \dfrac{\phi\delta - \chi a_0}{\phi} = \varphi & 0 & 0 & 0 \\
s^0 & \dfrac{\varphi a_0 - \phi 0}{\varphi} = a_0 & 0 & 0 & 0
\end{array}$$

The locations of the roots of the polynomial with respect to the imaginary axis can then be determined by the first column of the Routh table as follows.

Routh–Hurwitz Criterion

The roots of the polynomial $\varphi(s) = a_n s^n + a_{n-1} s^{n-1} + \cdots + a_1 s + a_0$ are all in the open left half of the complex plane if and only if all the elements of the first column of the Routh table are of the same sign. If there are changes of signs of the elements of the first column of the Routh table, then the number of sign changes is equal to the number of roots outside the open left half of the complex plane.

Since we usually assume that the first coefficient is positive, $a_n > 0$, all the elements of the first column of the Routh table must be positive to ensure that the system with characteristic polynomial $\varphi(s) = a_n s^n + a_{n-1} s^{n-1} + \cdots + a_1 s + a_0$ is stable. If the system is not stable, then the number of unstable roots is equal to the number of sign changes in the first column.

The proof of the Routh–Hurwitz criterion involves first constructing the following $n \times n$ Hurwitz matrix

$$\begin{bmatrix} a_{n-1} & a_n & 0 & 0 & \cdots & 0 \\ a_{n-3} & a_{n-2} & a_{n-1} & a_n & \cdots & 0 \\ a_{n-5} & a_{n-4} & a_{n-3} & a_{n-2} & \cdots & 0 \\ \cdots & \cdots & \cdots & \cdots & \cdots & \cdots \\ 0 & 0 & 0 & 0 & \cdots & a_2 \\ 0 & 0 & 0 & 0 & \cdots & a_0 \end{bmatrix}$$

For example, for $n = 3$ and 4, the Hurwitz matrices are

$$\begin{bmatrix} a_2 & a_3 & 0 \\ a_0 & a_1 & a_2 \\ 0 & 0 & a_0 \end{bmatrix} \quad \text{and} \quad \begin{bmatrix} a_3 & a_4 & 0 & 0 \\ a_1 & a_2 & a_3 & a_4 \\ 0 & a_0 & a_1 & a_2 \\ 0 & 0 & 0 & a_0 \end{bmatrix}$$

respectively. From the Hurwitz matrix, we can find its principal determinants, called the Hurwitz determinants, as follows.

$$D_1 = a_{n-1}$$

$$D_2 = \begin{vmatrix} a_{n-1} & a_n \\ a_{n-3} & a_{n-2} \end{vmatrix}$$

$$D_3 = \begin{vmatrix} a_{n-1} & a_n & 0 \\ a_{n-3} & a_{n-2} & a_{n-1} \\ a_{n-5} & a_{n-4} & a_{n-3} \end{vmatrix}$$

...

$$D_n = \begin{vmatrix} a_{n-1} & a_n & 0 & 0 & \dots & 0 \\ a_{n-3} & a_{n-2} & a_{n-1} & a_n & \dots & 0 \\ a_{n-5} & a_{n-4} & a_{n-3} & a_{n-2} & \dots & 0 \\ \dots & \dots & \dots & \dots & \dots & \dots \\ 0 & 0 & 0 & 0 & \dots & a_2 \\ 0 & 0 & 0 & 0 & .. & a_0 \end{vmatrix}$$

Hurwitz showed that the roots of the polynomial $\varphi(s)$ are all in the open left half of the complex plane if and only if all the Hurwitz determinants are positive. Routh then showed that this is equivalent to the Routh–Hurwitz criterion.

Example 3.3

For the polynomial

$$\varphi(s) = s^4 + 6s^3 + 13s^2 + 12s + 4$$

we construct its Routh table as follows.

$$\begin{array}{llll} s^4 & 1 & 13 & 4 \\ s^3 & 6 & 12 & 0 \\ s^2 & \dfrac{6 \times 13 - 1 \times 12}{6} = 11 & \dfrac{6 \times 4 - 1 \times 0}{6} = 4 & 0 \\ s^1 & \dfrac{11 \times 12 - 6 \times 4}{11} = 9.8182 & 0 & 0 \\ s^0 & \dfrac{9.8182 \times 4 - 11 \times 0}{9.8182} = 4 & 0 & 0 \end{array}$$

Since all the elements of the first column are positive, all the roots are all in the open left half of the complex plane. Indeed, we have $\varphi(s) = (s+1)^2(s+2)^2$. The corresponding system is (globally asymptotically) stable.

Example 3.4

Consider the polynomial

$$\varphi(s) = s^3 - 4s^2 + s + 6$$

Its Routh table is

$$\begin{array}{lll} s^3 & 1 & 1 \\ s^2 & -4 & 6 \\ s^1 & \dfrac{-4\times 1-1\times 6}{-4}=2.5 & 0 \\ s^0 & \dfrac{2.5\times 6-(-4)\times 0}{2.5}=6 & 0 \end{array}$$

There are two sign changes in the elements of the first column: $s^3 \leftrightarrow s^2$ and $s^2 \leftrightarrow s^1$. Therefore, there are two roots outside the open left half of the complex plane. Indeed, we have $\varphi(s)=(s+1)(s-2)(s-3)$.

Example 3.5

For the polynomial

$$\varphi(s)=2s^4+s^3+3s^2+5s+10$$

we construct its Routh table as follows.

$$\begin{array}{llll} s^4 & 2 & 3 & 10 \\ s^3 & 1 & 5 & 0 \\ s^2 & \dfrac{1\times 3-2\times 5}{1}=-7 & \dfrac{1\times 10-2\times 0}{1}=10 & 0 \\ s^1 & \dfrac{-7\times 5-1\times 10}{-7}=6.4286 & 0 & 0 \\ s^0 & \dfrac{6.4286\times 10-(-7)\times 0}{6.4286}=10 & 0 & 0 \end{array}$$

There are two sign changes in the elements of the first column: $s^3 \leftrightarrow s^2$ and $s^2 \leftrightarrow s^1$. Therefore, there are two roots outside the open left half of the complex plane.

Example 3.6

Consider a general third-order polynomial

$$\varphi(s)=a_3s^3+a_2s^2+a_1s+a_0$$

Its Routh table is

$$\begin{array}{lll} s^3 & a_3 & a_1 \\ s^2 & a_2 & a_0 \\ s^1 & \dfrac{a_2\times a_1-a_3\times a_0}{a_2} & 0 \\ s^0 & a_0 & 0 \end{array}$$

Without loss of generality, assume $a_3 > 0$. A necessary and sufficient condition for all three roots to be in the open left half of the complex plane is

$$a_3 > 0 \wedge a_2 > 0 \wedge \frac{a_2 \times a_1 - a_3 \times a_0}{a_2} > 0 \wedge a_0 > 0$$

$$\Leftrightarrow \quad a_3 > 0 \wedge a_2 > 0 \wedge (a_2 \times a_1 - a_3 \times a_0) > 0 \wedge a_0 > 0$$

$$\Leftrightarrow a_3 > 0 \wedge a_2 > 0 \wedge a_1 > 0 \wedge a_0 > 0 \wedge (a_2 \times a_1 - a_3 \times a_0) > 0$$

Therefore, for a general third-order system to be stable, all the coefficients must be positive and $a_2 \times a_1 - a_3 \times a_0 > 0$.

For first- and second-order systems, necessary and sufficient conditions for stability are also obvious and they can be summarized in Table 3.1.

The Routh table can be used for most systems. However, there are some special cases where the Routh table needs to be modified.

The first special case is when the first element, but not all the elements, of a row in the Routh table is zero. In this case, we will have difficulty in constructing the next row, because we cannot divide a number by zero. In this case, what we need to do is to replace the zero by a small number ε.

Example 3.7

Consider the following polynomial

$$\varphi(s) = s^3 - 3s + 2$$

If we try to construct the Routh table, we will get

$$\begin{array}{lcc} s^3 & 1 & -3 \\ s^2 & 0 & 2 \\ s^1 & \dfrac{0 \times (-3) - 1 \times 2}{0} = ? & \\ s^0 & & \end{array}$$

Table 3.1 Necessary and sufficient conditions for stability of 1st, 2nd, and 3rd order systems.

System	Characteristic polynomial	Stability condition
1st order	$a_1 s + a_0$	$a_1 > 0 \wedge a_0 > 0$
2nd order	$a_2 s^2 + a_1 s + a_0$	$a_2 > 0 \wedge a_1 > 0 \wedge a_0 > 0$
3rd order	$a_3 s^3 + a_2 s^2 + a_1 s + a_0$	$a_3 > 0 \wedge a_2 > 0 \wedge a_1 > 0 \wedge a_0 > 0$ $\wedge (a_2 \times a_1 - a_3 \times a_0) > 0$

So, let us replace the first zero in the second row by ε. We assume that ε is positive: $\varepsilon > 0$.

$$\begin{array}{lcc}
s^3 & 1 & -3 \\
s^2 & \varepsilon > 0 & 2 \\
s^1 & \dfrac{\varepsilon \times (-3) - 1 \times 2}{\varepsilon} \approx \dfrac{-2}{\varepsilon} < 0 & 0 \\
s^0 & 2 & 0
\end{array}$$

There are two sign changes in the elements of the first column: $s^2 \leftrightarrow s^1$ and $s^1 \leftrightarrow s^0$. Therefore, there are two roots outside the open left half of the complex plane. Indeed, $\varphi(s) = (s-1)^2(s+2)$. Note that the assumption $\varepsilon > 0$ is made without loss of generality. If we assume $\varepsilon < 0$, then

$$\begin{array}{lcc}
s^3 & 1 & -3 \\
s^2 & \varepsilon < 0 & 2 \\
s^1 & \dfrac{\varepsilon \times (-3) - 1 \times 2}{\varepsilon} \approx \dfrac{-2}{\varepsilon} > 0 & 0 \\
s^0 & 2 & 0
\end{array}$$

Still, there are two sign changes in the elements of the first column: $s^3 \leftrightarrow s^2$ and $s^2 \leftrightarrow s^1$.

The second special case is when all the elements of a row in the Routh table are zero. In this case, we need to go back to the previous row, find the auxiliary polynomial, take its derivative, and then put the coefficients of the derivative in place of the row of zeros.

Example 3.8

Consider the following polynomial

$$\varphi(s) = s^4 + s^3 - 3s^2 - s + 2$$

If we try to construct the Routh table, we will have

$$\begin{array}{lccc}
s^4 & 1 & -3 & 2 \\
s^3 & 1 & -1 & 0 \\
s^2 & \dfrac{1 \times (-3) - 1 \times (-1)}{1} = -2 & \dfrac{1 \times 2 - 1 \times 0}{1} = 2 & 0 \\
s^1 & \dfrac{-2 \times (-1) - 1 \times 2}{-2} = 0 & 0 & 0 \\
s^0 & & &
\end{array}$$

Since the elements of the row of s^1 are all zeros, we go back to the row of s^2 and find the corresponding auxiliary polynomial

$$\phi(s) = -2s^2 + 2$$

The derivative of $\phi(s) = -2s^2 + 2$ is

$$\frac{d\phi(s)}{ds} = -4s$$

Using its coefficients for the row of s^1, we continue the construction of the Routh table as follows.

$$\begin{array}{llll}
s^4 & 1 & -3 & 2 \\
s^3 & 1 & -1 & 0 \\
s^2 & \dfrac{1\times(-3)-1\times(-1)}{1} = -2 & \dfrac{1\times 2-1\times 0}{1} = 2 & 0 \\
s^1 & -4 & 0 & 0 \\
s^0 & 2 & 0 & 0
\end{array}$$

There are two sign changes in the elements of the first column: $s^3 \leftrightarrow s^2$ and $s^1 \leftrightarrow s^0$. Therefore, there are two roots outside the open left half of the complex plane. Indeed, $\varphi(s) = (s-1)^2(s+2)(s+1)$.

With the wide availability of computer programs such as MATLAB, it is straightforward to calculate the roots of polynomials of numerical coefficients. Therefore, in all the numerical examples presented above, we can bypass the Routh–Hurwitz criterion and find the stability of systems directly by solving the roots numerically. In other words, the Routh–Hurwitz criterion is not so useful in stability analysis of these systems. However, the Routh–Hurwitz criterion is still very useful in control synthesis, as illustrated in the following example.

Example 3.9

Consider the feedback control system in Figure 3.2. K is the controller gain that can be adjusted. We want to determine the range of K such that the closed-loop system is stable. We first find the characteristic equation from the transfer function of the closed-loop system:

$$\frac{Y(s)}{U(s)} = \frac{\dfrac{400000K}{s(s^2+1040s+48500)}}{1+\dfrac{400000K}{s(s^2+1040s+48500)}} = \frac{400000K}{s^3+1040s^2+48500s+400000K}$$

The characteristic equation of the closed-loop system is

$$\varphi(s) = s^3 + 1040s^2 + 48500s + 400000K$$

Clearly, we cannot determine the stability of the system by using MATLAB to find the roots of the characteristic equation. Here we must use

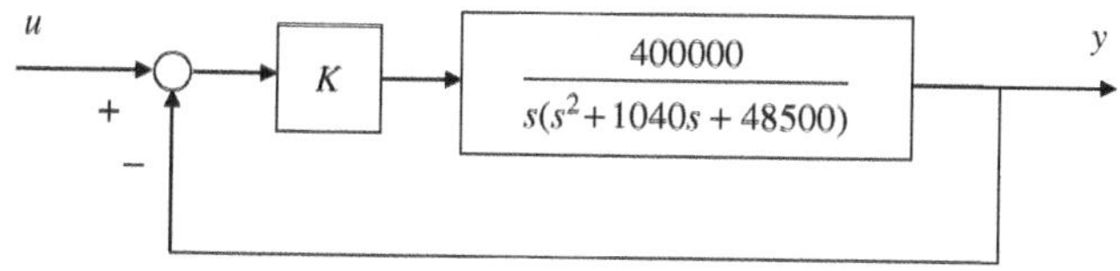

Figure 3.2 Determine stable K using the Routh–Hurwitz criterion.

the Routh–Hurwitz criterion. From Table 3.1, we know that this third-order system is stable if and only if

$$a_3 > 0 \wedge a_2 > 0 \wedge a_1 > 0 \wedge a_0 > 0$$
$$\wedge (a_2 \times a_1 - a_3 \times a_0) > 0$$

In other words

$$400000K > 0 \wedge (1040 \times 48500 - 400000K) > 0$$
$$\Leftrightarrow K > 0 \wedge K < \frac{1040 \times 48500}{400000} = 126.1$$

Intuitively, the reason for $K > 0$ is that if K is negative, then the system will have a positive feedback. As we know, positive feedback usually leads to instability. The reason for requiring $K < 126.1$ is that if K is too large, then the system will have very large gain, which also leads to instability.

The Routh–Hurwitz criterion is used to determine the pole locations with respect to the imaginary axis. But sometimes, we would like to know the pole locations with respect to a line parallel to the imaginary axis. For example, for robustness, we may want all the poles to be located at the left of $-a$, for some $a > 0$. We cannot apply the Routh–Hurwitz criterion directly to check if the poles are at the left of $-a$, but we can do a variable change and then apply the Routh–Hurwitz criterion. Figure 3.3 illustrates this variable change from s to s'. The imaginary axis of the s'-plane is at $-a$. The relationship between s and s' is given by $s = s' - a$. Furthermore, for any roots λ_i

$$\mathrm{Re}(\lambda_i) < -a \Leftrightarrow \lambda_i \text{ is in the open left half of the } s'\text{-plane.}$$

Hence, we can apply the Routh–Hurwitz criterion in the s'-plane.

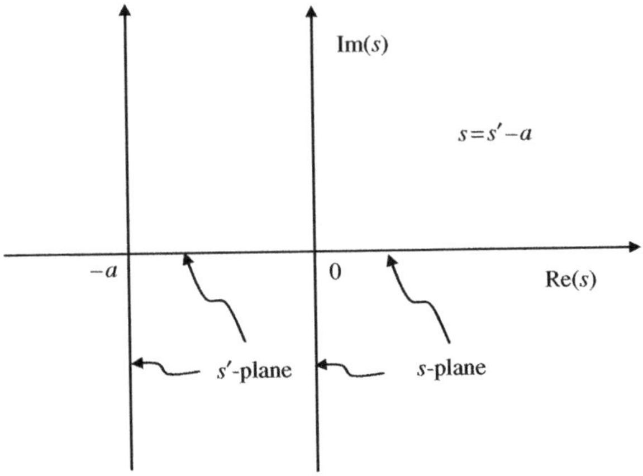

Figure 3.3 Locate poles at the left of $-a$ using the Routh–Hurwitz criterion.

Example 3.10

For the feedback control system in Figure 3.4, we need to determine the range of K such that the real parts of the poles of the closed-loop system are less than -2. The transfer function of the closed-loop system is

$$\frac{Y(s)}{U(s)} = \frac{\dfrac{(s^2+13s+14)K}{s^3+6s^2+14s+16}}{1+\dfrac{(s^2+13s+14)K}{s^3+6s^2+14s+16}} = \frac{(s^2+13s+14)K}{s^3+6s^2+14s+16+(s^2+13s+14)K}$$

The characteristic equation of the closed-loop system is

$$\varphi(s) = s^3+6s^2+14s+16+(s^2+13s+14)K$$

Substitute s by $s'-2$, we have

$$\begin{aligned}\varphi(s') &= (s'-2)^3+6(s'-2)^2+14(s'-2)+16+((s'-2)^2+13(s'-2)+14)K \\ &= s'^3+3Ks'^2+(K+2)s'+4\end{aligned}$$

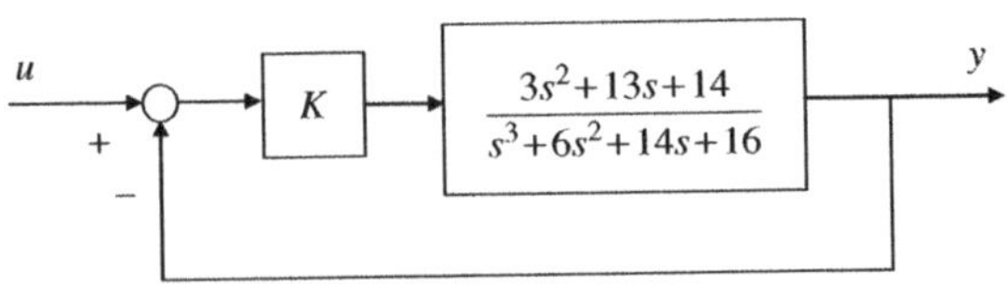

Figure 3.4 Use K to move the poles.

From Table 3.1, we know that all poles are in the open left half of the s'-plane if and only if

$$\begin{aligned}
&3K > 0 \wedge K+2 > 0 \wedge 3K(K+2)-4 > 0\\
\Leftrightarrow &3K > 0 \wedge K+2 > 0 \wedge 3K^2+6K-4 > 0\\
\Leftrightarrow &K > 0 \wedge K > -2 \wedge (K < -2.5275 \vee K > 0.5275)\\
\Leftrightarrow &K > 0 \wedge (K < -2.5275 \vee K > 0.5275)\\
\Leftrightarrow &(K > 0 \wedge K < -2.5275) \vee (K > 0 \wedge K > 0.5275)\\
\Leftrightarrow &K > 0.5275
\end{aligned}$$

3.4 NYQUIST CRITERION

The Nyquist criterion is another way to check the stability of a linear time-invariant system by determining the locations of roots of its characteristic polynomial with respect to the imaginary axis. Unlike the Routh–Hurwitz criterion, the Nyquist criterion is a frequency domain method based on the frequency response of a linear time-invariant system.

The Nyquist criterion is based on a fundamental theorem of complex analysis, Cauchy's argument principle. To present Cauchy's argument principle, let us consider a complex function $F: C \rightarrow C$; that is, F maps a point s in the s-plane to a point $F(s)$ in the $F(s)$-plane, as shown in Figure 3.5. As the point moves, F will map closed path Ω_s in the s-plane to a closed path $\Omega_{F(s)}$ in the $F(s)$-plane.

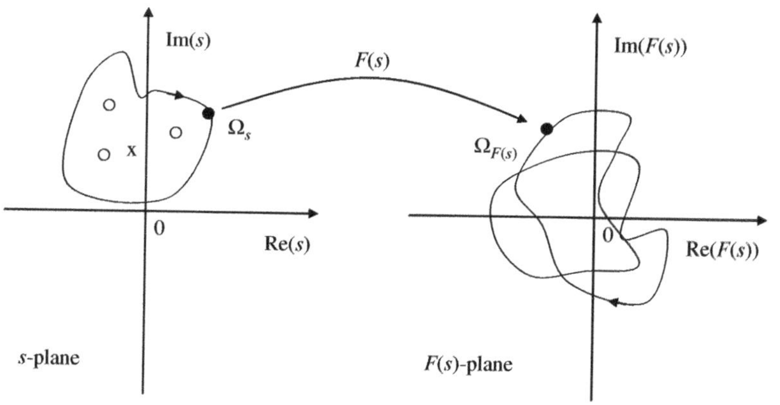

Figure 3.5 Mapping from the s-plane to the $F(s)$-plane.

Assume that the function F has only finite poles and zeros. Assume that Ω_s is an arbitrary closed path in the s-plane that does not go through any poles or zeros of F. Cauchy's argument principle says that the corresponding closed path $\Omega_{F(s)}$ in the $F(s)$-plane shall encircle the origin as many times as the difference between the number of zeros and the number of poles that are encircled by Ω_s in the s-plane. In other words

$$N = Z - P$$

where N is the number of the encirclements of the origin by the closed path $\Omega_{F(s)}$ in the $F(s)$-plane; Z is the number of zeros of $F(s)$ encircled by the closed path Ω_s in the s-plane; and P is the number of poles of $F(s)$ encircled by the closed path Ω_s in the s-plane.

We would like to apply Cauchy's argument principle to investigate the stability of the closed-loop system in Figure 3.6.

We know that the characteristic equation of the closed-loop system is given by

$$1 + G(s)H(s) = 0$$

The necessary and sufficient condition for the system to be stable is that $1 + G(s)H(s)$ has no zeros in the right half of the s-plane. To use Cauchy's argument principle, let us construct a closed path Ω_s in the s-plane as illustrated in Figure 3.7. Ω_s is a right half circle whose centre is at the origin and radius is R. Obviously if $R \to \infty$, Ω_s will encircle the entire right half of the s-plane. We call Ω_s the Nyquist path. The corresponding closed path in the $1 + G(s)H(s)$-plane (or the $G(s)H(s)$-plane) is called the Nyquist plot and is shown in Figure 3.8. The difference between the $1 + G(s)H(s)$-plane and the $G(s)H(s)$-plane is only a simple horizontal shift: the origin in the $1 + G(s)H(s)$-plane is the $(-1, j0)$ point in the $G(s)H(s)$-plane.

To check the stability of the closed-loop system, we apply Cauchy's argument principle to $F(s) = 1 + G(s)H(s)$; that is

$$N = Z - P$$

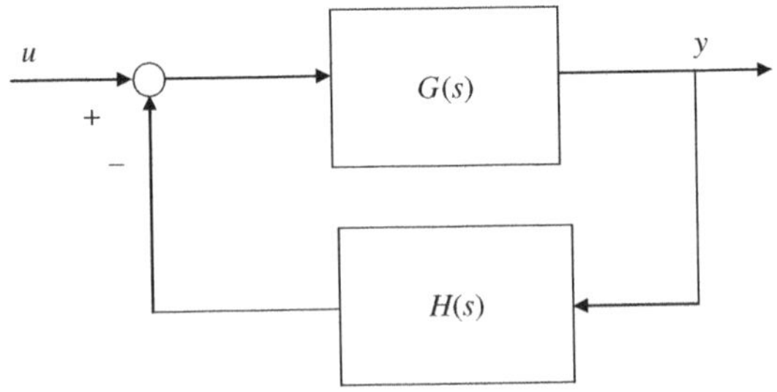

Figure 3.6 Check closed-loop stability using the Nyquist criterion.

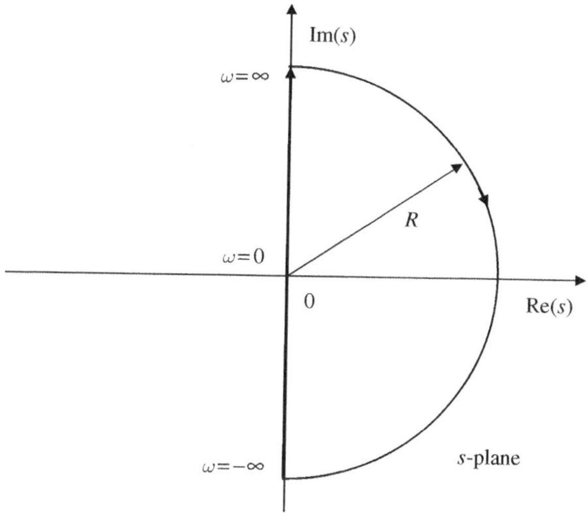

Figure 3.7 Closed path to encircle the right half of the s-plane.

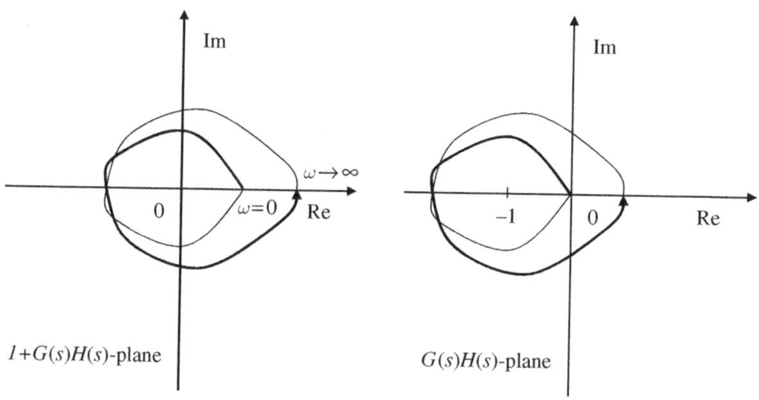

Figure 3.8 Nyquist Plot in $1+G(s)H(s)$-plane and $G(s)H(s)$-plane.

where N is the number of encirclements of the origin by the Nyquist plot in the $1+G(s)H(s)$-plane, Z is the number of zeros of $1+G(s)H(s)$ encircled by the Nyquist path in the s-plane, and P is the number of poles of $1+G(s)H(s)$ encircled by the Nyquist path in the s-plane.

For the closed-loop system with the characteristic equation $1+G(s)H(s)=0$ to be stable, Z, the number of zeros encircled by the Nyquist path in the s-plane, must be zero. $Z=0$ implies $N=-P$; that is, the number of the encirclements of the origin by the Nyquist plot in the $1+G(s)H(s)$-plane must equal the number of poles $1+G(s)H(s)$ encircled by the Nyquist path in the s-plane. The minus sign in the equation $N = -P$ means the encirclements of the origin by the Nyquist plot must be made counter-clockwise. Since

The number of encirclements of the origin
by the Nyquist plot in the $1 + G(s)H(s)$-plane
= The number of encirclements of the$(-1, j0)$ point
by the Nyquist plot in the $G(s)H(s)$-plane

and

The number of poles of $1 + G(s)H(s)$
encircled by the Nyquist path in the s-plane
= The number of poles of $1 + G(s)H(s)$
in the right half of the s-plane
= The number of poles of $G(s)H(s)$
in the right half of the s-plane

we obtain the following criterion.

Nyquist Criterion

For the closed-loop system with the characteristic equation $1 + G(s)H(s) = 0$ to be stable, the Nyquist plot of $G(s)H(s)$ must encircle the $(-1, j0)$ point as many times as the number of poles of $G(s)H(s)$ that are in the right half of the s-plane. The encirclements, if any, must be made counter-clockwise.

In many applications, $G(s)H(s)$ has no zeros or poles in the right half of the s-plane or on the imaginary axis, excluding the origin. Such a system is called a minimum-phase system.

Nyquist Criterion for Minimum-Phase Systems

For a minimum-phase system with the characteristic equation $1 + G(s)H(s) = 0$ to be stable, the Nyquist plot of $G(s)H(s)$ must not encircle the $(-1, j0)$ point.

To use the Nyquist criterion, we need to draw the Nyquist plot of $G(s)H(s)$. Note that the Nyquist plot is symmetric with respect to the real axis: the plot from $\omega = 0$ to $\omega = -\infty$ is the complex conjugate of the plot from $\omega = 0$ to $\omega = \infty$. Therefore, we only need to draw the Nyquist plot from $\omega = 0$ to $\omega = \infty$.

Example 3.11

Consider the system

$$G(s)H(s) = \frac{K(s+20)}{(s+10)(s-10)}$$

where $K > 0$ is the feedback gain to be determined. $G(s)H(s)$ has one pole in the right half of the s-plane, $\lambda = 10$. So, by the Nyquist criterion, for the closed-loop system to be stable, the Nyquist plot must encircle the $(-1, j0)$ point once. Let us sketch the Nyquist plot by first considering the two limits $\omega \to 0$ and $\omega \to -\infty$ of $G(j\omega)H(j\omega)$. For $\omega \to 0$

$$G(j\omega)H(j\omega) = \frac{K(j\omega+20)}{(j\omega+10)(j\omega-10)} \to \frac{K \times 20}{10 \times (-10)} = -\frac{K}{5}$$

For $\omega \to -\infty$

$$G(j\omega)H(j\omega) = \frac{K(j\omega+20)}{(j\omega+10)(j\omega-10)} \approx \frac{K \times j\omega}{(j\omega)^2} = \frac{K}{j\omega} = -\frac{jK}{\omega} \to -j0$$

Let us also estimate the phase of $G(j\omega)H(j\omega)$

$$\begin{aligned}\angle G(j\omega)H(j\omega) &= \angle(j\omega+20) - \angle(j\omega+10) - \angle(j\omega-10)\\ &\approx (0^\circ \sim 90^\circ) - (0^\circ \sim 90^\circ) - (90^\circ \sim 180^\circ)\\ &\approx -90^\circ \sim -180^\circ\end{aligned}$$

Therefore, the Nyquist plot is in the third quadrant; starting at $-K/5$, and approaching the origin from the $-j$ direction. It is sketched in Figure 3.9.

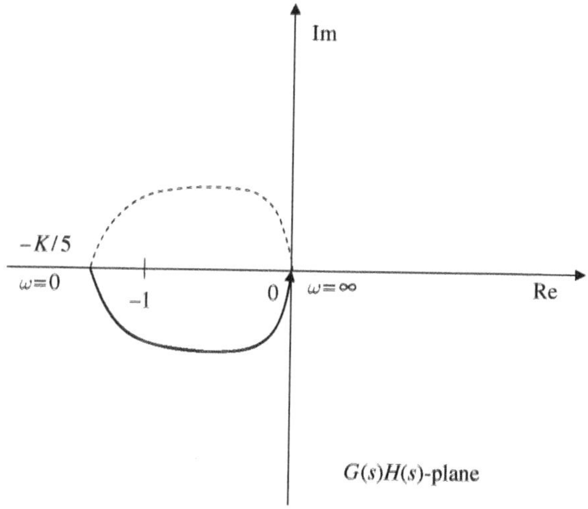

Figure 3.9 The Nyquist plot for the system in Example 3.11.

For the system to be stable, the Nyquist plot must encircle the $(-1, j0)$ point once. Hence, we must have

$$-\frac{K}{5} < -1 \Leftrightarrow K > 5$$

Example 3.12

For

$$G(s)H(s) = \frac{K}{s(s+a)}$$

where $K > 0$ and $a > 0$, we want to determine its stability. The system is a minimum-phase system. We sketch the Nyquist plot of $G(j\omega)H(j\omega)$ as follows. For $\omega \to 0$

$$G(j\omega)H(j\omega) = \frac{K}{j\omega(j\omega+a)} \approx \frac{K}{j\omega \times a} = -\frac{jK}{a\omega} \to -j\infty$$

For $\omega \to -\infty$

$$G(j\omega)H(j\omega) = \frac{K}{j\omega(j\omega+a)} \approx \frac{K}{(j\omega)^2} = -\frac{K}{\omega^2} \to -0$$

Also

$$\begin{aligned}\angle G(j\omega)H(j\omega) &= -\angle j\omega - \angle(j\omega+a) \\ &\approx -90° - (0° \sim 90°) \\ &\approx -90° \sim -180°\end{aligned}$$

Therefore, the Nyquist plot is in the third quadrant; it approaches the origin from the -1 direction and approaches ∞ from the $-j$ direction. It is sketched in Figure 3.10.

For this minimum-phase system to be stable, its Nyquist plot must not encircle the $(-1, j0)$ point. From Figure 3.8, we know that this is the case as long as $K > 0$ and $a > 0$.

Example 3.13

Consider the system

$$G(s)H(s) = \frac{K}{s(s+1)(s+2)}$$

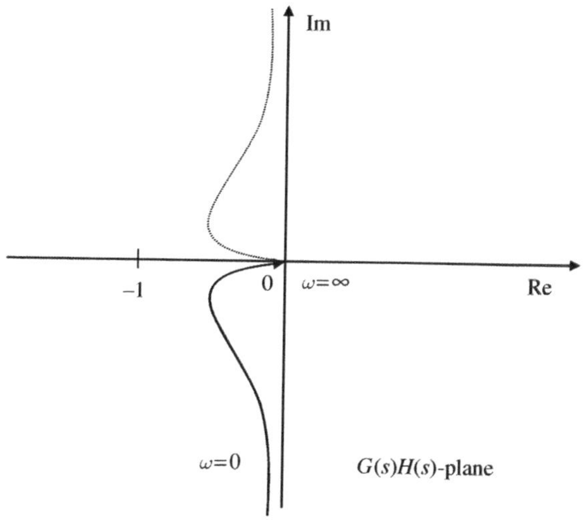

Figure 3.10 The Nyquist plot for the system in Example 3.12.

where $K > 0$. We want to determine the range of K to ensure stability. We sketch the Nyquist plot of $G(j\omega)H(j\omega)$ as follows. For $\omega \to 0$

$$G(j\omega)H(j\omega) = \frac{K}{j\omega(j\omega+1)(j\omega+2)} \approx \frac{K}{j\omega \times 2} = -\frac{jK}{2\omega} \to -j\infty$$

For $\omega \to -\infty$

$$G(j\omega)H(j\omega) = \frac{K}{j\omega(j\omega+1)(j\omega+2)} \approx \frac{K}{(j\omega)^3} = \frac{jK}{\omega^3} \to j0$$

Also

$$\begin{aligned}\angle G(j\omega)H(j\omega) &= -\angle j\omega - \angle(j\omega+1) - \angle(j\omega+1)\\ &\approx -90^\circ - (0^\circ \sim 90^\circ) - (0^\circ \sim 90^\circ)\\ &\approx -90^\circ \sim -270^\circ\end{aligned}$$

Therefore, the Nyquist plot is in the second and third quadrants; it approaches the origin from the j direction and approaches ∞ from the $-j$ direction. It is sketched in Figure 3.11.

The system is a minimum-phase system. For stability, its Nyquist plot must not encircle the $(-1, j0)$ point. For small K, the plot will not encircle the $(-1, j0)$ point as shown in Figure 3.9. However, if we increase K, then the plot will enlarge and eventually encircle the $(-1, j0)$ point. To find when this will happen, we need to calculate the intersection A of the plot

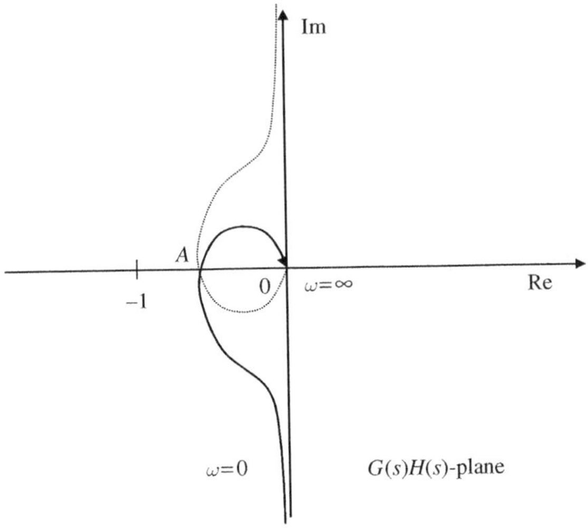

Figure 3.11 The Nyquist plot for the system in Example 3.13.

and the real axis. At the intersection, $G(j\omega)H(j\omega)$ is real. So we can calculate the corresponding frequency ω_c as follows.

$$G(j\omega_c)H(j\omega_c) = \frac{K}{j\omega_c(j\omega_c+1)(j\omega_c+2)}$$
$$= \frac{K}{j\omega_c(j\omega_c+1)(j\omega_c+2)}$$
$$= \frac{K}{j\omega_c(2-\omega_c^2)-3\omega_c^2}$$

Clearly, $G(j\omega)H(j\omega)$ = real implies $\omega_c(2-\omega_c^2)=0$. Therefore, the frequency at which the Nyquist plot intersects the real axis is given by $\omega_c = \sqrt{2}$.

ω_c is called the crossover frequency. The intersection is

$$A = G(j\omega_c)H(j\omega_c)|_{\omega_c=\sqrt{2}} = \left.\frac{K}{j\omega_c(2-\omega_c^2)-3\omega_c^2}\right|_{\omega_c=\sqrt{2}} = \frac{K}{-3\times 2} = -\frac{K}{6}$$

For the Nyquist plot not encircling the $(-1, j0)$ point, it is required that

$$A > -1 \Leftrightarrow -\frac{K}{6} > -1 \Leftrightarrow K < 6$$

Both Routh–Hurwitz and Nyquist criteria can be used to check the stability of a linear system. The Routh–Hurwitz criterion is often used in time-domain analysis and synthesis of control systems, while the Nyquist criterion is often used in frequency-domain analysis and synthesis. However, these topics are outside the scope of this book.

3.5 STABILIZABILITY AND DETECTABILITY

Stabilizability is related to both stability and controllability. From Chapter 2, we know that if a system is controllable, then we can use a state feedback to move its poles or eigenvalues to any locations in the s-plane. Therefore, we can always use state feedback to stabilize a controllable system: we just need to move the eigenvalues to the open left half of the s-plane. However, if the system is not controllable, then we may not be able to stabilize a system using state feedback. Let us consider two situations illustrated in Figure 3.12. In the figure, we use $\times$ to denote the eigenvalues and

to denote that the eigenvalue can be moved by state feedback. Both systems in (a) and (b) are not controllable, because both have two eigenvalues that can be moved and three eigenvalues that cannot. However, the system in (a) is very different from the system in (b) from the control point of view. For the system in (a), although eigenvalues λ_3 and λ_4 are unstable, they can be moved to the stable region by state feedback. So, the fact that they are unstable is not a big deal. However, for the system in (a), the unstable eigenvalues λ_3 and λ_4 will have a big problem because they cannot be

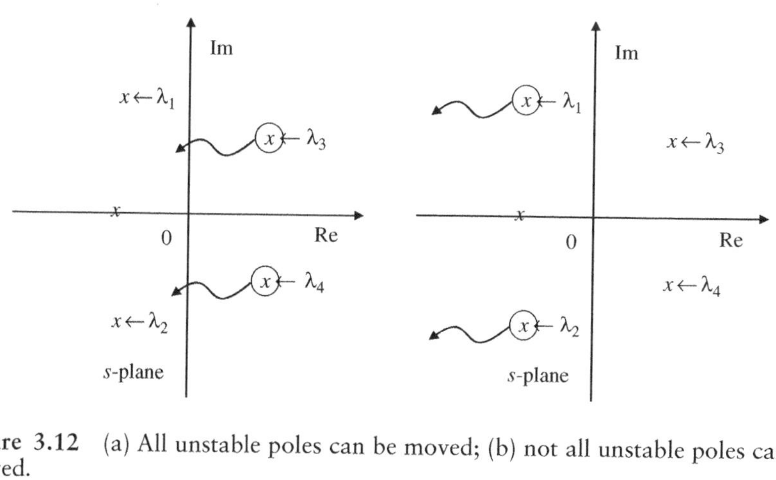

Figure 3.12 (a) All unstable poles can be moved; (b) not all unstable poles can be moved.

moved. We say that an eigenvalue λ_i is not controllable if it cannot be moved by state feedback.

Formally, $\lambda_i \in \lambda(A)$ is unstable if $\mathrm{Re}(\lambda_i) \geq 0$. $\lambda_i \in \lambda(A)$ is not controllable if $(\forall K)\lambda_i \in \lambda(A+BK)$.

Stabilizability

A linear time-invariant system

$$\dot{x} = Ax + Bu$$
$$y = Cx + Du$$

is stabilizable if all unstable eigenvalues are controllable.

The following two sufficient conditions can be easily obtained from the definition of stabilizability:

1. If a system is stable, then it is stabilizable.
2. If a system is controllable, then it is stabilizable.

Necessary and sufficient conditions for checking stabilizability are more complex, we need first to find all the eigenvalues of the system and then to determine if these eigenvalues are controllable or not.

Theorem 3.3

An eigenvalue $\lambda_i \in \lambda(A) = \lambda(A^T)$ of a linear time-invariant system (A, B) is controllable if and only if its corresponding eigenvector v_i of A^{T} satisfies the condition $v_i^{\mathrm{T}} B \neq 0$.

Proof

We prove only for the case when all eigenvalues $\lambda_1\, \lambda_2 \ldots \lambda_n$ of A are real and distinct. In this case, we know that there exist a transformation matrix T such that

$$\widetilde{A} = T^{-1}AT = \begin{bmatrix} \lambda_1 & 0 & \ldots & 0 \\ 0 & \lambda_2 & \ldots & 0 \\ \ldots & & & \\ 0 & 0 & \ldots & \lambda_n \end{bmatrix}$$

and

$$\widetilde{B} = T^{-1}B$$

On the other hand, from the discussion in Chapter 2, we know that for

$$V = [\, v_1 \;\; v_2 \;\; \ldots \;\; v_n \,]$$

we have

$$V^{-1}A^{\mathrm{T}}V = \begin{bmatrix} \lambda_1 & 0 & \ldots & 0 \\ 0 & \lambda_2 & \ldots & 0 \\ \ldots & & & \\ 0 & 0 & \ldots & \lambda_n \end{bmatrix}$$

Take the transpose

$$V^{\mathrm{T}}AV^{\mathrm{T}^{-1}} = \begin{bmatrix} \lambda_1 & 0 & \ldots & 0 \\ 0 & \lambda_2 & \ldots & 0 \\ \ldots & & & \\ 0 & 0 & \ldots & \lambda_n \end{bmatrix} = \widetilde{A}$$

Hence, we can let $T = V^{\mathrm{T}^{-1}}$ or $T^{-1} = V^{\mathrm{T}}$. Now

$$\widetilde{B} = T^{-1}B = \begin{bmatrix} v_1^{\mathrm{T}} \\ v_2^{\mathrm{T}} \\ \ldots \\ v_n^{\mathrm{T}} \end{bmatrix} B = \begin{bmatrix} v_1^{\mathrm{T}}B \\ v_2^{\mathrm{T}}B \\ \ldots \\ v_n^{\mathrm{T}}B \end{bmatrix}.$$

Since $\widetilde{A}$ is in the Jordan canonical form, it is clear that λ_i is controllable if and only if $v_i^{\mathrm{T}}B \neq 0$.

Q.E.D.

Checking stabilizability is more complex than checking controllability. Since controllability implies stabilizability, we can first check controllability. If the system is controllable, then we know it is stabilizable. If it is not, then we can check stabilizability as illustrated in the following example.

Example 3.14

Consider the following system

$$\dot{x} = \begin{bmatrix} -10 & -9 & -6 & -3 \\ 22 & 21 & 16 & 8 \\ -14 & -14 & -12 & -4 \\ -2 & -2 & -2 & -5 \end{bmatrix} x + \begin{bmatrix} 1 \\ -2 \\ 2 \\ -1 \end{bmatrix} u$$

$$y = [7\,6\,4\,2]\,x$$

The eigenvalues of A^{T} (or A) are $\lambda_1 = -1$, $\lambda_2 = 2$, $\lambda_3 = -3$, and $\lambda_4 = -4$. The corresponding eigenvectors of A^{T} are

$$v_1 = \begin{bmatrix} -0.7303 \\ -0.5477 \\ -0.3651 \\ -0.1826 \end{bmatrix} \quad v_2 = \begin{bmatrix} 0.6255 \\ 0.6255 \\ 0.4170 \\ 0.2085 \end{bmatrix} \quad v_3 = \begin{bmatrix} -0.5547 \\ -0.5547 \\ -0.5547 \\ -0.2774 \end{bmatrix} \quad v_4 = \begin{bmatrix} 0.5000 \\ 0.5000 \\ 0.5000 \\ 0.5000 \end{bmatrix}$$

For the system to be stabilizable, the unstable eigenvalue $\lambda_2 = 2$ must be controllable; that is, $v_2^{\mathrm{T}} B \neq 0$. However

$$v_2^{\mathrm{T}} B = \begin{bmatrix} 0.6255 & 0.6255 & 0.4170 & 0.2085 \end{bmatrix} \begin{bmatrix} 1 \\ -2 \\ 2 \\ -1 \end{bmatrix} = 0$$

Hence, the system is not stabilizable.

Detectability is dual to stabilizability. For detectability, we consider observability of an eigenvalue $\lambda_i \in \lambda(A)$. Formally, λ_i is not observable if $(\forall G)\lambda_i \in \lambda(A + GC)$.

Detectability

A linear time-invariant system

$$\dot{x} = Ax + Bu$$
$$y = Cx + Du$$

is detectable if all unstable eigenvalues are observable.

Dual to Theorem 3.3, we have the following theorem.

Theorem 3.4

An eigenvalue $\lambda_i \in \lambda(A)$ of a linear time-invariant system (A, C) is observable if and only if its corresponding eigenvector v_i of A satisfies the condition $Cv_i \neq 0$.

Proof

We prove only for the case when all eigenvalues $\lambda_1 \ \lambda_2 \ \ldots \ \lambda_n$ of A are real and distinct. In this case, we know that the transformation matrix

$$T = \begin{bmatrix} v_1 & v_2 & \ldots & v_n \end{bmatrix}$$

can transform (A, C) into its Jordan canonical form

$$\tilde{A} = T^{-1}AT = \begin{bmatrix} \lambda_1 & 0 & \dots & 0 \\ 0 & \lambda_2 & \dots & 0 \\ \dots & & & \\ 0 & 0 & \dots & \lambda_n \end{bmatrix}$$

Since

$$\tilde{C} = CT = C\begin{bmatrix} v_1 & v_2 & \dots & v_n \end{bmatrix} = \begin{bmatrix} Cv_1 & Cv_2 & \dots & Cv_n \end{bmatrix}$$

it is clear that λ_i is observable if and only if $Cv_i \neq 0$.

Q.E.D.

Checking detectability requires checking if all unstable eigenvalues are observable.

Example 3.15

Consider the system

$$\dot{x} = \begin{bmatrix} -10 & -9 & -6 & -3 \\ 22 & 21 & 16 & 8 \\ -14 & -14 & -12 & -4 \\ -2 & -2 & -2 & -5 \end{bmatrix} x + \begin{bmatrix} 1 \\ -2 \\ 2 \\ -1 \end{bmatrix} u$$

$$y = \begin{bmatrix} 7 & 6 & 4 & 2 \end{bmatrix} x$$

We know that eigenvalues of A are $\lambda_1 = -1$, $\lambda_2 = 2$, $\lambda_3 = -3$, and $\lambda_4 = -4$. The eigenvector of A corresponding to $\lambda_2 = 2$ is

$$v_2 = \begin{bmatrix} 0.4082 \\ -0.8165 \\ 0.4082 \\ 0 \end{bmatrix}$$

Since

$$Cv_2 = \begin{bmatrix} 7 & 6 & 4 & 2 \end{bmatrix} \begin{bmatrix} 0.4082 \\ -0.8165 \\ 0.4082 \\ 0 \end{bmatrix} = -0.4082 \neq 0$$

the system is detectable.

3.6 NOTES AND REFERENCES

In this chapter, we have discussed the issues related to stability of systems. We have defined stability and presented various criteria for checking stability. For nonlinear systems, the most common tool for checking stability is the Lyapunov stability theorem. This theorem is standard and can be found in many books. Here we mention the books by Vidyasagar [174] and by Antsaklis and Michel [7]. For linear systems, two main criteria for checking stability are the Routh–Hurwitz criterion and the Nyquist criterion. Again, many books present these two criteria, for example, the books by Kuo and Golnaraghi [92] and by Antsaklis and Michel [7]. We have also discussed stabilizability and detectability; they are properties weaker than controllability and observability. References of these properties can be found in books by Antsaklis and Michel [7], Belanger [26], Chui and Chen [44], and Rugh [140].

3.7 PROBLEMS

3.1 Consider the following nonlinear system

$$\dot{x}_1 = x_2$$
$$\dot{x}_2 = -x_1 - x_1^2 x_2$$

(a) Find the equilibrium of the system.
(b) Determine the stability of the system using the Lyapunov method if possible.

3.2 A nonlinear system is described by

$$\dot{x}_1 = x_2 - x_1(x_1^2 + x_2^2)$$
$$\dot{x}_2 = -x_1 - x_2(x_1^2 + x_2^2)$$

(a) Find the equilibrium of the system.
(b) Determine the stability of the system using the Lyapunov method if possible.

3.3 Consider the following linear time-invariant system

$$\dot{x} = \begin{bmatrix} -1 & 7 & 8 \\ 0 & -5 & 0 \\ 0 & 3 & -2 \end{bmatrix} x$$

Prove that the system is stable by finding a Lyapunov function.

3.4 Check stability of the systems with the following characteristic equation using the Routh–Hurwitz criterion. If it is unstable, find how many unstable roots.

(a) $4s^5+6s^4+s^3+7s^2+2s+9=0$
(b) $4s^4+7s^2+2s+5=0$
(c) $s^7+2s^5+5s^3+2s=0$

3.5 Consider two closed-loop systems whose loop transfer functions $G(s)H(s)$ are given by

$$\frac{s+1}{s(s+10)(s+15)} \quad \text{and} \quad \frac{s+5}{(s+2)^2(s+3)^2}$$

Determine the stability of the systems using the Routh–Hurwitz criterion.

3.6 The closed-loop systems have the characteristic equations given below. Determine the ranges of K where the systems are stable using Routh–Hurwitz criterion.

(a) $1+\dfrac{K(s+30)}{s(s+10)(s+15)}=0$

(b) $1+\dfrac{K}{s(s+10)(s^2+10s+50)}=0$

(c) $1+\dfrac{K(s+12)}{s(s+10)(s^2+10s+50)}=0$

3.7 For systems with the following characteristic equation, use the Routh–Hurwitz criterion to determine if all the poles of the systems are at the left of -3.

(a) $s^4+4s^3+7s^2+2s+5=0$
(b) $s^5+7s^4+2s^3+s^2+s+9=0$
(c) $s^4+8s^3+4s^2+2s+9=0$

3.8 Sketch the Nyquist plot for

$$G(s)H(s)=\frac{K(s+2)}{(s+10)(s-5)}$$

(a) Determine the range of K such that the closed-loop system is stable using the Nyquist criterion.
(b) Check the result of (a) using the Routh–Hurwitz criterion.

3.9 Sketch the Nyquist plot for

$$G(s)H(s) = \frac{K}{s(s+10)(s+15)}$$

(a) Determine the range of K such that the closed-loop system is stable using the Nyquist criterion.
(b) Check the result of (a) using the Routh–Hurwitz criterion.

3.10 Answer 'true' or 'false'. Explain your answers.

(a) If a system is controllable, then it is stabilizable.
(b) If a system is stable, then it is controllable.
(c) If a system is stable, then it is detectable.
(d) If a system is observable, then it is stable.

3.11 Check if the following systems are controllable. If not, find their uncontrollable modes and determine if the systems are stabilizable.

(a)

$$\dot{x} = \begin{bmatrix} -2 & 0 & 0 \\ 0 & 1 & 0 \\ 0 & 3 & -1 \end{bmatrix} x + \begin{bmatrix} 0 \\ 2 \\ -1 \end{bmatrix} u$$

(b)

$$\dot{x} = \begin{bmatrix} -2 & -7 & 4 \\ 8 & 1 & -3 \\ 0 & 0 & 3 \end{bmatrix} x + \begin{bmatrix} 2 \\ -3 \\ 0 \end{bmatrix} u$$

4

Optimal Control and Optimal Observers

In this chapter, we discuss how to design an optimal control and how to design an optimal observer. We derive the results of optimal control from a basic idea called the principle of optimality. We apply the principle of optimality to general nonlinear systems and obtain the Hamilton–Jacobi–Bellman equation for solving an optimal control problem. The Hamilton–Jacobi–Bellman equation will be used in Chapters 5 and 6 to prove robust stability of controlled systems. The Hamilton–Jacobi–Bellman equation can also be used to derive the Riccati equation for solving a linear quadratic regulator problem. Unlike the Hamilton–Jacobi–Bellman equation, the Riccati equation can always be solved if the system is stabilizable. Hence, the solution to the linear quadratic regulator problem exists as long as the system is stabilizable. The optimal observation problem is dual to the linear quadratic regulator problem. An optimal observer is often called the Kalman or Kalman–Bucy filter. Deriving results for the Kalman filter often requires knowledge of stochastic processes; to avoid this, a new method to derive the Kalman filter will be presented in this chapter.

4.1 OPTIMAL CONTROL PROBLEM

We first consider optimal control problems for general nonlinear time-invariant systems of the form

Robust Control Design: An Optimal Control Approach F. Lin

$$\dot{x} = f(x, u)$$

where $x \in R^n$ and $u \in R^m$ are the state variables and control inputs, respectively, and $f(.,.)$ is a nonlinear function that satisfies the usual condition for the existence of the solution to the differential equation.

Our goal is to find a control that minimizes the following cost functional

$$J(x, t) = \int_t^{t_f} L(x, u)\mathrm{d}\tau$$

where t is the current time, t_f is the terminating time, $x = x(t)$ is the current state, and $L(x, u)$ characterizes the cost objective.

The above cost functional is very general and can cover a large class of practical problems in everyday control applications. Let us look at some examples.

Minimal Tracking Error Problem

If our objective is to drive the state variable of the system to a desired value x_d, then we can take $L(x, u)$ to be of the form $L(x, u) = \|x - x_d\|$, $L(x, u) = (x - x_d)^T(x - x_d)$, or $L(x, u) = (x - x_d)^T Q(x - x_d)$, where $Q = Q^T \geq 0$ is a symmetric, positive semidefinite matrix describing the relative weights of state variables in x.

There are many such problems in practice. For example, in cruise control of an automobile, the goal is usually to keep the speed of the automobile at a constant, say 70 miles per hour.

Minimal Energy Problem

If our objective is to use minimal energy to control the system, then we can take $L(x, u)$ to be $L(x, u) = \|u\|$, $L(x, u) = u^T u$, or $L(x, u) = u^T Ru$ for some symmetric, positive definite matrix $R = R^T > 0$. This is because the input is usually related to the energy consumed by the system. So minimizing energy used requires minimizing the input.

For example, in a resistive circuit, if the input is a voltage source, then the power consumed by the circuit is proportional to the square of the voltage.

Combined Minimization Problem

We can also combine the above two objectives of minimizing tracking error and energy by letting, for example, $L(x, u) = (x - x_d)^T Q(x - x_d) + u^T Ru$.

Here Q and R are the relative weights on state and control. If we want to accomplish very precise tracking, then we choose matrix Q to have large values: if we want to save energy, we choose matrix R to have large values.

Example 4.1

Figure 4.1 shows an inverted pendulum mounted on a cart. In the figure, M is the mass of the cart; m is the mass of the pendulum; L is the length of the pendulum; y is the displacement of the cart, θ is the angle of the pendulum; and u is the force acting on the cart, which is the input to the system.

We assume that the mass of the pendulum is concentrated at the end of the pendulum. We also do not consider friction. Under these assumptions, we can derive the equations describing the dynamics of the system as follows (see Appendix A).

$$\ddot{y} = \frac{u + mL\dot{\theta}^2 \sin\theta - mg \sin\theta \cos\theta}{M + m\sin^2\theta}$$

$$\ddot{\theta} = \frac{-u\cos\theta - mL\dot{\theta}^2 \sin\theta \cos\theta + (M+m)g\sin\theta}{L(M + m\sin^2\theta)}$$

From the above equations, we can derive the state equations of the system. Let us define the state variables as: $x_1 = y$, $x_2 = \dot{y}$, $x_3 = \theta$, and $x_4 = \dot{\theta}$. The state equations are

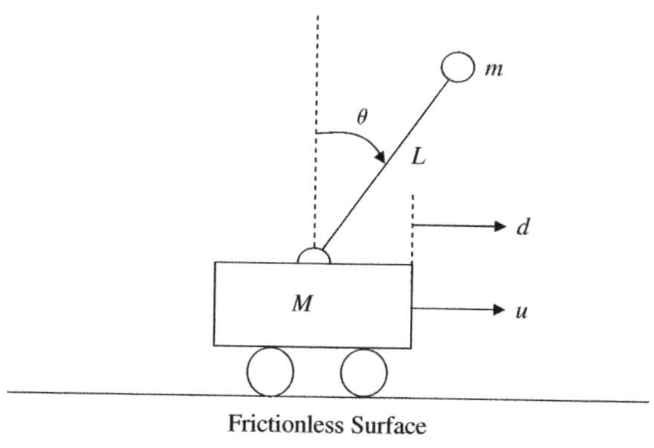

Figure 4.1 An inverted pendulum.

$$\dot{x}_1 = x_2$$

$$\dot{x}_2 = \frac{u + mLx_4^2 \sin x_3 - mg \sin x_3 \cos x_3}{M + m \sin^2 x_3}$$

$$\dot{x}_3 = x_4$$

$$\dot{x}_4 = \frac{-u \cos x_3 - mLx_4^2 \sin x_3 \cos x_3 + (M+m)g \sin x_3}{L(M + m \sin^2 x_3)}$$

If our goal is to find a control that keeps θ as close to 0 as possible with minimal energy, then we can take the cost functional as

$$J(x, t) = \int_t^{t_f} (qx_3^2 + ru^2)\mathrm{d}\tau$$

where q and r are the relative weights on state and control. If we really want θ close to 0, then we can pick a large q. On the other hand, if we really want to save energy, we can pick a large r. Obviously, the corresponding Q and R are

$$Q = \begin{bmatrix} 0 & 0 & 0 & 0 \\ 0 & 0 & 0 & 0 \\ 0 & 0 & q & 0 \\ 0 & 0 & 0 & 0 \end{bmatrix}$$

$$R = r$$

An optimal control problem for a nonlinear system is often difficult to solve. Therefore, we often need to restrict ourselves to a special class of problems. In particular, as we will show, the following problem can always be solved.

Linear Quadratic Regulator (LQR) Problem

If the system is linear time-invariant

$$\dot{x} = Ax + Bu$$

with desired state value $x_\mathrm{d} = 0$, and the cost function is quadratic

$$J(x, t) = \int_t^{t_f} (x^\mathrm{T} Qx + u^\mathrm{T} Ru)\mathrm{d}\tau$$

then, the optimal control problem is called a linear quadratic regulator (LQR) problem. Let us discuss how to solve the general nonlinear optimal control problem as well as how to solve the LQR problem.

4.2 PRINCIPLE OF OPTIMALITY

One way to solve an optimal control problem is to apply Bellman's principle of optimality. Let us illustrate this principle using Figure 4.2. The figure shows a trajectory of a second-order system. The trajectory starts in state x_0 at time t_0. It is driven by input $u(t), t \in [t_0, t_f]$. It ends in state x_f at time t_f. Denote this trajectory by $\eta(x_0, t_0, u[t_0, t_f])$. Let x_m be the state reached by the trajectory at some time $t_m \in [t_0, t_f]$.

If $\eta^*(x_0, t_0, u^*[t_0, t_f])$ is the optimal trajectory from x_0, t_0 under the optimal control $u^*[t_0, t_f]$, then the remaining trajectory $\eta^*(x_m, t_m, u^*[t_m, t_f])$ must be the optimal trajectory from x_m, t_m. The optimal control is $u^*[t_m, t_f]$, which is identical to the optimal control $u^*[t_0, t_f]$ over the remaining time interval $[t_m, t_f]$. The reason is obvious. If $u^*[t_m, t_f]$ is not the optimal control and $\eta^*(x_m, t_m, u^*[t_m, t_f])$ is not the optimal trajectory from x_m, t_m, then there must exist a different control $u'[t_m, t_f]$ and the corresponding trajectory $\eta'(x_m, t_m, u'[t_m, t_f])$ that are optimal and hence better than $u^*[t_m, t_f]$. Let $u''[t_0, t_f]$ be a control that is identical to $u^*[t_0, t_f]$ over the time interval $[t_0, t_m]$, but identical to $u'[t_m, t_f]$ over the time interval $[t_m, t_f]$. Then the new control $u''[t_0, t_f]$ over the time interval $[t_0, t_f]$ and the corresponding trajectory is better than the optimal control $u^*[t_0, t_f]$ and the corresponding trajectory, which is a contradiction.

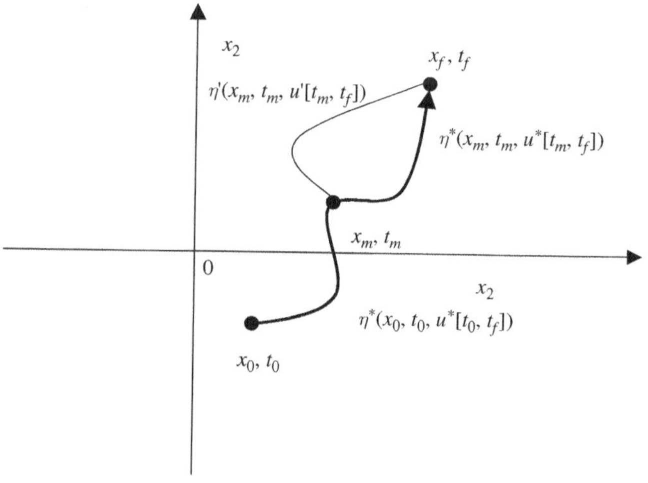

Figure 4.2 Principle of optimality.

Principle of Optimality

If a control is optimal from some initial state, then it must satisfy the following property: after any initial period, the control for the remaining period must also be optimal with regard to the state resulting from the control of the initial period.

The following example illustrates the application of the principle of optimality.

Example 4.2

Figure 4.3 shows a map of various cities and the cost of travel among these cities. We would like to find optimal (that is, least expensive) paths from City A to City B. The control here is interpreted as the decision on which city to go next.

If we enumerate all possible paths, there are a total of 34 paths from City A to City B. The number of paths will increase exponentially as the number of cities increases. Obviously, it is not wise to solve this problem by an exhaustive search.

To avoid an exhaustive search, we apply the principle of optimality as follows. Starting from the destination City B, we calculate backwards the minimal cost to reach the destination City B from each city. This is done recursively. The initial condition is shown in Figure 4.4, where the minimal cost to reach the destination City B from City B is obviously equal to zero.

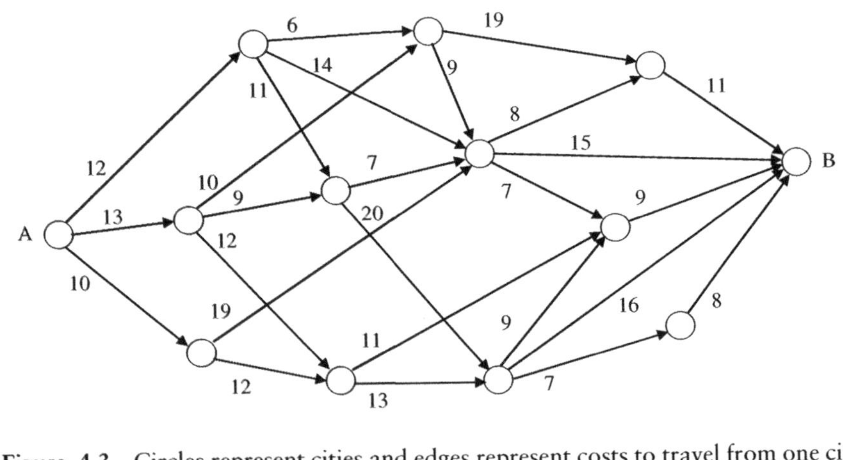

Figure 4.3 Circles represent cities and edges represent costs to travel from one city to another.

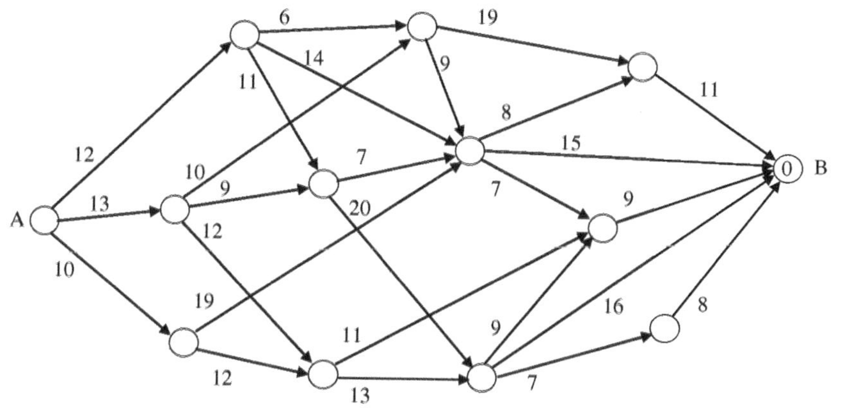

Figure 4.4 Initial condition of applying the principle of optimality.

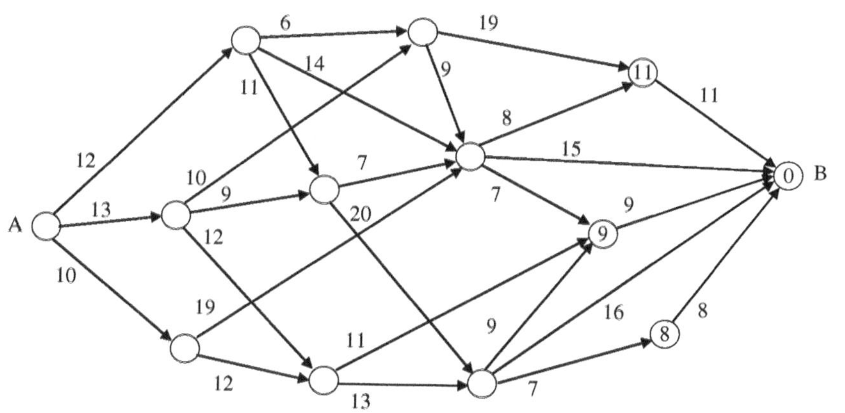

Figure 4.5 Step 1 of applying the principle of optimality.

In the next step, we calculate the minimal costs of three cities that can reach the destination City B directly. The results are obvious and shown in Figure 4.5.

In Step 2, the minimal costs of two more cities are calculated as shown in Figure 4.6. The first city has three paths leaving the city, all going to cities whose minimal costs have been calculated previously. The cost to travel along the first path is $8+11=19$. The cost to travel along the second path is 15. The cost to travel along the third path is $9+7=16$. The minimum of $\{19, 15, 16\}$ is 15, which is recorded. Similarly, the minimal cost of the second city is also $15(=7+8)$.

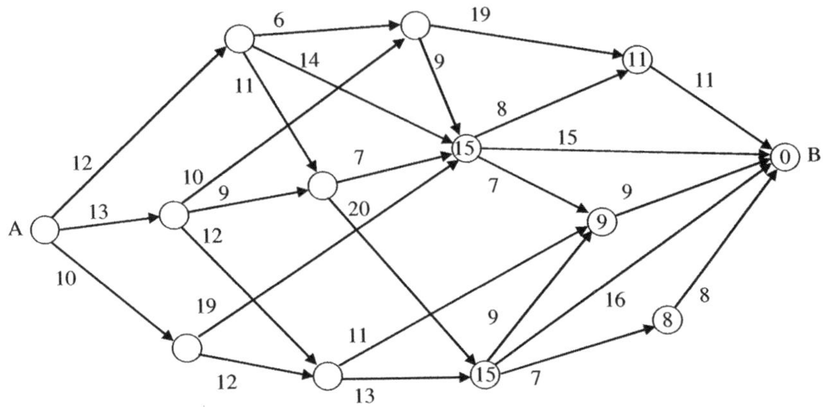

Figure 4.6 Step 2 of applying the principle of optimality.

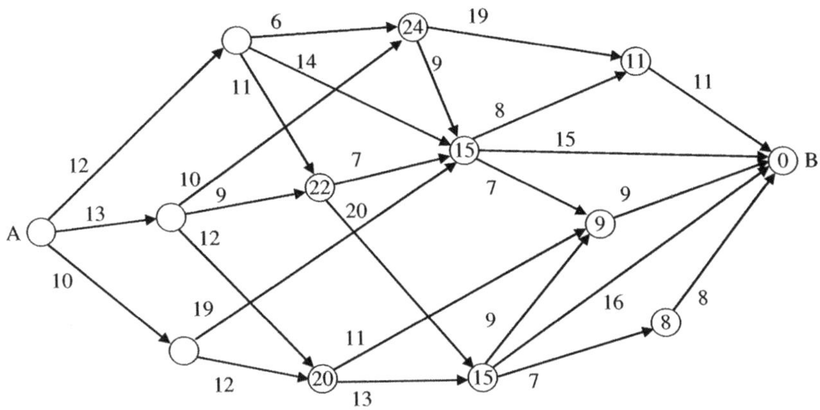

Figure 4.7 Step 3 of applying the principle of optimality.

In Step 3, the minimal costs of three more cities are calculated. The first is 24, which is obtained by minimizing $\{19+11=30, 9+15=24\}$. The second is 22, which is obtained by minimizing $\{7+15=22, 20+15=35\}$. The third is 20, which is obtained by minimizing $\{11+9=20, 13+15=28\}$. The results are shown in Figure 4.7. The next two steps of calculation are shown in Figures 4.8 and 4.9 respectively.

Therefore, the minimal cost to travel from City A to the destination City B is 41. From the minimal costs calculated, we can also find the corresponding optimal paths. For example, the optimal path from City A to City B is shown in Figure 4.10. Note that we have used the principle of optimality in the above procedure. An optimal control has the property that no matter

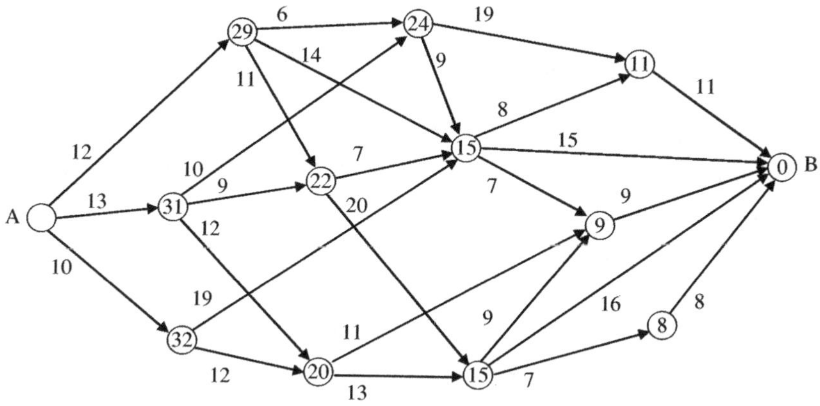

Figure 4.8 Step 4 of applying the principle of optimality.

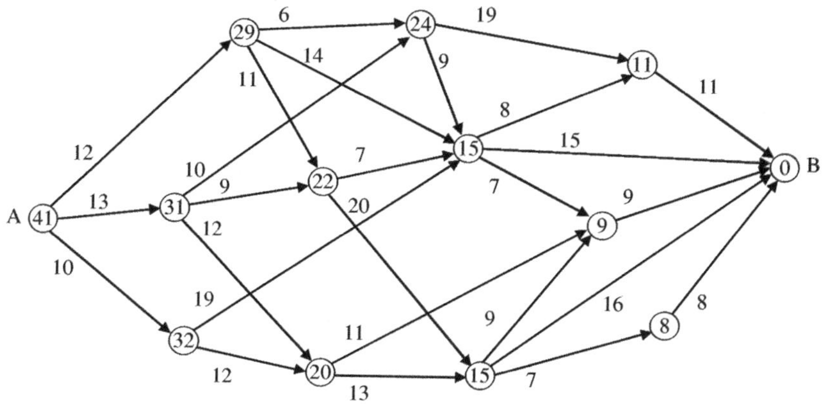

Figure 4.9 Step 5 of applying the principle of optimality.

what the previous controls have been, the remaining control must constitute an optimal control with regard to the state resulting from the previous controls. For example, since path A $\rightarrow$ C $\rightarrow$ D $\rightarrow$ B is the optimal path from A to B, path C $\rightarrow$ D $\rightarrow$ B is therefore the optimal path from C to B, by the principle of optimality. This is the reason that we can calculate minimal costs recursively.

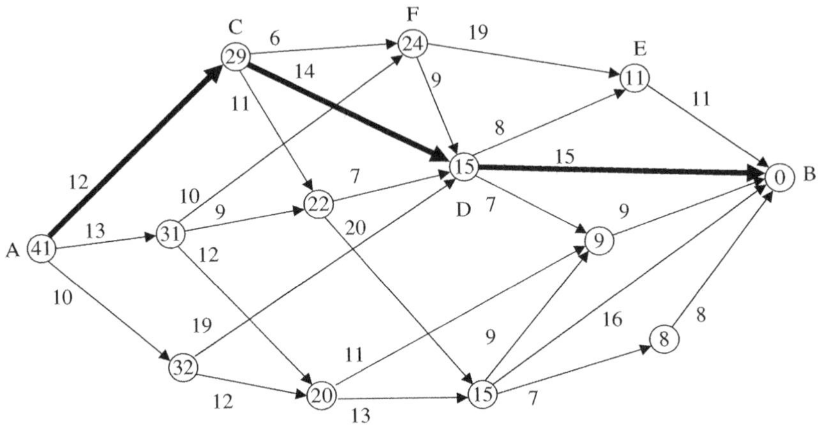

Figure 4.10 The optimal path.

4.3 HAMILTON–JACOBI–BELLMAN EQUATION

Let us now apply the principle of optimality to the optimal control of nonlinear systems. To this end, let us consider the current time t and a future time $t+\Delta t$ closed to t and the control during the interval $[t, t+\Delta t)$.

Clearly the cost $J(x, t)$ can be written as

$$J(x, t) = \int_t^{t_f} L(x, u)\mathrm{d}\tau = \int_t^{t+\Delta t} L(x, u)\mathrm{d}\tau + \int_{t+\Delta t}^{t_f} L(x, u)\mathrm{d}\tau$$
$$= \int_t^{t+\Delta t} L(x, u)\mathrm{d}\tau + J(x+\Delta x, t+\Delta t),$$

where $x+\Delta x$ is the state at $t+\Delta t$ and Δx can be approximated as $\Delta x = f(x, u)\Delta t$. Let * to denote the minimal cost under optimal control, then by the principle of optimality,

$$J^*(x, t) = \min_{u(\tau)\in R^m,\ t\leq\tau<t+\Delta t} \left\{ \int_t^{t+\Delta t} L(x, u)\mathrm{d}\tau + J^*(x+\Delta x, t+\Delta t) \right\}$$

In the above equation, $\int_t^{t+\Delta t} L(x, u)\mathrm{d}\tau$ can be approximated as $L(x, u)\Delta t$ and $J^*(x+\Delta x, t+\Delta t)$ can be approximated by its Taylor expansion:

$$J^*(x+\Delta x, t+\Delta t) = J^*(x, t) + \left(\frac{\partial J^*}{\partial x}\right)^T \Delta x + \frac{\partial J^*}{\partial t}\Delta t$$

Therefore

$$J^*(x, t) = \min_{u(\tau)\in R^m,\ t\leq\tau<t+\Delta t} \left\{ L(x, u)\Delta t + J^*(x, t) + \left(\frac{\partial J^*}{\partial x}\right)^T \Delta x + \frac{\partial J^*}{\partial t}\Delta t \right\}$$

Since $J^*(x, t)$ and $(\partial J^*/\partial t)\Delta t$ are independent of $u(\tau) \in R^m, t \leq \tau < t+\Delta t$, the above equation can be written as

$$J^*(x,t) = J^*(x,t) + \frac{\partial J^*}{\partial t}\Delta t + \min_{u(\tau)\in R^m,\ t\leq\tau<t+\Delta t}\left\{L(x,u)\Delta t + \left(\frac{\partial J^*}{\partial x}\right)^T \Delta x\right\}$$

or

$$-\frac{\partial J^*}{\partial t}\Delta t = \min_{u(\tau)\in R^m,\ t\leq\tau<t+\Delta t}\left\{L(x,u)\Delta t + \left(\frac{\partial J^*}{\partial x}\right)^T \Delta x\right\}$$

$$-\frac{\partial J^*}{\partial t} = \min_{u(\tau)\in R^m,\ t\leq\tau<t+\Delta t}\left\{L(x,u) + \left(\frac{\partial J^*}{\partial x}\right)^T \frac{\Delta x}{\Delta t}\right\}$$

Let $\Delta t \to 0$, then $\dfrac{\Delta x}{\Delta t} \to \dot{x} = f(x,u)$. Therefore, we obtain the following Hamilton–Jacobi–Bellman equation:

$$-\frac{\partial J^*}{\partial t} = \min_{u(t)\in R^m}\left\{L(x,u) + \left(\frac{\partial J^*}{\partial x}\right)^T f(x,u)\right\}$$

In this book, we consider mainly time-invariant systems with an infinite horizon ($t_f = \infty$). For such systems, $J^*(x, t)$ is independent of t. Hence, the Hamilton–Jacobi–Bellman equation reduces to

$$\min_{u(t)\in R^m}\left\{L(x,u) + \left(\frac{\partial J^*}{\partial x}\right)^T f(x,u)\right\} = 0$$

Example 4.3

Consider the following nonlinear system

$$\dot{x}_1 = -2x_1 + x_2^2 u$$
$$\dot{x}_2 = 2x_1^3 - x_2 + 2u$$

The cost functional is given by

$$J = \int_t^{t_f} (x_1^4 + 2x_2^2 + u^2)\, d\tau$$

The Hamilton–Jacobi–Bellman equation is then given below:

$$-\frac{\partial J^*}{\partial t} = \min_u\left\{x_1^4 + 2x_2^2 + u^2 + \frac{\partial J^*}{\partial x_1}(-2x_1 + x_2^2 u) + \frac{\partial J^*}{\partial x_2}(2x_1^3 - x_2 + 2u)\right\}$$

For general nonlinear systems, it is not always easy to solve the Hamilton–Jacobi–Bellman equations.

4.4 LINEAR QUADRATIC REGULATOR PROBLEM

Let us now consider the LQR problem: for a linear time-invariant system

$$\dot{x} = Ax + Bu$$

find an optimal control that minimizes the quadratic cost function

$$J(x, t) = \int_t^{t_f} (x^T Qx + u^T Ru)d\tau$$

where $Q = Q^T \geq 0$ is a symmetric and positive semi-definite matrix and $R = R^T > 0$ is a symmetric and positive definite matrix.

To find the solution to the LQR problem, we assume the minimal cost to be quadratic:

$$J^*(x, t) = x^T S(t)x$$

where $S(t) = S(t)^T \geq 0$ is a symmetric and positive semidefinite matrix function of t. By the Hamilton–Jacobi–Bellman equation, the optimal control u^* satisfies

$$-\frac{\partial J^*}{\partial t} = \min_{u(t)\in R^m} \left\{ L(x, u) + \left(\frac{\partial J^*}{\partial x}\right)^T f(x, u) \right\}$$

or

$$-x^T \dot{S}(t)x = \min_{u(t)\in R^m} \left\{ x^T Qx + u^T Ru + 2x^T S(t)(Ax + Bu) \right\}$$

To calculate $\min_{u(t)\in R^m} \left\{ x^T Qx + u^T Ru + 2x^T S(t)(Ax + Bu) \right\}$, we set the derivative of $x^T Qx + u^T Ru + 2x^T S(t)(Ax + Bu)$ with respect to u to be zero:

$$2Ru^* + 2B^T S(t)x = 0$$

Therefore, the optimal control is given by

$$u^* = -R^{-1}B^T S(t)x$$

Furthermore, $\min_{u(t)\in R^m} \left\{ x^T Qx + u^T Ru + 2x^T S(t)(Ax + Bu) \right\}$ can be calculated as

$$\begin{aligned}
&\min_{u(t)\in R^m} \{x^T Qx + u^T Ru + 2x^T S(t)(Ax + Bu)\} \\
&= x^T Qx + u^{*T} Ru^* + 2x^T S(t)(Ax + Bu^*) \\
&= x^T Qx + (-R^{-1}B^T S(t)x)^T R \left(-R^{-1}B^T S(t)x\right) + 2x^T S(t)\{Ax + B(-R^{-1}B^T S(t)x)\}
\end{aligned}$$

$$= x^{\mathrm{T}}Qx + x^{\mathrm{T}}S(t)BR^{-1}B^{\mathrm{T}}S(t)x + 2x^{\mathrm{T}}S(t)Ax - 2x^{\mathrm{T}}S(t)BR^{-1}B^{\mathrm{T}}S(t)x$$
$$= x^{\mathrm{T}}Qx - x^{\mathrm{T}}S(t)BR^{-1}B^{\mathrm{T}}S(t)x + 2x^{\mathrm{T}}S(t)Ax$$
$$= x^{\mathrm{T}}Qx - x^{\mathrm{T}}S(t)BR^{-1}B^{\mathrm{T}}S(t)x + x^{\mathrm{T}}S(t)Ax + x^{\mathrm{T}}A^{\mathrm{T}}S(t)x$$
$$= x^{\mathrm{T}}(Q - S(t)BR^{-1}B^{\mathrm{T}}S(t) + S(t)A + A^{\mathrm{T}}S(t))x$$

By the Hamilton–Jacobi–Bellman equation

$$-x^{\mathrm{T}}\dot{S}(t)x = x^{\mathrm{T}}(Q - S(t)BR^{-1}B^{\mathrm{T}}S(t) + S(t)A + A^{\mathrm{T}}S(t))x$$

In other words, $S(t)$ satisfies

$$\dot{S}(t) = -(S(t)A + A^{\mathrm{T}}S(t) + Q - S(t)BR^{-1}B^{\mathrm{T}}S(t))$$

The above equation is called the Riccati equation. The Riccati equation is much simpler than the Hamilton–Jacobi–Bellman equation. Let us show the above result by an example.

Example 4.4

Consider the following second-order system

$$\dot{x} = \begin{bmatrix} 0 & 1 \\ 2 & 3 \end{bmatrix} x + \begin{bmatrix} -1 \\ 1 \end{bmatrix} u$$

The cost functional is given by

$$J(x,t) = \int_t^{t_f} (x^T \begin{bmatrix} 1 & 1 \\ 1 & 2 \end{bmatrix} x + 3u^2) d\tau$$

To derive the Riccati equation, we denote

$$S(t) = \begin{bmatrix} S_1(t) & S_2(t) \\ S_2(t) & S_3(t) \end{bmatrix}$$

then the Riccati equation is as follows.

$$\begin{bmatrix} \dot{S}_1(t) & \dot{S}_2(t) \\ \dot{S}_2(t) & \dot{S}_3(t) \end{bmatrix} = -\begin{bmatrix} S_1(t) & S_2(t) \\ S_2(t) & S_3(t) \end{bmatrix}\begin{bmatrix} 0 & 1 \\ 2 & 3 \end{bmatrix} - \begin{bmatrix} 0 & 2 \\ 1 & 3 \end{bmatrix}\begin{bmatrix} S_1(t) & S_2(t) \\ S_2(t) & S_3(t) \end{bmatrix} - \begin{bmatrix} 1 & 1 \\ 1 & 2 \end{bmatrix}$$
$$+ \begin{bmatrix} S_1(t) & S_2(t) \\ S_2(t) & S_3(t) \end{bmatrix}\begin{bmatrix} -1 \\ 1 \end{bmatrix}\frac{1}{3}[-1\ 1]\begin{bmatrix} S_1(t) & S_2(t) \\ S_2(t) & S_3(t) \end{bmatrix}$$

However, solving the Riccati equation is still not easy. We can further simplify the Riccati equation if we consider the LQR problem where the horizon is infinite; that is

$$J(x) = \int_0^{\infty} (x^{\mathrm{T}} Qx + u^{\mathrm{T}} Ru)\mathrm{d}\tau$$

When this is the case, $S(t) = S$ is a constant matrix. Therefore, $\dot{S}(t) = 0$ and the above Riccati equation reduces to the following algebraic Riccati equation.

$$SA + A^{\mathrm{T}} S + Q - SBR^{-1} B^{\mathrm{T}} S = 0$$

Finally, some conditions must be satisfied in order for the optimal control to exist. For the LQR problem with an infinite horizon, the system must be stabilizable. Because otherwise, $x(t) \nrightarrow 0$ and hence

$$\int_0^{\infty} (x^T Qx + u^T Ru) d\tau \to \infty$$

that is, the optimal control does not exist. The following theorem summarizes our results.

Theorem 4.1

For the LQR problem with

$$\dot{x} = Ax + Bu$$

$$J(x) = \int_0^{\infty} (x^{\mathrm{T}} Qx + u^{\mathrm{T}} Ru)\mathrm{d}\tau$$

if $Q \geq 0$, $R > 0$, and (A, B) is stabilizable, then the solution to the problem exists and is given by

$$u^* = -R^{-1} B^{\mathrm{T}} Sx$$

where S is the unique positive, definite solution to the following algebraic Riccati equation.

$$SA + A^{\mathrm{T}} S + Q - SBR^{-1} B^{\mathrm{T}} S = 0$$

Example 4.5

Consider the following second-order system

$$\dot{x} = \begin{bmatrix} 0 & 1 \\ 0 & 0 \end{bmatrix} x + \begin{bmatrix} 0 \\ 1 \end{bmatrix} u$$

We would like to find an optimal control that minimizes the following cost functional

$$J(x) = \int_0^{\infty} (x_1^2 + \rho u^2)\mathrm{d}\tau$$

where $\rho > 0$ is a parameter. Therefore, we have

$$Q = \begin{bmatrix} 1 & 0 \\ 0 & 0 \end{bmatrix}$$

$$R = \rho$$

We denote

$$S = \begin{bmatrix} S_1 & S_2 \\ S_2 & S_3 \end{bmatrix}$$

then the algebraic Riccati equation becomes

$$\begin{bmatrix} S_1 & S_2 \\ S_2 & S_3 \end{bmatrix}\begin{bmatrix} 0 & 1 \\ 0 & 0 \end{bmatrix} + \begin{bmatrix} 0 & 0 \\ 1 & 0 \end{bmatrix}\begin{bmatrix} S_1 & S_2 \\ S_2 & S_3 \end{bmatrix} + \begin{bmatrix} 1 & 0 \\ 0 & 0 \end{bmatrix} - \begin{bmatrix} S_1 & S_2 \\ S_2 & S_3 \end{bmatrix}\begin{bmatrix} 0 \\ 1 \end{bmatrix}\rho^{-1}\begin{bmatrix} 0 & 1 \end{bmatrix}$$
$$\begin{bmatrix} S_1 & S_2 \\ S_2 & S_3 \end{bmatrix} = \begin{bmatrix} 0 & 0 \\ 0 & 0 \end{bmatrix}$$

This leads to the following three equations:

$$\rho - S_2^2 = 0$$

$$\rho S_1 - S_2 S_3 = 0$$

$$2\rho S_2 - S_3^2 = 0$$

The only positive, definite solution of S is given by

$$S_1 = 2^{\frac{1}{2}}\rho^{\frac{1}{4}}$$

$$S_2 = \rho^{\frac{1}{2}}$$

$$S_3 = 2^{\frac{1}{2}}\rho^{\frac{3}{4}}$$

or

$$S = \begin{bmatrix} 2^{\frac{1}{2}}\rho^{\frac{1}{4}} & \rho^{\frac{1}{2}} \\ \rho^{\frac{1}{2}} & 2^{\frac{1}{2}}\rho^{\frac{3}{4}} \end{bmatrix}.$$

The corresponding optimal control is

$$u^* = -\rho^{-1}[0 \quad 1]\begin{bmatrix} 2^{\frac{1}{2}}\rho^{\frac{1}{4}} & \rho^{\frac{1}{2}} \\ \rho^{\frac{1}{2}} & 2^{\frac{1}{2}}\rho^{\frac{3}{4}} \end{bmatrix}x = -\left[\rho^{-\frac{1}{2}} \quad 2^{\frac{1}{2}}\rho^{-\frac{1}{4}}\right]x$$

Example 4.6

Consider the following LQR problem

$$\dot{x} = \begin{bmatrix} 1 & 0 & -5 & 3 \\ 2 & -4 & 0 & 0 \\ 0 & 3 & -7 & 0 \\ -6 & 9 & 0 & -2 \end{bmatrix} x + \begin{bmatrix} 1 \\ -4 \\ 3 \\ -2 \end{bmatrix} u$$

$$J(x) = \int_0^\infty (x^T \begin{bmatrix} 9 & 0 & 2 & 0 \\ 0 & 6 & -1 & 0 \\ 2 & -1 & 7 & -3 \\ 0 & 0 & -3 & 4 \end{bmatrix} x + u^T u)d\tau$$

Using MATLAB, we can calculate the solution to the LQR problem using the command 'lqr', as illustrated in Figure 4.11. The solution to the Riccati equation

$$SA + A^T S + Q - SBR^{-1}B^T S = 0$$

is given by

$$S = \begin{bmatrix} 3.8655 & 0.1559 & -1.2756 & 0.9400 \\ 0.1559 & 0.5306 & 0.0715 & 0.0831 \\ -1.2756 & 0.0715 & 0.9702 & -0.5679 \\ 0.9400 & 0.0831 & -0.5679 & 0.7431 \end{bmatrix}$$

The state feedback control is

$$u^* = -R^{-1}B^T Sx = \left[2.4647 \quad 1.9183 \quad -2.4847 \quad 2.5822\right]x$$

In the standard LQR problem, the goal is to drive the state of a system to zero. However, in some practical applications, the goal is to drive the

```
A =
     1     0    -5     3
     2    -4     0     0
     0     3    -7     0
    -6     9     0    -2
B =
     1
    -4
     3
    -2
Q =
     9     0     2     0
     0     6    -1     0
     2    -1     7    -3
     0     0    -3     4
R =
     1
>> [K, S] = lqr (A, B, Q, R)
K =
    -2.4647   -1.9183   2.4847    -2.5822
S =
    3.8655    0.1559    -1.2756   0.9400
    0.1559    0.5306    0.0715    0.0831
    -1.2756   0.0715    0.9702    -0.5679
    0.9400    0.0831    -0.5679   0.7431
```

Figure 4.11 MATLAB results of Example 4.6.

output of the system to some constant. For example, in an automotive cruise control, the goal is to drive an automobile at a constant speed, say 70 miles per hour. For such applications, we need to modify the LQR problem as follows.

For a linear time-invariant system

$$\dot{x} = Ax + Bu$$

$$y = Cx + Du$$

the goal is to find an optimal control so that the output $\lim_{t\to\infty} y(t) = y_d$ for some desired output y_d. To solve this problem, we first need to find the corresponding desired state $\lim_{t\to\infty} x(t) = x_d$ and input $\lim_{t\to\infty} u(t) = u_d$ that achieve y_d. Obviously, x_d, u_d, and y_d must satisfy the state and output equations; that is

$$\dot{x}_{\mathrm{d}} = Ax_{\mathrm{d}} + Bu_{\mathrm{d}}$$
$$y_{\mathrm{d}} = Cx_{\mathrm{d}} + Du_{\mathrm{d}}$$

Since x_{d} is a constant, $\dot{x}_{\mathrm{d}} = 0$, we have

$$0 = Ax_{\mathrm{d}} + Bu_{\mathrm{d}}$$
$$y_{\mathrm{d}} = Cx_{\mathrm{d}} + Du_{\mathrm{d}}$$

Given y_{d}, we will solve the above equations for x_{d} and u_{d}. If there is no solution for x_{d} and u_{d}, then the goal of $\lim_{t\to\infty} y(t) = y_d$ is not achievable. So, let us assume the solution exists; that is, the linear equation

$$\begin{bmatrix} A & B \\ C & D \end{bmatrix} \begin{bmatrix} x_{\mathrm{d}} \\ u_{\mathrm{d}} \end{bmatrix} = \begin{bmatrix} 0 \\ y_{\mathrm{d}} \end{bmatrix}$$

has a solution. For single-input–single-output system, x_{d} and u_{d} can be solved as

$$\begin{bmatrix} x_{\mathrm{d}} \\ u_{\mathrm{d}} \end{bmatrix} = \begin{bmatrix} A & B \\ C & D \end{bmatrix}^{-1} \begin{bmatrix} 0 \\ y_{\mathrm{d}} \end{bmatrix}$$

Next we define new variables as

$$\Delta x(t) = x(t) - x_{\mathrm{d}}$$
$$\Delta u(t) = u(t) - u_{\mathrm{d}}$$
$$\Delta y(t) = y(t) - y_{\mathrm{d}}$$

Derive the state and output equations for Δx, Δu, and Δy as follows.

$$\Delta\dot{x} = \dot{x} - \dot{x}_{\mathrm{d}} = \dot{x} = Ax + Bu = A\Delta x + B\Delta u + Ax_{\mathrm{d}} + Bu_{\mathrm{d}} = A\Delta x + B\Delta u$$
$$\Delta y = y - y_{\mathrm{d}} = Cx + Du - Cx_{\mathrm{d}} - Du_{\mathrm{d}} = C\Delta x + D\Delta u$$

Therefore, the state and output equations for Δx, Δu, and Δy are given by the same (A, B, C, D):

$$\Delta\dot{x} = A\Delta x + B\Delta u$$
$$\Delta y = C\Delta x + D\Delta u$$

We can find a control that minimizes the cost functional

$$J(\Delta x, t) = \int_t^{t_f} (\Delta x^{\mathrm{T}} Q \Delta x + \Delta u^{\mathrm{T}} R \Delta u) \mathrm{d}\tau$$

After finding the optimal control Δu^*, we take $u^* = \Delta u^* + u_{\mathrm{d}}$, which is the optimal control of the original system.

Example 4.7

Consider the following system modelling a DC motor with load torque = 0

$$\begin{bmatrix} \dot{\theta} \\ \dot{\omega} \\ \dot{i} \end{bmatrix} = \begin{bmatrix} 0 & 1 & 0 \\ 0 & 0 & 4.438 \\ 0 & -12 & -24 \end{bmatrix} \begin{bmatrix} \theta \\ \omega \\ i \end{bmatrix} + \begin{bmatrix} 0 \\ 0 \\ 20 \end{bmatrix} v$$

$$\theta = \begin{bmatrix} 1 & 0 & 0 \end{bmatrix} \begin{bmatrix} \theta \\ \omega \\ i \end{bmatrix}$$

where the states θ, ω, i are the angle position, angle velocity, and current respectively; the input v is the voltage applied. Our goal is to drive the motor to $\theta_d = 10$ while minimizing

$$J(x) = \int_0^\infty (9(\theta - \theta_d)^2 + v^2) d\tau$$

We first find x_d and u_d as

$$\begin{bmatrix} \theta_d \\ \omega_d \\ i_d \\ v_d \end{bmatrix} = \begin{bmatrix} A & B \\ C & D \end{bmatrix}^{-1} \begin{bmatrix} 0 \\ 0 \\ 0 \\ 10 \end{bmatrix} = \begin{bmatrix} 0 & 1 & 0 & 0 \\ 0 & 0 & 4.438 & 0 \\ 0 & -12 & -24 & 20 \\ 1 & 0 & 0 & 0 \end{bmatrix}^{-1} \begin{bmatrix} 0 \\ 0 \\ 0 \\ 10 \end{bmatrix} = \begin{bmatrix} 10 \\ 0 \\ 0 \\ 0 \end{bmatrix}$$

The modified LQR problem is

$$\begin{bmatrix} \Delta\dot{\theta} \\ \Delta\dot{\omega} \\ \Delta\dot{i} \end{bmatrix} = \begin{bmatrix} 0 & 1 & 0 \\ 0 & 0 & 4.438 \\ 0 & -12 & -24 \end{bmatrix} \begin{bmatrix} \Delta\theta \\ \Delta\omega \\ \Delta i \end{bmatrix} + \begin{bmatrix} 0 \\ 0 \\ 20 \end{bmatrix} \Delta v$$

$$Q = \begin{bmatrix} 9 & 0 & 0 \\ 0 & 0 & 0 \\ 0 & 0 & 0 \end{bmatrix}$$

$$R = 1$$

Its solution is obtained using MATLAB as

$$S = \begin{bmatrix} 4.4387 & 0.9145 & 0.1500 \\ 0.9145 & 0.2550 & 0.0440 \\ 0.1500 & 0.0440 & 0.0076 \end{bmatrix}$$

$$\Delta v^* = \begin{bmatrix} -3.0000 & -0.8796 & -0.1529 \end{bmatrix} \begin{bmatrix} \Delta\theta \\ \Delta\omega \\ \Delta i \end{bmatrix}$$

Hence, the optimal control of the original problem is

$$
\begin{aligned}
v^* &= \Delta v^* + v_d \\
&= \begin{bmatrix} -3.0000 & -0.8796 & -0.1529 \end{bmatrix} \begin{bmatrix} \Delta\theta \\ \Delta\omega \\ \Delta i \end{bmatrix} \\
&= \begin{bmatrix} -3.0000 & -0.8796 & -0.1529 \end{bmatrix} \begin{bmatrix} \theta - \theta_d \\ \omega - \omega_d \\ i - i_d \end{bmatrix} \\
&= \begin{bmatrix} -3.0000 & -0.8796 & -0.1529 \end{bmatrix} \begin{bmatrix} \theta \\ \omega \\ i \end{bmatrix} \\
&\quad - \begin{bmatrix} -3.0000 & -0.8796 & -0.1529 \end{bmatrix} \begin{bmatrix} \theta_d \\ \omega_d \\ i_d \end{bmatrix} \\
&= \begin{bmatrix} -3.0000 & -0.8796 & -0.1529 \end{bmatrix} \begin{bmatrix} \theta \\ \omega \\ i \end{bmatrix} \\
&\quad - \begin{bmatrix} -3.0000 & -0.8796 & -0.1529 \end{bmatrix} \begin{bmatrix} 10 \\ 0 \\ 0 \end{bmatrix} \\
&= 30 + \begin{bmatrix} -3.0000 & -0.8796 & -0.1529 \end{bmatrix} \begin{bmatrix} \theta \\ \omega \\ i \end{bmatrix}
\end{aligned}
$$

4.5 KALMAN FILTER

The problem dual to the optimal control design is the problem of optimal observer design. An optimal observer is often called the Kalman or Kalman–Bucy filter. To present the results on the Kalman filter, let us first consider a linear time-invariant system

$$
\begin{aligned}
\dot{x} &= Ax + Bu \\
y &= Cx + Du
\end{aligned}
$$

Let us assume that there is some noise in the system, both in the state equation and in the output equation:

$$\dot{x} = Ax + Bu + \psi$$
$$y = Cx + Du + \lambda$$

where ψ is the plant noise and λ is the measurement noise.

We assume that ψ is an uncorrelated, zero-mean, and Gaussian white-noise random vector:

$$\begin{aligned} E[\psi(t)] &= 0 \\ E[\psi(t)\psi(\tau)^T] &= \Psi\delta(t-\tau) \end{aligned} \tag{4.1}$$

where E denotes the expectation, Ψ is a $n \times n$ matrix describing the 'strength' of the noise and δ is the Dirac delta function. Similarly, λ is an uncorrelated, zero-mean, and Gaussian white-noise random vector:

$$\begin{aligned} E[\lambda(t)] &= 0 \\ E[\lambda(t)\lambda(\tau)^T] &= \Gamma\delta(t-\tau) \end{aligned} \tag{4.2}$$

where Γ is an $m \times m$ matrix. We further assume that ψ and λ are uncorrelated; that is

$$E[\psi(t)\lambda(\tau)^{\mathrm{T}}] = 0 \tag{4.3}$$

The above assumptions on stochastic features of the noise are all reasonable from a practical point of view.

Because of the noise, the state estimates given by an observer as described in Chapter 2 are no longer accurate. What an optimal observer can do is to minimize the expected estimation error, knowing the stochastic features of the noises. To this end, let us assume that an observer has the form

$$\dot{\hat{x}} = A\hat{x} + Bu - G(y - \hat{y})$$

The estimation error $\tilde{x} = x - \hat{x}$ satisfies the following differential equation:

$$\begin{aligned}
\dot{\tilde{x}} &= \dot{x} - \dot{\hat{x}} \\
&= Ax + Bu + \psi - A\hat{x} - Bu + G(y - \hat{y}) \\
&= Ax + \psi - A\hat{x} + G(Cx + Du + \lambda - C\hat{x} - Du) \\
&= Ax + \psi - A\hat{x} + GCx + G\lambda - GC\hat{x}) \\
&= (A + GC)x - (A + GC)\hat{x} + \psi + G\lambda \\
&= (A + GC)\tilde{x} + \psi + G\lambda
\end{aligned}$$

From Equation (2.4), the solution of $\widetilde{x}$ when $t_0 = 0$ is

$$\widetilde{x}(t) = e^{(A+GC)t}\widetilde{x}(0) + \int_0^t e^{(A+GC)(t-\tau)}(\psi(\tau) + G\lambda(\tau))\mathrm{d}\tau$$

The first term $e^{(A+GC)t}\widetilde{x}(0)$ depends on the initial state estimates and is not stochastic. We also know that if $A+GC$ is stable, then $e^{(A+GC)t}\widetilde{x}(0) \to 0$. Hence, to minimize the expected estimation error, we need to minimize the error due to the second term, denoted by

$$e(t) = \int_0^t e^{(A+GC)(t-\tau)}(\psi(\tau) + G\lambda(\tau))\mathrm{d}\tau$$

Let us change the variable from τ to $\alpha = t - \tau$

$$\begin{aligned} e(t) &= \int_0^t e^{(A+GC)(t-\tau)}(\psi(\tau) + G\lambda(\tau))\ \mathrm{d}\tau \\ &= \int_0^t e^{(A+GC)\alpha}(\psi(t-\alpha) + G\lambda(t-\alpha))\ \mathrm{d}\alpha \end{aligned}$$

The expected estimation error due to the second term is

$$\begin{aligned} &E[e(t)e(t)^{\mathrm{T}}] \\ &= E\left[\int_0^t e^{(A+GC)\alpha}(\psi(t-\alpha) + G\lambda(t-\alpha))d\alpha\left(\int_0^t e^{(A+GC)\beta}(\psi(t-\beta)\right.\right. \\ &\quad \left.\left. +G\lambda(t-\beta))\ \mathrm{d}\beta\right)^{\mathrm{T}}\right] \\ &= E\left[\int_0^t\int_0^t e^{(A+GC)\alpha}(\psi(t-\alpha) + G\lambda(t-\alpha))(\psi(t-\beta) + G\lambda(t-\beta))^{\mathrm{T}}\right. \\ &\quad \left. \times e^{(A+GC)^{\mathrm{T}}\beta}\ \mathrm{d}\alpha\mathrm{d}\beta\right] \\ &= \int_0^t\int_0^t e^{(A+GC)\alpha} E\left[(\psi(t-\alpha) + G\lambda(t-\alpha))(\psi(t-\beta) + G\lambda(t-\beta))^{\mathrm{T}}\right] \\ &\quad \times e^{(A+GC)^{\mathrm{T}}\beta}\ \mathrm{d}\alpha\mathrm{d}\beta \end{aligned}$$

Let us calculate the expectation value of E inside the integral.

$$\begin{aligned} &E\left[(\psi(t-\alpha) + G\lambda(t-\alpha))(\psi(t-\beta) + G\lambda(t-\beta))^{\mathrm{T}}\right] \\ &= E\left[\psi(t-\alpha)\psi(t-\beta)^{\mathrm{T}}\right] + E\left[G\lambda(t-\alpha)\lambda(t-\beta)^{\mathrm{T}}G^{\mathrm{T}}\right] \\ &\quad + E\left[\psi(t-\alpha)\lambda(t-\beta)^{\mathrm{T}}G^{\mathrm{T}}\right] + E\left[G\lambda(t-\alpha)\psi(t-\beta)^{\mathrm{T}}\right] \end{aligned}$$

By Equations (4.1–4.3)

$$E\left[(\psi(t-\alpha)+G\lambda(t-\alpha))(\psi(t-\beta)+G\lambda(t-\beta))^{\mathrm{T}}\right]$$
$$=\Psi\delta(\beta-\alpha)+G\Gamma G^{\mathrm{T}}\delta(\beta-\alpha)$$
$$=(\Psi+G\Gamma G^{\mathrm{T}})\delta(\beta-\alpha)$$

Hence

$$E[e(t)e(t)^{\mathrm{T}}]$$
$$=\int_0^t\int_0^t e^{(A+GC)\alpha}(\Psi+G\Gamma G^{\mathrm{T}})\delta(\beta-\alpha)e^{(A+GC)^{\mathrm{T}}\beta}\mathrm{d}\alpha\mathrm{d}\beta$$
$$=\int_0^t\int_0^t e^{(A+GC)\alpha}(\Psi+G\Gamma G^{\mathrm{T}})e^{(A+GC)^{\mathrm{T}}\beta}\delta(\beta-\alpha)\mathrm{d}\alpha\mathrm{d}\beta$$
$$=\int_0^t e^{(A+GC)\beta}(\Psi+G\Gamma G^{\mathrm{T}})e^{(A+GC)^{\mathrm{T}}\beta}\mathrm{d}\beta$$

Our problem is to select G to minimize

$$\int_0^t e^{(A+GC)\beta}(\Psi+G\Gamma G^{\mathrm{T}})e^{(A+GC)^{\mathrm{T}}\beta}\,\mathrm{d}\beta$$

Or in the case of infinite horizon

$$\int_0^\infty e^{(A+GC)\beta}(\Psi+G\Gamma G^{\mathrm{T}})e^{(A+GC)^{\mathrm{T}}\beta}\,\mathrm{d}\beta \tag{4.4}$$

It is difficult to solve this problem directly. So we will convert this problem to a problem we know how to solve. Let us use feedback control $u=Kx$ in the LQR problem

$$\dot{x}=Ax+Bu$$
$$J(x)=\int_0^t (x^{\mathrm{T}}Qx+u^{\mathrm{T}}Ru)\,\mathrm{d}\tau$$

The closed-loop system is described by

$$\dot{x}=Ax+Bu=Ax+BKx=(A+BK)x$$

Its response at time τ is

$$x(\tau)=e^{(A+BK)\tau}x(0)$$

The cost can be calculated as

$$\begin{aligned} J(x) &= \int_0^t (x^{\mathrm{T}}Qx + (Kx)^{\mathrm{T}}RKx)\,\mathrm{d}\tau \\ &= \int_0^t x^{\mathrm{T}}(Q + K^{\mathrm{T}}RK)x\,\mathrm{d}\tau \\ &= \int_0^t (e^{(A+BK)\tau}x(0))^{\mathrm{T}}(Q + K^{\mathrm{T}}RK)e^{(A+BK)\tau}x(0)\,\mathrm{d}\tau \\ &= x(0)^{\mathrm{T}}\Big(\int_0^t e^{(A+BK)^{\mathrm{T}}\tau}(Q + K^{\mathrm{T}}RK)e^{(A+BK)\tau}\,\mathrm{d}\tau\Big)x(0) \end{aligned}$$

Minimizing $J(x)$ is equivalent to minimizing

$$\int_0^t e^{(A+BK)^{\mathrm{T}}\tau}(Q + K^{\mathrm{T}}RK)e^{(A+BK)\tau}\,\mathrm{d}\tau$$

or in the case of an infinite horizon,

$$\int_0^\infty e^{(A+BK)^{\mathrm{T}}\tau}(Q + K^{\mathrm{T}}RK)e^{(A+BK)\tau}\,\mathrm{d}\tau \tag{4.5}$$

From Theorem 4.1, the minimum is achieved if

$$K = -R^{-1}B^{\mathrm{T}}S,$$

where S is the unique positive definite solution to the following algebraic Riccati equation

$$SA + A^{\mathrm{T}}S + Q - SBR^{-1}B^{\mathrm{T}}S = 0$$

Compare Equations (4.4) and (4.5); we see the following duality.

$$\begin{array}{ccc} A+BK & \leftrightarrow & A^{\mathrm{T}}+C^{\mathrm{T}}G^{\mathrm{T}} \\ Q+K^{\mathrm{T}}RK & \leftrightarrow & \Psi+G\Gamma G^{\mathrm{T}} \\ A & \leftrightarrow & A^{\mathrm{T}} \\ B & \leftrightarrow & C^{\mathrm{T}} \\ K & \leftrightarrow & G^{\mathrm{T}} \\ Q & \leftrightarrow & \Psi \\ R & \leftrightarrow & \Gamma \end{array}$$

From the above duality, we can obtain the following theorem.

Theorem 4.2

For a linear time-invariant system

$$\dot{x} = Ax + Bu + \psi$$
$$y = Cx + Du + \lambda$$

with noise satisfying Equations (4.1)–(4.3), if (A, C) is detectable, then the optimal observer or Kalman filter is given by

$$\dot{\hat{x}} = A\hat{x} + Bu - G(y - \hat{y})$$

where $G = -(\Gamma^{-1}CU)^{\mathrm{T}} = -UC^{\mathrm{T}}\Gamma^{-1}$ and U is the unique positive, definite solution to the following algebraic Riccati equation:

$$UA^{\mathrm{T}} + AU + \Psi - UC^{\mathrm{T}}\Gamma^{-1}CU = 0$$

Example 4.8

Consider the following linear time-invariant system

$$\dot{x} = \begin{bmatrix} -2 & 0 & 0 \\ 0 & 0 & -2 \\ 0 & 0 & -1 \end{bmatrix} x + \begin{bmatrix} 2 \\ 0 \\ -1 \end{bmatrix} u + \psi$$
$$y = \begin{bmatrix} 2 & 1 & 1 \end{bmatrix} x + \lambda$$

where the noises ψ and λ satisfy Equations (4.1)–(4.3) with

$$\Psi = \begin{bmatrix} 4 & 6 & 2 \\ 6 & 9 & 3 \\ 2 & 3 & 1 \end{bmatrix}$$
$$\Gamma = 1$$

To find the Kalman filter, we solve the Riccati equation

$$UA^{T} + AU + \Psi - UC^{T}\Gamma^{-1}CU = 0$$

to obtain

$$U = \begin{bmatrix} 0.3768 & 0.6203 & 0.2049 \\ 0.6203 & 1.1616 & 0.3526 \\ 0.2049 & 0.3526 & 0.1150 \end{bmatrix}$$

Hence

$$G = -UC^{\mathrm{T}}\Gamma^{-1} = -\begin{bmatrix} 1.5788 \\ 2.7549 \\ 0.8774 \end{bmatrix}$$

The Kalman filter is given by

$$\dot{\hat{x}} = A\hat{x} + Bu - G(y - \hat{y}) = \begin{bmatrix} -2 & 0 & 0 \\ 0 & 0 & -2 \\ 0 & 0 & -1 \end{bmatrix}\hat{x} + \begin{bmatrix} 2 \\ 0 \\ -1 \end{bmatrix}u + \begin{bmatrix} 1.5788 \\ 2.7549 \\ 0.8774 \end{bmatrix}(y - \hat{y})$$

Example 4.9

Consider the linear time-invariant system

$$\dot{x} = \begin{bmatrix} 7 & 0 & 0 & -2 \\ 0 & 3 & 0 & -5 \\ -1 & 9 & 0 & 0 \\ 3 & 0 & 0 & -5 \end{bmatrix}x + \begin{bmatrix} 2 & 0 \\ 3 & 0 \\ 0 & 0 \\ 0 & -5 \end{bmatrix}u + \psi$$

$$y = \begin{bmatrix} 1 & 0 & -1 & 2 \\ 0 & -3 & 0 & 0 \end{bmatrix}x + \lambda$$

where the noises ψ and λ satisfy Equations (4.1)–(4.3) with

$$\Psi = \begin{bmatrix} 9 & -3 & 2 & 0 \\ -3 & 4 & 8 & 0 \\ 2 & 8 & 7 & 1 \\ 0 & 0 & 1 & 5 \end{bmatrix}$$

$$\Gamma = \begin{bmatrix} 2 & 4 \\ 4 & 9 \end{bmatrix}$$

Using MATLAB command $[G, U] = lqr(A', C', \Psi, \Gamma)$, we obtain

$$U = \begin{bmatrix} 11.1565 & 3.5347 & 15.8112 & 2.7756 \\ 3.5347 & 5.5952 & 12.3399 & 0.6219 \\ 15.8112 & 12.3399 & 38.7903 & 4.8480 \\ 2.7756 & 0.6219 & 4.8480 & 1.2995 \end{bmatrix}$$

$$G = -\begin{bmatrix} 25.2423 & -12.3970 \\ -0.4550 & -1.6629 \\ 14.2654 & -10.4535 \\ 6.1011 & -2.9189 \end{bmatrix}$$

The Kalman filter is given by

$$\dot{\hat{x}} = \begin{bmatrix} 7 & 0 & 0 & -2 \\ 0 & 3 & 0 & -5 \\ -1 & 9 & 0 & 0 \\ 3 & 0 & 0 & -5 \end{bmatrix} \hat{x} + \begin{bmatrix} 2 & 0 \\ 3 & 0 \\ 0 & 0 \\ 0 & -5 \end{bmatrix} u + \begin{bmatrix} 25.2423 & -12.3970 \\ -0.4550 & -1.6629 \\ 14.2654 & -10.4535 \\ 6.1011 & -2.9189 \end{bmatrix} (y - \hat{y})$$

4.6 NOTES AND REFERENCES

In this chapter, we have discussed the problems of optimal control and optimal observer, also called the Kalman filter. We have defined the optimal control problem for general nonlinear systems. We have derived the Hamilton–Jacobi–Bellman equation from the principle of optimality. For the linear quadratic regulator problem, the Hamilton–Jacobi–Bellman equation is reduced to the Riccati equation or the algebraic Riccati equation, which can be solved easily. We have also studied the problem of designing the Kalman filter, which is dual to the linear quadratic regulator problem. The optimal control problem and the linear quadratic regulator problem have been discussed in many books, including those by Bryson and Ho [29], Chui and Chen [44], Lewis and Syrmos [101], and Sage and White [141]. Discussions on the Kalman filter can be found without proof, for example, in Belanger [26]. Our proof, however, does not exist in any of the references.

4.7 PROBLEMS

4.1 The costs to travel from one place to another are indicated in Figure 4.12.

(a) Calculate the optimal cost from A to B.
(b) Indicate the optimal path.

4.2 Consider the following system

$$\dot{x}_1 = -2x_1 + x_2^3$$
$$\dot{x}_2 = x_1 - 3x_2 + u$$

We want to find a control that minimizes

$$J = \int_t^{t_f} (x_1^2 + 2u^2)\mathrm{d}t$$

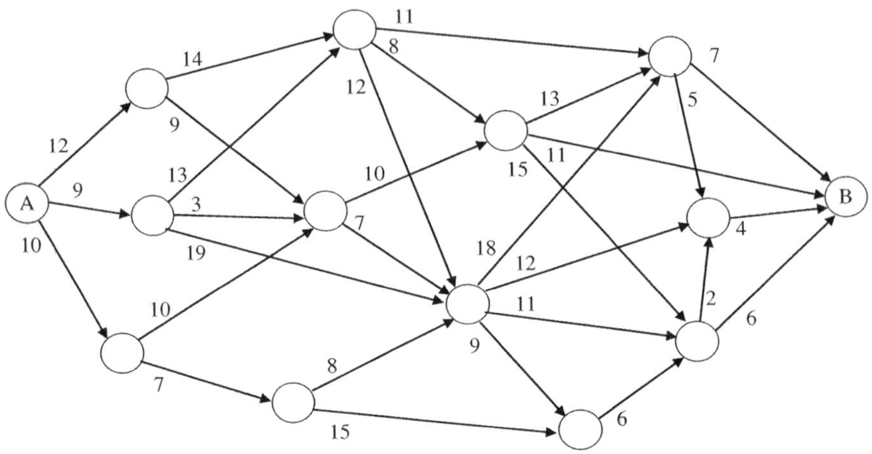

Figure 4.12 Figure of Problem 4.1.

Write the Hamilton–Jacobi–Bellman equation.

4.3 For the following nonlinear system

$$\dot{x}_1 = -4x_1 + x_2^3 + x_2 x_3$$
$$\dot{x}_2 = x_1 - 3x_2 - x_3^3 + 5u$$
$$\dot{x}_3 = -5x_1 + x_2^4 + x_3^2 - 4u$$

with the cost functional

$$J = \int_t^{t_f} (x_1^2 + 2x_2^2 + 9x_3^2 + 2u^2)\mathrm{d}t$$

write the Hamilton–Jacobi–Bellman equation.

4.4 Consider the following system

$$\dot{x} = \begin{bmatrix} 0 & 1 \\ 0 & 2 \end{bmatrix} x + \begin{bmatrix} 0 \\ 1 \end{bmatrix} u$$

with the cost functional

$$J = \int_0^{\infty} \left(x^{\mathrm{T}} \begin{bmatrix} 1 & 0 \\ 0 & 0 \end{bmatrix} x + 4u^2\right)\mathrm{d}t$$

(a) Write the corresponding algebraic Riccati equation.
(b) Solve the algebraic Riccati equation.
(c) Find the optimal control.

4.5 Consider the following system

$$\dot{x} = \begin{bmatrix} 4 & -2 & 0 & 9 \\ 3 & 7 & 0 & -8 \\ -1 & 0 & 3 & 0 \\ 0 & 4 & -7 & 1 \end{bmatrix} x + \begin{bmatrix} 0 & 2 \\ -6 & 0 \\ 0 & 0 \\ 1 & -9 \end{bmatrix} u$$

with the cost functional

$$J = \int_0^\infty (x^T \begin{bmatrix} 2 & 3 & 0 & 0 \\ 3 & 5 & 0 & 0 \\ 0 & 0 & 9 & -2 \\ 0 & 0 & -2 & 1 \end{bmatrix} x + u^T \begin{bmatrix} 4 & 2 \\ 2 & 6 \end{bmatrix} u) dt$$

(a) Write the algebraic Riccati equation.
(b) Solve the algebraic Riccati equation using MATLAB.
(c) Find the optimal control.

4.5 Using SIMULINK to simulate the closed-loop system obtained in Problem 4.4.

4.6 For the system

$$\dot{x} = \begin{bmatrix} 0 & 1 \\ 0 & 0 \end{bmatrix} x + \begin{bmatrix} 0 \\ 1 \end{bmatrix} u$$

find a state feedback to minimize

$$J = \int_0^\infty (x^T \begin{bmatrix} 1 & b \\ b & a \end{bmatrix} x + 4u^2) dt$$

4.7 Design an optimal control for the system defined by

$$\dot{x} = \begin{bmatrix} 0 & 1 \\ 0 & 0 \end{bmatrix} x + \begin{bmatrix} 0 \\ 1 \end{bmatrix} u$$

such that the following cost functional is minimized

$$J = \frac{1}{2} \int_0^\infty (x^T \begin{bmatrix} 1 & 0 \\ 0 & \mu \end{bmatrix} x + u^2) dt$$

4.8 For the following LQR problem, write the Riccati equation and the corresponding feedback. Under what condition (on α), does the optimal control exist?

$$\dot{x} = \begin{bmatrix} 1 & -3 \\ 0 & 2 \end{bmatrix} x + \begin{bmatrix} 2 \\ \alpha \end{bmatrix} u$$

$$J = \frac{1}{2}\int_0^\infty (x^T x + u^T Ru) d\tau$$

4.9 Design an optimal observer for the system

$$\dot{x} = ax + u + \psi$$

$$y = x + \lambda$$

(a) Express the observer gain in terms of a and correlations Ψ and Γ. Note that $\Psi \geq 0$, $\Gamma > 0$, and a may be positive or negative.
(b) Calculate the transfer function $\hat{X}(s)/Y(s)$.

4.10 Consider the following linear time-invariant system

$$\dot{x} = \begin{bmatrix} 2 & -4 \\ 1 & -3 \end{bmatrix} x + \begin{bmatrix} 1 \\ -3 \end{bmatrix} u + \psi$$

$$y = \begin{bmatrix} 1 & 3 \end{bmatrix} x + \lambda$$

where the noise terms ψ and λ satisfy Equations (4.1)–(4.3) with

$$\Psi = \begin{bmatrix} 3 & 2 \\ 2 & 7 \end{bmatrix}$$

$$\Gamma = 17$$

(a) Write the corresponding algebraic Riccati equation.
(b) Solve the algebraic Riccati equation.
(c) Write the equations for the Kalman filter.

4.11 Consider the following linear time-invariant system

$$\dot{x} = \begin{bmatrix} 0 & 4 & 8 & 0 \\ 9 & -2 & 5 & 0 \\ 5 & 0 & 1 & -7 \\ 4 & -9 & 0 & 3 \end{bmatrix} x + \begin{bmatrix} 4 \\ -8 \\ -1 \\ 3 \end{bmatrix} u + \psi$$

$$y = \begin{bmatrix} 7 & 3 & -8 & 2- \end{bmatrix} x + \lambda$$

where the noise terms ψ and λ satisfy Equations (4.1–4.3) with

$$\Psi = \begin{bmatrix} 7 & 1 & 0 & 2 \\ 1 & 9 & 3 & 0 \\ 0 & 3 & 6 & 1 \\ 2 & 0 & 1 & 8 \end{bmatrix}$$

$$\Gamma = 8$$

(a) Write the corresponding algebraic Riccati equation.
(b) Solve the algebraic Riccati equation using MATLAB.
(c) Write the equations for the Kalman filter.

5

Robust Control of Linear Systems

Although our optimal control approach to robust control problems can be used for both linear and nonlinear systems, we will start with linear systems in this chapter. There are two reasons for this. First, it is conceptually easier to present our approach of translating a robust control problem into an optimal control problem in linear systems. Second, for linear systems, the resulting optimal control problem is a linear quadratic regulator (LQR) problem, whose solution can be easily obtained.

We first consider the case that there is no uncertainty in B and the matching condition is satisfied. We show that not only robust stabilization, but also robust pole assignment to any arbitrary left half plane can be achieved in this case by solving the corresponding optimal control problem which reduces to an LQR problem.

We then relax the matching condition and assume arbitrary uncertainty in A. In this case, we decompose the uncertainty into a matched component and an unmatched component. An augmented control is introduced for the unmatched uncertainty. This augmented control will be discarded in the control implementation. Because of the unmatched uncertainty, a computable condition on the upper bound of the augmented control (and hence on the unmatched uncertainty) needs to be satisfied in order to guarantee the robust stability. This condition, however, is only sufficient and depends on three design parameters that can be chosen by designers. Interestingly, this sufficient condition will always be violated if we require,

Robust Control Design: An Optimal Control Approach F. Lin

instead of stabilization, pole placement to a remote left half plane with a sufficient distance from the imaginary axis.

We also extend the results to allow uncertainty in the input matrix B. In this case, we need to select a lower bound on the uncertainty in B, and design the robust control based on the lower bound.

5.1 INTRODUCTION

In this chapter, we discuss robust control of linear time-invariant systems of the form

$$\dot{x} = Ax + Bu$$

where $x \in R^n$ and $u \in R^m$ are state variables and control inputs respectively. Matrices A and B have uncertainties. The following example shows various types of uncertainties in A and B.

Example 5.1

Consider the circuit in Figure 5.1 with one voltage source, two inductors and three resistors.

Applying Kirchhoff's voltage law to the two meshes, we have

$$v_{in} = R_1 i_1 + L_1 \frac{di_1}{dt} + R_2(i_1 - i_2)$$

$$R_2(i_1 - i_2) = L_2 \frac{di_2}{dt} + R_3 i_2$$

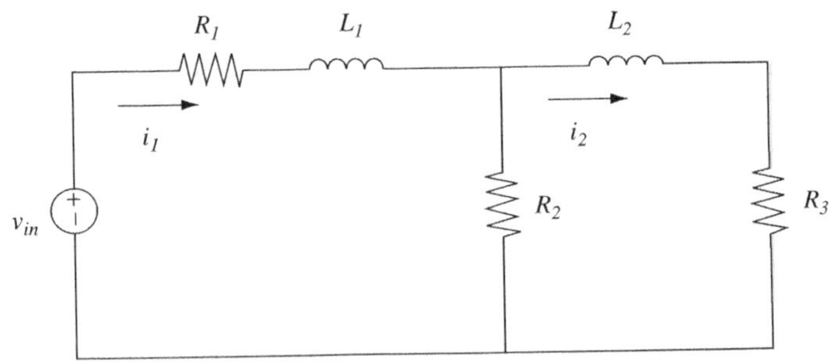

Figure 5.1 Circuit diagram of the system to illustrate various types of uncertainty.

Defining state variables to be i_1 and i_2, we obtain the following state equations:

$$\begin{bmatrix} \dot{i}_1 \\ \dot{i}_2 \end{bmatrix} = \begin{bmatrix} -(R_1+R_2)/L_1 & R_2/L_1 \\ R_2/L_2 & -(R_2+R_3)/L_2 \end{bmatrix} \begin{bmatrix} i_1 \\ i_2 \end{bmatrix} + \begin{bmatrix} 1/L_1 \\ 0 \end{bmatrix} v_{in}$$

If R_1 is uncertain, then the difference of A between the actual value R_1 and the nominal value R_{1o}, called the uncertainty in A, can be expressed as

$$\begin{aligned} & A(R_1) - A(R_{1o}) \\ & = \begin{bmatrix} -(R_1+R_2)/L_1 & R_2/L_1 \\ R_2/L_2 & -(R_2+R_3)/L_2 \end{bmatrix} - \begin{bmatrix} -(R_{1o}+R_2)/L_1 & R_2/L_1 \\ R_2/L_2 & -(R_2+R_3)/L_2 \end{bmatrix} \\ & = \begin{bmatrix} -\Delta R_1/L_1 & 0 \\ 0 & 0 \end{bmatrix} = \begin{bmatrix} 1/L_1 \\ 0 \end{bmatrix} [-\Delta R_1 \quad 0] \end{aligned}$$

where $\Delta R_1 = R_1 - R_{1o}$ is the deviation of R_1 from its nominal value. Note that for R_1, the uncertainty is in the range of B (that is, the uncertainty can be written in the form $B\phi(R_1)$ for some $\phi(R_1)$). When this is the case, we say that the matching condition is satisfied.

If R_2 is uncertain, then the uncertainty in A is

$$\begin{aligned} & A(R_2) - A(R_{2o}) \\ & = \begin{bmatrix} -(R_1+R_2)/L_1 & R_2/L_1 \\ R_2/L_2 & -(R_2+R_3)/L_2 \end{bmatrix} - \begin{bmatrix} -(R_1+R_{2o})/L_1 & R_{2o}/L_1 \\ R_{2o}/L_2 & -(R_{2o}+R_3)/L_2 \end{bmatrix} \\ & = \begin{bmatrix} -\Delta R_2/L_1 & \Delta R_2/L_1 \\ \Delta R_2/L_2 & -\Delta R_2/L_2 \end{bmatrix} \end{aligned}$$

where $\Delta R_2 = R_2 - R_{2o}$. Note that for R_2, the uncertainty is not in the range of B, that is, the matching condition is not satisfied. In fact, we can decompose the above uncertainty into a matched component and an unmatched component as follows.

$$\begin{aligned} & A(R_2) - A(R_{2o}) \\ & = \begin{bmatrix} -\Delta R_2/L_1 & \Delta R_2/L_1 \\ \Delta R_2/L_1 & -\Delta R_2/L_1 \end{bmatrix} \\ & = \begin{bmatrix} -\Delta R_2/L_1 & \Delta R_2/L_1 \\ 0 & 0_1 \end{bmatrix} + \begin{bmatrix} 0_1 & 0 \\ \Delta R_2/L_1 & -\Delta R_2/L_1 \end{bmatrix} \end{aligned}$$

where

$$\begin{bmatrix} -\Delta R_2/L_1 & \Delta R_2/L_1 \\ 0 & 0_1 \end{bmatrix} = \begin{bmatrix} 1/L_1 \\ 0 \end{bmatrix} [-\Delta R_2 \quad \Delta R_2]$$

is the matched component and

$$\begin{bmatrix} 0_1 & 0 \\ \Delta R_2/L_1 & -\Delta R_2/L_1 \end{bmatrix}$$

is the unmatched component.

If R_3 is uncertain, then the uncertainty in A is

$$\begin{aligned} &A(R_3)-A(R_{3o}) \\ &=\begin{bmatrix} -(R_1+R_2)/L_1 & R_2/L_1 \\ R_2/L_2 & -(R_2+R_3)/L_2 \end{bmatrix}-\begin{bmatrix} -(R_1+R_2)/L_1 & R_2/L_1 \\ R_2/L_2 & -(R_2+R_{3o})/L_2 \end{bmatrix} \\ &=\begin{bmatrix} 0_1 & 0 \\ 0_1 & -\Delta R_3/L_2 \end{bmatrix} \end{aligned}$$

where $\Delta R_3 = R_3 - R_{3o}$. Note that for R_3, the uncertainty is totally unmatched; that is, the matched component is 0.

If L_1 is uncertain, then there are uncertainties in both A and B. The uncertainty in B is

$$B(L_1)-B(L_{1o})=\begin{bmatrix} 1/L_1 \\ 0 \end{bmatrix}-\begin{bmatrix} 1/L_{1o} \\ 0 \end{bmatrix}=\begin{bmatrix} \Delta L_1 \\ 0 \end{bmatrix}$$

where $\Delta L_1 = 1/L_1 - 1/L_{1o}$. The uncertainty in A is

$$\begin{aligned} &A(L_1)-A(L_{1o}) \\ &=\begin{bmatrix} -(R_1+R_2)/L_1 & R_2/L_1 \\ R_2/L_2 & -(R_2+R_3)/L_2 \end{bmatrix}-\begin{bmatrix} -(R_1+R_2)/L_{1o} & R_2/L_{1o} \\ R_2/L_2 & -(R_2+R_3)/L_2 \end{bmatrix} \\ &=\begin{bmatrix} -(R_1+R_2)\Delta L_1 & R_2\Delta L_1 \\ 0 & 0 \end{bmatrix} \\ &=\begin{bmatrix} 1/L_1 \\ 0 \end{bmatrix}\left[-(R_1+R_2)L_1\Delta L_1 \quad R_2 L_1 \Delta L_1\right] \end{aligned}$$

Note that for L_1, the uncertainty in A is in the range of B (matched uncertainty).

If L_2 is uncertain, then the uncertainty in A is

$$\begin{aligned} &A(L_2)-A(L_{2o}) \\ &=\begin{bmatrix} -(R_1+R_2)/L_1 & R_2/L_1 \\ R_2/L_2 & -(R_2+R_3)/L_2 \end{bmatrix}-\begin{bmatrix} -(R_1+R_2)/L_1 & R_2/L_1 \\ R_2/L_{2o} & -(R_2+R_3)/L_{2o} \end{bmatrix} \\ &=\begin{bmatrix} 0 & 0 \\ R_2\Delta L_2 & -(R_2+R_3)\Delta L_2 \end{bmatrix} \end{aligned}$$

where $\Delta L_2 = 1/L_2 - 1/L_{2_o}$. Note that for L_2, the uncertainty in A is totally unmatched.

With this example in mind, let us first consider systems satisfying the matching condition.

5.2 MATCHED UNCERTAINTY

The system to be controlled is described by

$$\dot{x} = A(p)x + Bu$$

where $p \in P$ is an uncertain parameter vector. We first study the case where the matching condition is satisfied; that is, the uncertainty is in the range of B. In other words, the uncertainty in A can be written as $A(p) - A(p_o) = B\phi(p)$ for some $\phi(p)$, where $p_o \in P$ is the nominal value of p. Since we will translate a robust control problem into an optimal control problem, we would like to guarantee that the solution to the optimal control problem exists. For the above linear time-invariant system, the optimal control problem is actually a linear quadratic regulator (LQR) problem. As we discussed in Chapter 4, the solution to an LQR problem exists if the system is stabilizable. Therefore, we make the following assumptions.

Assumption 5.1

There exists a nominal value $p_o \in P$ such that $(A(p_o), B)$ is stabilizable.

Assumption 5.2

For any $p \in P$, there exists a $m \times n$ matrix $\phi(p)$ such that

$$A(p) - A(p_o) = B\phi(p) \tag{5.1}$$

and $\phi(p)$ is bounded.

It is not difficult to show that under Assumption 5.1, $(A(p), B)$ is stabilizable for all $p \in P$. Under Assumption 5.2, the system dynamics can be rewritten as

$$\dot{x} = A(p_o)x + Bu + B\phi(p)x$$

Our first goal is to solve the following robust control problem of stabilizing the system under uncertainty.

Robust Control Problem 5.1

Find a feedback control law $u = Kx$ such that the closed-loop system

$$\dot{x} = A(p_o)x + Bu + B\phi(p)x = A(p_o)x + BKx + B\phi(p)x$$

is asymptotically stable for all $p \in P$.

We will not attempt to solve the above robust control problem directly as its solution may not be straightforward. Our approach is to solve it indirectly by translating it into an optimal control problem. Since we consider linear systems here, the optimal control problem becomes an LQR problem.

LQR Problem 5.2

For the nominal system

$$\dot{x} = A(p_o)x + Bu$$

find a feedback control law $u = Kx$ that minimizes the cost functional

$$\int_0^\infty (x^T Fx + x^T x + u^T u)dt$$

where F is an upper bound on the uncertainty $\phi(p)^T\phi(p)$; that is, for all $p \in P$,

$$\phi(p)^T\phi(p) \le F \tag{5.2}$$

The existence of this upper bound is guaranteed by Assumption 5.2 on the boundedness of $\phi(p)$. Any F such that Equation (5.2) is satisfied can be used in the LQR problem.

To solve the LQR problem, we first solve the algebraic Riccati equation (note that $R = R^{-1} = I$)

$$A(p_o)^T S + SA(p_o) + F + I - SBB^T S = 0$$

for S. Then the solution to the LQR problem is given by $u = -B^T Sx$.

The following theorem shows that we can solve the robust control problem by solving the LQR problem.

Theorem 5.1

Robust Control Problem 5.1 is solvable under Assumptions 5.1 and 5.2. Furthermore, the solution to LQR Problem 5.2 is a solution to Robust Control Problem 5.1.

Proof

Since $(A(p_o), B)$ is stabilizable and $F \geq 0$, by Theorem 2.2, the solution to LQR Problem 5.2 exists. Let the solution be $u = Kx$. We would like to prove that it is also a solution to Robust Control Problem 5.1; that is

$$\dot{x} = A(p_o)x + BKx + B\phi(p)x \tag{5.3}$$

is asymptotically stable for all $p \in P$.

To prove this, we define

$$V(x_o) = \min_{u \in R^m} \int_0^\infty (x^T Fx + x^T x + u^T u)dt$$

to be the minimum cost of the optimal control of the nominal system from some initial state x_o. We would like to show that $V(x)$ is a Lyapunov function for system (5.3). By definition, $V(x)$ must satisfy the Hamilton–Jacobi–Bellman equation

$$\min_{u(t) \in R^m} \{L(x, u) + \left(\frac{\partial J^*}{\partial x}\right)^T f(x, u)\} = 0$$

which reduces to

$$\min_{u \in R^m} (x^T Fx + x^T x + u^T u + V_x^T (A(p_o)x + Bu)) = 0$$

where $V_x = (\partial V / \partial x)$. Since $u = Kx$ is the optimal control, it must make: (1) the above minimum zero; and (2) the derivative of $x^T Fx + x^T x + u^T u + V_x^T (A(p_o)x + Bu)$ (with respect to u) zero.

$$x^T Fx + x^T x + x^T K^T Kx + V_x^T (A(p_o)x + BKx) = 0 \tag{5.4}$$

$$2x^T K^T + V_x^T B = 0 \tag{5.5}$$

With the aid of the above two equations, we can show that $V(x)$ is a Lyapunov function for System (5.3). Clearly,

$$V(x) > 0 \qquad x \neq 0$$
$$V(x) = 0 \qquad x = 0$$

To show $\dot{V}(x) < 0$ for all $x \neq 0$, we first use Equation (5.3)

$$\begin{aligned}\dot{V}(x) &= V_x^T \dot{x} \\ &= V_x^T (A(p_o)x + BKx + B\phi(p)x) \\ &= V_x^T (A(p_o)x + BKx) + V_x^T B\phi(p)x\end{aligned}$$

By Equation (5.4)

$$V_x^{\mathrm{T}}(A(p_o)x + BKx) = -(x^{\mathrm{T}}Fx + x^{\mathrm{T}}x + x^{\mathrm{T}}K^{\mathrm{T}}Kx)$$

By Equation (5.5)

$$V_x^{\mathrm{T}}B\phi(p)x = -2x^{\mathrm{T}}K^{\mathrm{T}}\phi(p)x$$

Hence

$$\begin{aligned}
\dot{V}(x) &= -x^{\mathrm{T}}Fx - x^{\mathrm{T}}x - x^{\mathrm{T}}K^{\mathrm{T}}Kx - 2x^{\mathrm{T}}K^{\mathrm{T}}\phi(p)x \\
&= -x^{\mathrm{T}}Fx - x^{\mathrm{T}}x - x^{\mathrm{T}}K^{\mathrm{T}}Kx - 2x^{\mathrm{T}}K^{\mathrm{T}}\phi(p)x - x^{\mathrm{T}}\phi(p)^{\mathrm{T}}\phi(p)x \\
&\quad + x^{\mathrm{T}}\phi(p)^{\mathrm{T}}\phi(p)x \\
&= -x^{\mathrm{T}}Fx + x^{\mathrm{T}}\phi(p)^{\mathrm{T}}\phi(p)x - x^{\mathrm{T}}x - x^{\mathrm{T}}K^{\mathrm{T}}Kx - 2x^{\mathrm{T}}K^{\mathrm{T}}\phi(p)x \\
&\quad - x^{\mathrm{T}}\phi(p)^{\mathrm{T}}\phi(p)x \\
&= -x^{\mathrm{T}}(F - \phi(p)^{\mathrm{T}}\phi(p))x - x^{\mathrm{T}}x - x^{\mathrm{T}}(K + \phi(p))^{\mathrm{T}}(K + \phi(p))x \\
&\leq -x^{\mathrm{T}}x
\end{aligned}$$

In other words

$$\begin{aligned}
\dot{V}(x) &< 0 \qquad x \neq 0 \\
\dot{V}(x) &= 0 \qquad x = 0
\end{aligned}$$

Therefore, by the Lyapunov Stability Theorem , System (5.3) is stable for all $p \in P$. In other words, $u = Kx$ is a solution to Robust Control Problem 5.1.

Q.E.D.

Example 5.2

Consider the following second-order system

$$\begin{bmatrix} \dot{x}_1 \\ \dot{x}_2 \end{bmatrix} = \begin{bmatrix} 0 & 1 \\ 1+p & p \end{bmatrix} \begin{bmatrix} x_1 \\ x_2 \end{bmatrix} + \begin{bmatrix} 0 \\ 1 \end{bmatrix} u$$

where $p \in [-10, \ \ 1]$ is the uncertainty. We would like to design a robust control $u = Kx$ so that the closed-loop system is stable for all $p \in [-10, \ \ 1]$.

To translate this problem into an LQR problem, let us pick $p_o = 0$ and check the controllability of $(A(p_o), B)$. The controllability matrix of $(A(p_o), B)$ is

$$C = \begin{bmatrix} B & A(p_o)B \end{bmatrix} = \begin{bmatrix} 0 & 1 \\ 1 & 0 \end{bmatrix}$$

Since C is of full rank, the nominal system is controllable. Because the state equation can be written as

$$\begin{bmatrix} \dot{x}_1 \\ \dot{x}_2 \end{bmatrix} = \begin{bmatrix} 0 & 1 \\ 1 & 0 \end{bmatrix} \begin{bmatrix} x_1 \\ x_2 \end{bmatrix} + \begin{bmatrix} 0 \\ 1 \end{bmatrix} u + \begin{bmatrix} 0 \\ 1 \end{bmatrix} \begin{bmatrix} p & p \end{bmatrix} \begin{bmatrix} x_1 \\ x_2 \end{bmatrix}$$

the matching condition is satisfied with $\phi(p) = \begin{bmatrix} p & p \end{bmatrix}$.

Let us calculate F as follows.

$$\phi(p)^{\mathrm{T}} \phi(p) = \begin{bmatrix} p \\ p \end{bmatrix} \begin{bmatrix} p & p \end{bmatrix} = \begin{bmatrix} p^2 & p^2 \\ p^2 & p^2 \end{bmatrix} \leq \begin{bmatrix} 100 & 100 \\ 100 & 100 \end{bmatrix} = F$$

Therefore, the corresponding LQR problem is as follows. For the nominal system

$$\begin{bmatrix} \dot{x}_1 \\ \dot{x}_2 \end{bmatrix} = \begin{bmatrix} 0 & 1 \\ 1 & 0 \end{bmatrix} \begin{bmatrix} x_1 \\ x_2 \end{bmatrix} + \begin{bmatrix} 0 \\ 1 \end{bmatrix} u$$

find a feedback control law $u = Kx$ that minimizes the cost functional

$$\int_0^\infty (x^{\mathrm{T}} Fx + x^{\mathrm{T}} x + u^{\mathrm{T}} u) dt = \int_0^\infty (x^{\mathrm{T}} (F + I) x + u^{\mathrm{T}} u) dt$$

In other words, for the LQR problem

$$Q = F + I = \begin{bmatrix} 101 & 100 \\ 100 & 101 \end{bmatrix}$$

$$R = I = 1$$

We can use MATLAB to solve this LQR problem. From the MATLAB results, the solution to the algebraic Riccati equation

$$SA + A^{\mathrm{T}} S + Q - SBR^{-1} B^{\mathrm{T}} S = 0$$

is given by

$$S = \begin{bmatrix} 12.0995 & 11.0995 \\ 11.0995 & 11.0995 \end{bmatrix}$$

Table 5.1 Eigenvalues for different values of p.

p	λ_1	λ_2
1	−9.0995	−1.0000
0	−10.0995	−1.0000
−1	−11.0995	−1.0000
−5	−15.0995	−1.0000
−10	−20.0995	−1.0000

The corresponding control $u = -R^{-1}B^{\mathrm{T}}Sx$ is

$$u = \begin{bmatrix} -11.0995 & -11.0995 \end{bmatrix} x$$

Note that the MATLAB uses the convention $u = -Kx$.

To verify the results, we check the eigenvalues of the controlled system

$$\begin{bmatrix} \dot{x}_1 \\ \dot{x}_2 \end{bmatrix} = \begin{bmatrix} 0 & 1 \\ 1+p & p \end{bmatrix} \begin{bmatrix} x_1 \\ x_2 \end{bmatrix} + \begin{bmatrix} 0 \\ 1 \end{bmatrix} u$$

for different p. For $p = 1, 0, -1, -5, -10$, the corresponding eigenvalues λ_1 λ_2 are listed in Table 5.1. From the table, we can see that the controlled system is indeed robustly stable.

So far, our goal has been to stabilize the system. However, in some cases, we need not only to stabilize the system, but also to ensure some stability margin. This problem can be studied as follows. If we strengthen our assumption and require that $(A(p_o), B)$ be controllable, then we can not only stabilize the system, but also place the poles to the left of $-\gamma$, where γ is some arbitrary positive real number. In other words, we will solve the following robust pole placement problem.

Robust Pole Placement Problem 5.3

For an arbitrary positive real number γ, find a feedback control law $u = Kx$ such that the closed-loop system

$$\dot{x} = A(p_o)x + Bu + B\phi(p)x = A(p_o)x + BKx + B\phi(p)x$$

has all its poles on the left of $-\gamma$ for all $p \in P$.

This problem can be solved by solving the following LQR problem.

LQR Problem 5.4

For the auxiliary system

$$\dot{x} = A(p_o)x + \gamma x + Bu$$

find a feedback control law $u = Kx$ that minimizes the cost functional

$$\int_0^\infty (x^T Fx + x^T x + u^T u)dt$$

where F is an upper bound on the uncertainty $\phi(p)^T\phi(p)$ as defined in Equation (5.2).

The following theorem shows that the result obtained by translating the robust control problem into the LQR problem is indeed correct.

Theorem 5.2

Under the assumption that $(A(p_o), B)$ is controllable, Robust Pole Placement Problem 5.3 is solvable. Furthermore, the solution to LQR Problem 5.4 is a solution to Robust Pole Placement Problem 5.3.

Proof

Since $(A(p_o), B)$ and hence $(A(p_o) + \gamma I, B)$ is controllable (for all positive real γ) and $F \geq 0$, by Theorem 2.2, the solution to LQR Problem 5.4 exists. By Theorem 5.1, its solution $u = Kx$ has the following property: The system

$$\dot{x} = A(p_o)x + \gamma x + BKx + B\phi(p)x$$

is asymptotically stable for all $p \in P$; that is

$$(\forall p \in P)(\forall s, \mathrm{Re}(s) \geq 0)|sI - A(p_o) - \gamma I - BK - B\phi(p)| \neq 0$$
$$\Rightarrow (\forall p \in P)(\forall s, \mathrm{Re}(s) \geq 0)|(s - \gamma)I - A(p_o) - BK - B\phi(p)| \neq 0$$

Let $s' = s - \gamma$, then $\mathrm{Re}(s) = \mathrm{Re}(s' + \gamma) = \mathrm{Re}(s') + \gamma \geq 0 \Leftrightarrow \mathrm{Re}(s') \geq -\gamma$. Therefore

$$(\forall p \in P)(\forall s', \mathrm{Re}(s') \geq -\gamma)|s'I - A(p_o) - BK - B\phi(p)| \neq 0$$

which implies that $u = Kx$ is a solution to Robust Pole Placement Problem 5.3.

Q.E.D.

Example 5.3

Consider the system discussed in Example 5.2:

$$\begin{bmatrix}\dot{x}_1\\ \dot{x}_2\end{bmatrix}=\begin{bmatrix}0 & 1\\ 1+p & p\end{bmatrix}\begin{bmatrix}x_1\\ x_2\end{bmatrix}+\begin{bmatrix}0\\ 1\end{bmatrix}u$$

where $p \in [-10, \ \ 1]$ is the uncertainty. We would like to design a robust control $u = Kx$ so that the closed-loop system has all its poles on the left of -8 for all $p \in [-10, \ \ 1]$.

We pick $p_o = 0$ and we can check that $(A(p_o), B)$ is controllable.

The corresponding LQR problem is as follows: for the nominal system

$$\begin{bmatrix}\dot{x}_1\\ \dot{x}_2\end{bmatrix}=\begin{bmatrix}0 & 1\\ 1 & 0\end{bmatrix}\begin{bmatrix}x_1\\ x_2\end{bmatrix}+8\begin{bmatrix}x_1\\ x_2\end{bmatrix}+\begin{bmatrix}0\\ 1\end{bmatrix}u=\begin{bmatrix}8 & 1\\ 1 & 8\end{bmatrix}\begin{bmatrix}x_1\\ x_2\end{bmatrix}+\begin{bmatrix}0\\ 1\end{bmatrix}u$$

find a feedback control law $u = Kx$ that minimizes the cost functional

$$\int_0^\infty (x^T Fx + x^T x + u^T u)dt = \int_0^\infty (x^T(F+I)x + u^T u)dt$$

where $F=\begin{bmatrix}100 & 100\\ 100 & 100\end{bmatrix}$ is calculated in Example 5.1. In other words

$$Q = F + I = \begin{bmatrix}101 & 100\\ 100 & 101\end{bmatrix} \qquad R = 1$$

Solving the LQR problem using MATLAB, we obtained

$$u = \begin{bmatrix}-323.5 & -36.5\end{bmatrix}x$$

To verify the results, we check the eigenvalues of the controlled system

$$\begin{bmatrix}\dot{x}_1\\ \dot{x}_2\end{bmatrix}=\begin{bmatrix}0 & 1\\ 1+p & p\end{bmatrix}\begin{bmatrix}x_1\\ x_2\end{bmatrix}+\begin{bmatrix}0\\ 1\end{bmatrix}u$$

for different p. For $p = 1, 0, -1, -5, -10$, the corresponding eigenvalues λ_1 λ_2 are listed in Table 5.2.

Indeed, all the poles are at the left of -8, that is, the control $u = Kx$ solves the robust pole placement problem.

Table 5.2 Eigenvalues for different values of p.

p	λ_1	λ_2
1	−17.7474 + 2.5501*j*	−17.7474 − 2.5501*j*
0	−21.4863	−15.0086
−1	−24.0378	−13.4571
−5	−30.8952	−10.5997
−10	−37.6684	−8.8285

5.3 UNMATCHED UNCERTAINTY

We now relax the matching condition (5.1) in Assumption 5.2 of Section 5.2. Consider the following system

$$\dot{x} = A(p)x + Bu$$

The assumptions that we make are as follows.

Assumption 5.3

There exists a nominal value $p_o \in P$ of p such that $(A(p_o), B)$ is stabilizable.

Assumption 5.4

$A(p)$ is bounded.

Our goal is to solve the following robust control problem of stabilizing the system under uncertainty.

Robust Control Problem 5.5

Find a feedback control law $u = Kx$ such that the closed-loop system

$$\dot{x} = A(p)x + Bu = A(p)x + BKx$$

is asymptotically stable for all $p \in P$.

In order to solve this robust control problem, we first decompose the uncertainty $A(p) - A(p_o)$ into the sum of a matched component and an unmatched component. This can be done by using pseudo-inverse B^+ of B. If B is a tall matrix of full rank, then $B^+ = (B^T B)^{-1} B^T$. Let

$$A(p) - A(p_o) = BB^+(A(p) - A(p_o)) + (I - BB^+)(A(p) - A(p_o))$$

Then $BB^+(A(p) - A(p_o))$ is the matched component and $(I - BB^+)(A(p) - A(p_o))$ is the unmatched component. Note that if the matching condition is satisfied, then the unmatched part $(I - BB^+)(A(p) - A(p_o)) = 0$. Let $\phi(p) = B^+(A(p) - A(p_o))$, then

$$A(p) - A(p_o) = B\phi(p)$$

as in Section 5.2. Define F and H as the following upper bounds on the uncertainty: for all $p \in P$

$$(A(p) - A(p_o))^T B^{+T} B^{+} (A(p) - A(p_o)) \leq F \tag{5.6}$$

$$\alpha^{-2}(A(p) - A(p_o))^T (A(p) - A(p_o)) \leq H \tag{5.7}$$

where $\alpha \geq 0$ is a design parameter whose usefulness will be discussed shortly. Note that F defined in Equation (5.6) is same as F defined in Equation (5.2).

As in Section 5.2, our approach is to solve the above Robust Control Problem 5.5 indirectly by translating it into the following LQR problem.

LQR Problem 5.6

For the auxiliary system

$$\dot{x} = A(p_o)x + Bu + \alpha(I - BB^{+})v$$

find a feedback control law $u = Kx$, $v = Lx$ that minimizes the cost functional

$$\int_0^{\infty} (x^T(F + \rho^2 H + \beta^2 I)x + u^T u + \rho^2 v^T v)dt$$

where $\alpha \geq 0$, $\rho \geq 0$ and $\beta \geq 0$ are design parameters.

In this LQR problem, v is an augmented control that is used to deal with the unmatched uncertainty.

Note that if the matching condition is satisfied, then we can take the design parameters to be $\alpha = 0$, $\rho = 0$, $\beta = 1$. In this case, LQR Problem 5.6 reduces to LQR Problem 5.2. The design parameters will be selected so that a sufficient condition in the following theorem is satisfied.

The solution to the LQR problem is given by

$$\begin{bmatrix} u \\ v \end{bmatrix} = -\widetilde{R}^{-1}\widetilde{B}^T Sx$$

where S is the unique positive definite solution to the following algebraic Riccati equation.

$$S\widetilde{A} + \widetilde{A}^T S + \widetilde{Q} - S\widetilde{B}\widetilde{R}^{-1}\widetilde{B}^T S = 0$$

In our case

$$\widetilde{A} = A(p_o), \quad \widetilde{B} = \begin{bmatrix} B & \alpha(I - BB^{+}) \end{bmatrix}$$

$$\widetilde{Q} = F + \rho^2 H + \beta^2 I, \quad \widetilde{R} = \begin{bmatrix} I & 0 \\ 0 & \rho^2 I \end{bmatrix}$$

Since

$$\widetilde{B}\widetilde{R}^{-1}\widetilde{B}^{\mathrm{T}} = \begin{bmatrix} B & \alpha(I-BB^{+}) \end{bmatrix} \begin{bmatrix} I & 0 \\ 0 & \rho^{-2}I \end{bmatrix} \begin{bmatrix} B^{\mathrm{T}} \\ \alpha(I-BB^{+}) \end{bmatrix} = BB^{\mathrm{T}} + \alpha^2\rho^{-2}(I-BB^{+})^2$$

the Riccati equation becomes

$$SA(p_o) + A(p_o)^T S + F + \rho^2 H + \beta^2 I - S(BB^T + \alpha^2\rho^{-2}(I-BB^{+})^2)S = 0$$

The control is given by

$$\begin{bmatrix} u \\ v \end{bmatrix} = \begin{bmatrix} -B^{\mathrm{T}}S \\ -\alpha\rho^2(I-BB^{+})S \end{bmatrix} x = \begin{bmatrix} K \\ L \end{bmatrix} x$$

The following theorem states the relation between Robust Control Problem 5.5 and LQR Problem 5.6.

Theorem 5.3

If one can choose α, ρ and β such that the solution to LQR Problem 5.6, $u = Kx$, $v = Lx$, satisfies

$$\beta^2 I - 2\rho^2 L^T L > 0$$

then $u = Kx$ is a solution to Robust Control Problem 5.5.

Proof

Since $(A(p_o), B)$ is stabilizable and $F \geq 0$ $H \geq 0$, by Theorem 2.2, the solution to LQR Problem 5.6 exists. Denote the solution by $u = Kx$, $v = Lx$. We would like to prove that it is also a solution to Robust Control Problem 5.5; that is,

$$\dot{x} = A(p)x + BKx \tag{5.8}$$

is asymptotically stable for all $p \in P$.

To prove this, we define

$$V(x_o) = \min_{u \in R^m} \int_0^\infty (x^{\mathrm{T}}(F + \rho^2 H + \beta^2 I)x + u^{\mathrm{T}}u + \rho^2 v^{\mathrm{T}}v)dt$$

to be the minimum cost of the optimal control of the auxiliary system from some initial state x_o. We would like to show that $V(x)$ is a Lyapunov function for system (5.8). By definition, $V(x)$ must satisfy the Hamilton–Jacobi–Bellman equation, which reduces to

$$\min_{u,v}(x^{\mathrm{T}}(F+\rho^2 H+\beta^2 I)x+u^{\mathrm{T}}u+\rho^2 v^{\mathrm{T}}v+V_x^{\mathrm{T}}(A(p_o)x+Bu+\alpha(I-BB^+)v))=0.$$

Since $u=Kx$, $v=Lx$ are the optimal controls, they must satisfy

$$x^{\mathrm{T}}(F+\rho^2 H+\beta^2 I)x+x^{\mathrm{T}}K^{\mathrm{T}}Kx+\rho^2 x^{\mathrm{T}}L^{\mathrm{T}}Lx \qquad (5.9)$$
$$+V_x^{\mathrm{T}}(A(p_o)x+BKx+\alpha(I-BB^+)Lx)=0$$

$$2x^{\mathrm{T}}K^{\mathrm{T}}+V_x^{\mathrm{T}}B=0 \qquad (5.10)$$

$$2\rho^2 x^{\mathrm{T}}L^{\mathrm{T}}+V_x^{\mathrm{T}}\alpha(I-BB^+)=0 \qquad (5.11)$$

With the aid of the above three equations, we can show that $V(x)$ is a Lyapunov function for System (5.8). Clearly

$$V(x)>0 \qquad x\neq 0$$
$$V(x)=0 \qquad x=0$$

To show $\dot V(x)<0$ for all $x\neq 0$, we first use Equation (5.8)

$$\begin{aligned}\dot V(x)&=V_x^{\mathrm{T}}\dot x\\ &=V_x^{\mathrm{T}}(A(p)x+BKx)\\ &=V_x^{\mathrm{T}}(A(p_o)x+BKx+\alpha(I-BB^+)Lx)+V_x^{\mathrm{T}}(A(p)-A(p_o))x\\ &\quad -V_x^{\mathrm{T}}\alpha(I-BB^+)Lx\\ &=V_x^{\mathrm{T}}(A(p_o)x+BKx+\alpha(I-BB^+)Lx)+V_x^{\mathrm{T}}BB^+(A(p)-A(p_o))x\\ &\quad +V_x^{\mathrm{T}}(I-BB^+)(A(p)-A(p_o))x-V_x^{\mathrm{T}}\alpha(I-BB^+)Lx\end{aligned}$$

By Equation (5.9),

$$\begin{aligned}&V_x^{\mathrm{T}}(A(p_o)x+BKx+\alpha(I-BB^+)Lx)\\ &=-x^{\mathrm{T}}(F+\rho^2 H+\beta^2 I)x-x^{\mathrm{T}}K^{\mathrm{T}}Kx-\rho^2 x^{\mathrm{T}}L^{\mathrm{T}}Lx\end{aligned}$$

By Equation (5.10)

$$V_x^{\mathrm{T}}BB^+(A(p)-A(p_o))x=-2x^{\mathrm{T}}K^{\mathrm{T}}B^+(A(p)-A(p_o))x$$

By Equation (5.11)

$$V_x^{\mathrm{T}}\alpha(I-BB^+)Lx=-2\rho^2 x^{\mathrm{T}}L^{\mathrm{T}}Lx$$
$$V_x^{\mathrm{T}}(I-BB^+)(A(p)-A(p_o))x=-2\alpha^{-1}\rho^2 x^{\mathrm{T}}L^{\mathrm{T}}(A(p)-A(p_o))x$$

Therefore

$$\begin{aligned}\dot{V}(x) &= V_x^{\mathrm{T}}(A(p_o)x + BKx + \alpha(I - BB^+)Lx) + V_x^{\mathrm{T}} BB^+(A(p) - A(p_o))x \\ &\quad + V_x^{\mathrm{T}}(I - BB^+)(A(p) - A(p_o))x - V_x^{\mathrm{T}}\alpha(I - BB^+)Lx \\ &= -x^{\mathrm{T}}(F + \rho^2 H + \beta^2 I)x - x^{\mathrm{T}}K^{\mathrm{T}}Kx - \rho^2 x^{\mathrm{T}}L^{\mathrm{T}}Lx - 2x^{\mathrm{T}}K^{\mathrm{T}}B^+ \\ &\quad (A(p) - A(p_o))x \\ &\quad - 2\alpha^{-1}\rho^2 x^{\mathrm{T}}L^{\mathrm{T}}(A(p) - A(p_o))x + 2\rho^2 x^{\mathrm{T}}L^{\mathrm{T}}Lx\end{aligned}$$

By Equation (5.6)

$$\begin{aligned}&-x^{\mathrm{T}}K^{\mathrm{T}}Kx - 2x^{\mathrm{T}}K^{\mathrm{T}}B^+(A(p) - A(p_o))x \\ &= -x^{\mathrm{T}}(K - B^+(A(p) - A(p_o)))^{\mathrm{T}}(K - B^+(A(p) - A(p_o)))x \\ &\quad + x^{\mathrm{T}}(B^+(A(p) - A(p_o)))^{\mathrm{T}}(B^+(A(p) - A(p_o)))x \\ &\le x^{\mathrm{T}}(B^+(A(p) - A(p_o)))^{\mathrm{T}}(B^+(A(p) - A(p_o)))x \\ &\le x^{\mathrm{T}}Fx\end{aligned}$$

By Equation (5.7)

$$\begin{aligned}&-2\alpha^{-1}\rho^2 x^{\mathrm{T}}L^{\mathrm{T}}(A(p) - A(p_o))x \\ &\le \rho^2 x^{\mathrm{T}}L^{\mathrm{T}}Lx + \rho^2\alpha^{-2}x^{\mathrm{T}}(A(p) - A(p_o))^{\mathrm{T}}(A(p) - A(p_o))x \\ &\le \rho^2 x^{\mathrm{T}}L^{\mathrm{T}}Lx + \rho^2 x^{\mathrm{T}}Hx\end{aligned}$$

Hence

$$\begin{aligned}\dot{V}(x) &= -x^{\mathrm{T}}(F + \rho^2 H + \beta^2 I)x - x^{\mathrm{T}}K^{\mathrm{T}}Kx - \rho^2 x^{\mathrm{T}}L^{\mathrm{T}}Lx \\ &\quad - 2x^{\mathrm{T}}K^{\mathrm{T}}B^+(A(p) - A(p_o))x \\ &\quad - 2\alpha^{-1}\rho^2 x^{\mathrm{T}}L^{\mathrm{T}}(A(p) - A(p_o))x + 2\rho^2 x^{\mathrm{T}}L^{\mathrm{T}}Lx \\ &\le -x^{\mathrm{T}}(F + \rho^2 H + \beta^2 I)x - \rho^2 x^{\mathrm{T}}L^{\mathrm{T}}Lx + x^{\mathrm{T}}Fx \\ &\quad + \rho^2 x^{\mathrm{T}}L^{\mathrm{T}}Lx + \rho^2 x^{\mathrm{T}}Hx + 2\rho^2 x^{\mathrm{T}}L^{\mathrm{T}}Lx \\ &= -x^{\mathrm{T}}(\beta^2 I - 2\rho^2 L^{\mathrm{T}}L)x\end{aligned}$$

If the sufficient condition $\beta^2 I - 2\rho^2 L^{\mathrm{T}}L > 0$ is satisfied, then

$$\begin{aligned}\dot{V}(x) &< 0 \qquad x \neq 0 \\ \dot{V}(x) &= 0 \qquad x = 0\end{aligned}$$

Therefore, by the Lyapunov Stability Theorem, System (5.3) is stable for all $p \in P$. In other words, $u = Kx$ is a solution to Robust Control Problem 5.5.

Q.E.D.

Example 5.4

Consider the following second-order system

$$\begin{bmatrix} \dot{x}_1 \\ \dot{x}_2 \end{bmatrix} = \begin{bmatrix} p & 1+p \\ 1 & 0 \end{bmatrix} \begin{bmatrix} x_1 \\ x_2 \end{bmatrix} + \begin{bmatrix} 0 \\ 1 \end{bmatrix} u$$

where $p \in [-2, \ \ 2]$ is the uncertainty. We would like to design a robust control $u = Kx$ so that the closed-loop system is stable for all $p \in [-2, \ \ 2]$.

To translate this problem into a LQR problem, let us pick $p_o = 0$ and check the controllability of $(A(p_o), B)$. The controllability matrix of $(A(p_o), B)$ is

$$\mathrm{C} = \begin{bmatrix} B & A(p_o)B \end{bmatrix} = \begin{bmatrix} 0 & 1 \\ 1 & 0 \end{bmatrix}$$

Since C is of full rank, $(A(p_o), B)$ is controllable.

For this system, the matching condition is not satisfied. Let us decompose the uncertainty into the matched component and unmatched component as follows.

$$A(p) - A(p_o) = \begin{bmatrix} p & 1+p \\ 1 & 0 \end{bmatrix} - \begin{bmatrix} 0 & 1 \\ 1 & 0 \end{bmatrix} = \begin{bmatrix} p & p \\ 0 & 0 \end{bmatrix}$$

$$B^{+} = (B^{\mathrm{T}}B)^{-1}B^{\mathrm{T}} = \left(\begin{bmatrix} 0 & 1 \end{bmatrix} \begin{bmatrix} 0 \\ 1 \end{bmatrix} \right)^{-1} \begin{bmatrix} 0 & 1 \end{bmatrix} = \begin{bmatrix} 0 & 1 \end{bmatrix}$$

The matched component is

$$BB^{+}(A(p) - A(p_o)) = \begin{bmatrix} 0 \\ 1 \end{bmatrix} \begin{bmatrix} 0 & 1 \end{bmatrix} \begin{bmatrix} p & p \\ 0 & 0 \end{bmatrix} = \begin{bmatrix} 0 & 0 \\ 0 & 1 \end{bmatrix} \begin{bmatrix} p & p \\ 0 & 0 \end{bmatrix} = \begin{bmatrix} 0 & 0 \\ 0 & 0 \end{bmatrix}$$

In other words, all uncertainty is unmatched. The unmatched component is

$$(I - BB^{+})(A(p) - A(p_o)) = \begin{bmatrix} 1 & 0 \\ 0 & 0 \end{bmatrix} \begin{bmatrix} p & p \\ 0 & 0 \end{bmatrix} = \begin{bmatrix} p & p \\ 0 & 0 \end{bmatrix}$$

To define the corresponding LQR problem, we need to select the design parameters α, ρ and β. How to select these parameters is still an open problem. We select $\alpha = 0.05, \rho = 1$ and $\beta = 10$ based on our experience with these types of systems. The matrices F and H can be found as

$$(A(p) - A(p_o))^{\mathrm{T}} B^{+^{\mathrm{T}}} B^{+} (A(p) - A(p_o))$$

$$= \begin{bmatrix} p & 0 \\ p & 0 \end{bmatrix} \begin{bmatrix} 0 \\ 1 \end{bmatrix} \begin{bmatrix} 0 & 1 \end{bmatrix} \begin{bmatrix} p & p \\ 0 & 0 \end{bmatrix} = \begin{bmatrix} 0 & 0 \\ 0 & 0 \end{bmatrix} = F$$

$$\alpha^{-2} (A(p) - A(p_o))^{\mathrm{T}} (A(p) - A(p_o))$$

$$= 400 \begin{bmatrix} p & 0 \\ p & 0 \end{bmatrix} \begin{bmatrix} p & p \\ 0 & 0 \end{bmatrix} = \begin{bmatrix} 400p^2 & 400p^2 \\ 400p^2 & 400p^2 \end{bmatrix} \leq \begin{bmatrix} 1600 & 1600 \\ 1600 & 1600 \end{bmatrix} = H$$

Therefore, the LQR problem is as follows. For the auxiliary system

$$\begin{bmatrix} \dot{x}_1 \\ \dot{x}_2 \end{bmatrix} = \begin{bmatrix} 0 & 1 \\ 1 & 0 \end{bmatrix} \begin{bmatrix} x_1 \\ x_2 \end{bmatrix} + \begin{bmatrix} 0 \\ 1 \end{bmatrix} u + \begin{bmatrix} 0.05 & 0 \\ 0 & 0 \end{bmatrix} v$$

find a feedback control law $u = Kx, v = Lx$ that minimizes the cost functional

$$\int_0^\infty \left(x^{\mathrm{T}} \left(\begin{bmatrix} 0 & 0 \\ 0 & 0 \end{bmatrix} + \begin{bmatrix} 1600 & 1600 \\ 1600 & 1600 \end{bmatrix} + \begin{bmatrix} 100 & 0 \\ 0 & 100 \end{bmatrix} \right) x + u^{\mathrm{T}} u + v^{\mathrm{T}} v \right) dt$$

Combining the inputs u and v, we obtain the following matrices:

$$\widetilde{A} = \begin{bmatrix} 0 & 1 \\ 1 & 0 \end{bmatrix} \qquad \widetilde{B} = \begin{bmatrix} 0 & 0.05 & 0 \\ 1 & 0 & 0 \end{bmatrix}$$

$$\widetilde{Q} = \begin{bmatrix} 1700 & 1600 \\ 1600 & 1700 \end{bmatrix} \qquad \widetilde{R} = \begin{bmatrix} 1 & 0 & 0 \\ 0 & 1 & 0 \\ 0 & 0 & 1 \end{bmatrix}$$

Solving the above LQR problem using MATLAB, we obtain

$$\begin{bmatrix} u \\ v_1 \\ v_2 \end{bmatrix} = \begin{bmatrix} -41.7185 & -42.1792 \\ -6.5577 & -2.0859 \\ 0 & 0 \end{bmatrix} \begin{bmatrix} x_1 \\ x_2 \end{bmatrix}$$

In other words,

$$u = Kx = \begin{bmatrix} -41.7185 & -42.1792 \end{bmatrix} x$$

$$v = Lx = \begin{bmatrix} -6.5577 & -2.0859 \\ 0 & 0 \end{bmatrix} x$$

For $u = Kx$ to be the solution to the robust control problem, we need to check the sufficient condition $\beta^2 I - 2\rho^2 L^{\mathrm{T}} L > 0$. Clearly

$$\beta^2 I - 2\rho^2 L^{\mathrm{T}} L = 100 \times \begin{bmatrix} 1 & 0 \\ 0 & 1 \end{bmatrix} - 2 \times \begin{bmatrix} -6.5577 & 0 \\ -2.0859 & 0 \end{bmatrix} \begin{bmatrix} -6.5577 & -2.0859 \\ 0 & 0 \end{bmatrix}$$

$$= \begin{bmatrix} 56.9969 & -13.6788 \\ -13.6788 & 95.6489 \end{bmatrix} > 0$$

Table 5.3 Eigenvalues for different values of p.

p	λ_1	λ_2
−2	−1.0109	−43.1683
−1	−1.0000	−42.1792
0	−0.9885	−41.1907
1	−0.9765	−40.2027
2	−0.9638	−39.2154

Therefore, $u = Kx$ is the solution to the robust control problem. To verify the results, we check the eigenvalues of the controlled system

$$\begin{bmatrix} \dot{x}_1 \\ \dot{x}_2 \end{bmatrix} = \begin{bmatrix} p & 1+p \\ 1 & 0 \end{bmatrix} \begin{bmatrix} x_1 \\ x_2 \end{bmatrix} + \begin{bmatrix} 0 \\ 1 \end{bmatrix} u$$

for different p. For $p = -2, -1, 0, 1, 2$, the corresponding eigenvalues λ_1 λ_2 are listed in Table 5.3. Clearly, the results show that the controlled system is robust.

One interesting question is whether the result on the robust pole placement similar to Theorem 5.2 will hold for unmatched uncertainty. That is, if the matching condition is not satisfied, is it still possible to place all the poles to the left of $-\gamma$ for any positive real γ, under the assumption of controllability of $(A(p_o), B)$?

Our intuition seems to indicate that this is not possible. However, we cannot prove this intuitive conjecture and it remains an open problem. One reason for our conjecture arises if we consider the following LQR problem. For the auxiliary system

$$\dot{x} = A(p_o)x + \gamma x + Bu + \alpha(I - BB^{+})v$$

find a feedback control law $u = Kx$, $v = Lx$ that minimizes the cost functional

$$\int_0^\infty (x^{\mathrm{T}}(F + \rho^2 H + \beta^2 I)x + u^{\mathrm{T}}u + \rho^2 v^{\mathrm{T}} v)\mathrm{d}t$$

where $\alpha \geq 0$, $\rho \geq 0$ and $\beta \geq 0$ are design parameters and F and H are given in Equations (5.6) and (5.7). We can then show that the condition $\beta^2 I - 2\rho^2 L^T L > 0$ will be violated for a sufficiently large γ.

Theorem 5.4

Let $u = Kx$, $v = Lx$ be the solution to the LQR problem. If the matching condition is not satisfied, then for any choice of α, ρ and β

$$\beta^2 I - 2\rho^2 L^T L < 0$$

for a sufficiently large γ.

Proof

The solution to the LQR problem is given by

$$\begin{bmatrix} u \\ v \end{bmatrix} = -\widetilde{R}^{-1}\widetilde{B}^{\mathrm{T}} Sx$$

where S is the unique positive definite solution to the following algebraic Riccati equation.

$$S\widetilde{A} + \widetilde{A}^T S + \widetilde{Q} - S\widetilde{B}\widetilde{R}^{-1}\widetilde{B}^{\mathrm{T}} S = 0$$

where

$$\widetilde{A} = A(p_o) + \gamma I, \qquad \widetilde{B} = \begin{bmatrix} B & \alpha(I - BB^+) \end{bmatrix}$$
$$\widetilde{Q} = F + \rho^2 H + \beta^2 I, \quad \widetilde{R} = \begin{bmatrix} I & 0 \\ 0 & \rho^2 I \end{bmatrix}$$

Since

$$\widetilde{B}\widetilde{R}^{-1}\widetilde{B}^{\mathrm{T}} = \begin{bmatrix} B\,\alpha(I - BB^+) \end{bmatrix}\begin{bmatrix} I & 0 \\ 0 & \rho^{-2} I \end{bmatrix}\begin{bmatrix} B^{\mathrm{T}} \\ \alpha(I - BB^+) \end{bmatrix} = BB^{\mathrm{T}} + \alpha^2\rho^{-2}(I - BB^+)^2$$

the Riccati equation becomes

$$S(A(p_o) + \gamma I) + (A(p_o)^{\mathrm{T}} + \gamma I)S + F + \rho^2 H + \beta^2 I$$
$$- S(BB^{\mathrm{T}} + \alpha^2\rho^{-2}(I - BB^+)^2)S = 0.$$

Therefore, for a sufficiently large γ, the solution to the Riccati equation approaches

$$S \to 2\gamma(BB^{\mathrm{T}} + \alpha^2\rho^{-2}(I - BB^+)^2)^{-1}$$

The corresponding control

$$\begin{bmatrix} u \\ v \end{bmatrix} = -\widetilde{R}^{-1}\widetilde{B}^T Sx \to -\begin{bmatrix} I & 0 \\ 0 & \rho^{-2} I \end{bmatrix}\begin{bmatrix} B^{\mathrm{T}} \\ \alpha(I - BB^+) \end{bmatrix} 2\gamma(BB^{\mathrm{T}} + \alpha^2\rho^{-2}(I - BB^+)^2)^{-1}x$$
$$= \begin{bmatrix} -2\gamma B^{\mathrm{T}}(BB^{\mathrm{T}} + \alpha^2\rho^{-2}(I - BB^+)^2)^{-1} \\ -2\gamma\alpha\rho^{-2}(I - BB^+)(BB^{\mathrm{T}} + \alpha^2\rho^{-2}(I - BB^+)^2)^{-1} \end{bmatrix} x$$

That is, $L \rightarrow -2\gamma\alpha\rho^{-2}(I - BB^{+})(BB^{T} + \alpha^{2}\rho^{-2}(I - BB^{+})^{2})^{-1}$. Therefore, for a sufficiently large γ

$$\beta^2 I - 2\rho^2 L^T L < 0$$

no matter how we choose α, ρ and β.

Q.E.D.

This theorem is, however, not a proof of our conjecture, because the condition $\beta^2 I - 2\rho^2 L^T L > 0$ is sufficient, but not necessary.

5.4 UNCERTAINTY IN THE INPUT MATRIX

We now allow uncertainty in the input matrix. We consider three cases.

Case 1

Input uncertainty enters the system via B and uncertainty in $A(p)$ satisfies the matching condition. In other words, we consider the following system

$$\dot{x} = A(p)x + BD(p)u$$

where $D(p)$ is an $m \times m$ matrix representing the uncertainty in the input matrix. We first make the following assumptions.

Assumption 5.5

There exists a nominal value $p_o \in P$ of p such that $(A(p_o), B)$ is stabilizable.

Assumption 5.6

There exists a constant matrix D such that for all $p \in P$

$$0 < D \leq D(p)$$

Assumption 5.7

For any $p \in P$, there exists a $m \times n$ matrix $\phi(p)$ such that

$$A(p) - A(p_o) = BD\phi(p)$$

and $\phi(p)$ is bounded.

Under these assumptions, the system dynamics can be rewritten as

$$\dot{x} = A(p_o)x + BD(u + E(p)u) + BD\phi(p)x$$

where $E(p) = D^{-1}D(p) - I \geq 0$

Our goal is to solve the following robust control problem of stabilizing the system under uncertainty.

Robust Control Problem 5.7

Find a feedback control law $u = Kx$ such that the closed-loop system

$$\begin{aligned} \dot{x} &= A(p_o)x + BD(u + E(p)u) + BD\phi(p)x \\ &= A(p_o)x + BD(Kx + E(p)Kx) + BD\phi(p)x \end{aligned}$$

is asymptotically stable for all $p \in P$.

We would like to translate the above problem into the following LQR Problem.

LQR Problem 5.8

For the auxiliary system

$$\dot{x} = A(p_o)x + BDu$$

find a feedback control law $u = Kx$ that minimizes the cost functional

$$\int_0^\infty (x^T Fx + x^T x + u^T u)dt$$

where F is an upper bound on the uncertainty $\phi(p)^T\phi(p)$, that is, for all $p \in P$

$$\phi(p)^T\phi(p) \leq F$$

To solve the LQR problem, we first solve the algebraic Riccati equation (note that $R = R^{-1} = I$)

$$A(p_o)^{\mathrm{T}}S + SA(p_o) + F + I - SBDD^{\mathrm{T}}B^{\mathrm{T}}S = 0$$

for S. The solution to the LQR problem is then given by $u = -D^{\mathrm{T}}B^{\mathrm{T}}Sx$.

The following theorem shows that we can solve the robust control problem by solving the LQR problem.

Theorem 5.5

Robust Control Problem 5.7 is solvable. Furthermore, the solution to LQR Problem 5.8 is a solution to Robust Control Problem 5.7.

Proof

Since $(A(p_o), B)$ is stabilizable and $F \geq 0$, by Theorem 2.2, the solution to LQR Problem 5.8 exists. Let the solution be $u = Kx$. We would like to prove that it is also a solution to the robust control problem; that is

$$\dot{x} = A(p_o)x + BDKx + BDE(p)Kx + BD\phi(p)x \tag{5.12}$$

is asymptotically stable for all $p \in P$.

To prove this, we define

$$V(x_o) = \min_{u\in R^m} \int_0^\infty (x^{\mathrm{T}}Fx + x^{\mathrm{T}}x + u^{\mathrm{T}}u)dt$$

to be the minimum cost of the optimal control of the auxiliary system from some initial state x_o. We would like to show that $V(x)$ is a Lyapunov function for system (5.12). By definition, $V(x)$ must satisfy the Hamilton–Jacobi–Bellman equation, which reduces to

$$\min_{u\in R^m}(x^{\mathrm{T}}Fx + x^{\mathrm{T}}x + u^{\mathrm{T}}u + V_x^{\mathrm{T}}\ (A(p_o)x + BDu)) = 0$$

Since $u = Kx$ is the optimal control, it must satisfy the following two equations.

$$x^{\mathrm{T}}Fx + x^{\mathrm{T}}x + x^{\mathrm{T}}K^{\mathrm{T}}Kx + V_x^{\mathrm{T}}\ (A(p_o)x + BDKx) = 0 \tag{5.13}$$

$$2x^{\mathrm{T}}K^{\mathrm{T}} + V_x^{\mathrm{T}}\ BD = 0 \tag{5.14}$$

With the aid of the above two equations, we can show that $V(x)$ is a Lyapunov function for System (5.12). Clearly

$$V(x) > 0 \qquad x \neq 0$$
$$V(x) = 0 \qquad x = 0$$

To show $\dot{V}(x) < 0$ for all $x \neq 0$, we first use Equation (5.12)

$$\begin{aligned}\dot{V}(x) &= V_x^{\mathrm{T}}\,\dot{x}\\ &= V_x^{\mathrm{T}}\,(A(p_o)x + BDKx + BDE(p)Kx + BD\phi(p)x)\\ &= V_x^{\mathrm{T}}\,(A(p_o)x + BDKx) + V_x^{\mathrm{T}}\,BD(E(p)K + \phi(p))x.\end{aligned}$$

By Equation (5.13)

$$V_x^{\mathrm{T}}\,(A(p_o)x + BDKx) = -(x^{\mathrm{T}}Fx + x^{\mathrm{T}}x + x^{\mathrm{T}}K^{\mathrm{T}}Kx).$$

By Equation (5.14)

$$V_x^{\mathrm{T}}\,BD(E(p)K + \phi(p))x = -2x^{\mathrm{T}}K^{\mathrm{T}}E(p)Kx - 2x^{\mathrm{T}}K^{\mathrm{T}}\phi(p)x$$

Hence

$$\begin{aligned}\dot{V}(x) &= -x^{\mathrm{T}}Fx - x^{\mathrm{T}}x - x^{\mathrm{T}}K^{\mathrm{T}}Kx - 2x^{\mathrm{T}}K^{\mathrm{T}}\phi(p)x - 2x^{\mathrm{T}}K^{\mathrm{T}}E(p)Kx\\ &= -x^{\mathrm{T}}Fx - x^{\mathrm{T}}x - x^{\mathrm{T}}K^{\mathrm{T}}Kx - 2x^{\mathrm{T}}K^{\mathrm{T}}\phi(p)x - x^{\mathrm{T}}\phi(p)^{\mathrm{T}}\phi(p)x\\ &\quad + x^{\mathrm{T}}\phi(p)^{\mathrm{T}}\phi(p)x - 2x^{\mathrm{T}}K^{\mathrm{T}}E(p)Kx\\ &= -x^{\mathrm{T}}Fx + x^{\mathrm{T}}\phi(p)^{\mathrm{T}}\phi(p)x - x^{\mathrm{T}}x - x^{\mathrm{T}}K^{\mathrm{T}}Kx - 2x^{\mathrm{T}}K^{\mathrm{T}}\phi(p)x\\ &\quad - x^{\mathrm{T}}\phi(p)^{\mathrm{T}}\phi(p)x - 2x^{\mathrm{T}}K^{\mathrm{T}}E(p)Kx\\ &= -x^{\mathrm{T}}(F + \phi(p)^{\mathrm{T}}\phi(p))x - x^{\mathrm{T}}x - x^{\mathrm{T}}(K + \phi(p))^{\mathrm{T}}(K + \phi(p))x\\ &\quad - 2x^{\mathrm{T}}K^{\mathrm{T}}E(p)Kx\\ &\leq -x^{\mathrm{T}}x\end{aligned}$$

In other words

$$\dot{V}(x) < 0 \qquad x \neq 0$$
$$\dot{V}(x) = 0 \qquad x = 0$$

Therefore, by the Lyapunov stability theorem , System (5.12) is stable for all $p \in P$. In other words, $u = Kx$ is a solution to Robust Control Problem 5.7.

Q.E.D.

Example 5.5

Consider the following second-order system

$$\begin{bmatrix} \dot{x}_1 \\ \dot{x}_2 \end{bmatrix} = \begin{bmatrix} 0 & 1 \\ 1+p & p \end{bmatrix} \begin{bmatrix} x_1 \\ x_2 \end{bmatrix} + \begin{bmatrix} 0 \\ q \end{bmatrix} u$$

where $p \in [-10, \quad 1]$ and $q \in [1, \quad 10]$ are the uncertainties. We would like to design a robust control $u = Kx$ so that the closed-loop system is stable for all $p \in [-10, \quad 1]$ and $q \in [1, \quad 10]$. Let us pick $p_o = 0$.

Since $q > 1 > 0$, we can take $B = \begin{bmatrix} 0 \\ 1 \end{bmatrix}$ and $D = 1$. Then the LQR problem is the same as the LQR problem in Example 5.2; that is, for the nominal system

$$\begin{bmatrix} \dot{x}_1 \\ \dot{x}_2 \end{bmatrix} = \begin{bmatrix} 0 & 1 \\ 1 & 0 \end{bmatrix} \begin{bmatrix} x_1 \\ x_2 \end{bmatrix} + \begin{bmatrix} 0 \\ 1 \end{bmatrix} u$$

find a feedback control law $u = Kx$ that minimizes the cost functional

$$\int_0^\infty (x^T Fx + x^T x + u^T u)dt = \int_0^\infty (x^T (F+I)x + u^T u)dt$$

In other words,

$$Q = F + I = \begin{bmatrix} 101 & 100 \\ 100 & 101 \end{bmatrix}$$

$$R = I = 1$$

The solution is

$$u = [-11.0995 \quad -11.0995]x$$

To verify the results, we check the eigenvalues of the controlled system

$$\begin{bmatrix} \dot{x}_1 \\ \dot{x}_2 \end{bmatrix} = \begin{bmatrix} 0 & 1 \\ 1+p & p \end{bmatrix} \begin{bmatrix} x_1 \\ x_2 \end{bmatrix} + \begin{bmatrix} 0 \\ q \end{bmatrix} u$$

for different p and q as shown in Table 5.4. Clearly, the results show that the controlled system is robust.

Case 2

Input uncertainty enters the system via B and uncertainty in $A(p)$ does not satisfy the matching condition. That is, we consider the following system

$$\dot{x} = A(p)x + BD(p)u$$

We relax the previous assumptions and assume the following.

Table 5.4 Eigenvalues for different values of p and q.

p	q	λ_1	λ_2
−10	1	−1.0000	−20.0995
−10	5	−1.0000	−64.4975
−10	10	−1.0000	−119.9950
1	1	−1.0000	−9.0995
1	5	−1.0000	−53.4975
1	10	−1.0000	−108.9950

Assumption 5.8

There exists a nominal value $p_o \in P$ of p such that $(A(p_o), B)$ is stabilizable.

Assumption 5.9

There exists a constant matrix D such that for all $p \in P$

$$0 < D \leq D(p)$$

Assumption 5.10

$A(p)$ is bounded.

Under these assumptions, the system dynamics can be rewritten as

$$\dot{x} = A(p)x + BD(u + E(p)u)$$

where $E(p) = D^{-1}D(p) - I \geq 0$.

Our goal is to solve the following robust control problem of stabilizing the system under uncertainty.

Robust Control Problem 5.9

Find a feedback control law $u = Kx$ such that the closed-loop system

$$\dot{x} = A(p)x + BD(u + E(p)u) = A(p)x + BD(Kx + E(p)Kx)$$

is asymptotically stable for all $p \in P$.

In order to solve this robust control problem, we first decompose the uncertainty $A(p)-A(p_o)$ into the sum of a matched component and an unmatched component by projecting it into the range of BD; that is

$$A(p)-A(p_o)=(BD)(BD)^{+}(A(p)-A(p_o))+(I-(BD)(BD)^{+})(A(p)-A(p_o))$$

Define H as in Equation (5.7) and G as follows: for all $p \in P$

$$(A(p)-A(p_o))^{\mathrm{T}}(BD)^{+^{\mathrm{T}}}(BD)^{+}(A(p)-A(p_o)) \leq G \tag{5.15}$$

We would like to translate the above problem into the following LQR Problem.

LQR Problem 5.10

For the auxiliary system

$$\dot{x}=A(p_o)x+BDu+\alpha(I-(BD)(BD)^{+})v$$

find a feedback control law $u=Kx$, $v=Lx$ that minimizes the cost functional

$$\int_0^{\infty}(x^{\mathrm{T}}(G+\rho^2H+\beta^2I)x+u^{\mathrm{T}}u+\rho^2v^{\mathrm{T}}v)dt$$

where $\alpha \geq 0$, $\rho \geq 0$ and $\beta \geq 0$ are design parameters.

The solution to the LQR problem is given by

$$\begin{bmatrix} u \\ v \end{bmatrix} = -\widetilde{R}^{-1}\widetilde{B}^{T}Sx$$

where S is the unique positive definite solution to the following algebraic Riccati equation.

$$S\widetilde{A}+\widetilde{A}^{\mathrm{T}}S+\widetilde{Q}-S\widetilde{B}\widetilde{R}^{-1}\widetilde{B}^{\mathrm{T}}S=0$$

where

$$\widetilde{A}=A(p_o) \qquad \widetilde{B}=\begin{bmatrix} BD & \alpha(I-(BD)(BD)^{+}) \end{bmatrix}$$

$$\widetilde{Q}=G+\rho^2H+\beta^2I \qquad \widetilde{R}=\begin{bmatrix} I & 0 \\ 0 & \rho^2I \end{bmatrix}$$

Since

$$\widetilde{B}\widetilde{R}^{-1}\widetilde{B}^{\mathrm{T}} = \begin{bmatrix} BD & \alpha(I-(BD)(BD)^{+}) \end{bmatrix} \begin{bmatrix} I & 0 \\ 0 & \rho^{-2}I \end{bmatrix} \begin{bmatrix} (BD)^{\mathrm{T}} \\ \alpha(I-(BD)(BD)^{+}) \end{bmatrix}$$
$$= (BD)(BD)^{\mathrm{T}} + \alpha^2\rho^{-2}(I-(BD)(BD)^{+})^2$$

the Riccati equation becomes

$$SA(p_o) + A(p_o)^{\mathrm{T}}S + G + \rho^2 H + \beta^2 I - S((BD)(BD)^{\mathrm{T}}$$
$$+ \alpha^2\rho^{-2}(I-(BD)(BD)^{+})^2)S = 0$$

The control is given by

$$\begin{bmatrix} u \\ v \end{bmatrix} = \begin{bmatrix} -(BD)^{\mathrm{T}}S \\ -\alpha\rho^{-2}(I-(BD)(BD)^{+})S \end{bmatrix} x = \begin{bmatrix} K \\ L \end{bmatrix} x$$

The following theorem states the relationship between Robust Control Problem 5.9 and LQR Problem 5.10.

Theorem 5.6

If one can choose α, ρ and β such that the solution to LQR Problem 5.10 $u = Kx$, $v = Lx$ satisfies

$$\beta^2 I - 2\rho^2 L^T L > 0$$

then $u = Kx$ is a solution to Robust Control Problem 5.9.

Proof

Since $(A(p_o), B)$ is stabilizable and $G \geq 0$ $H \geq 0$, by Theorem 2.2, the solution to LQR Problem 5.10 exists. Denote the solution by $u = Kx$, $v = Lx$. We would like to prove that it is also a solution to the robust control problem; that is

$$\dot{x} = A(p)x + BD(Kx + E(p)Kx) \tag{5.16}$$

is asymptotically stable for all $p \in P$.

To prove this, we define

$$V(x_o) = \min_{u \in R^m} \int_0^\infty (x^{\mathrm{T}}(G + \rho^2 H + \beta^2 I)x + u^{\mathrm{T}}u + \rho^2 v^{\mathrm{T}}v)dt$$

to be the minimum cost of the optimal control of the auxiliary system from some initial state x_o. We would like to show that $V(x)$ is a Lyapunov function for system (5.16). By definition, $V(x)$ must satisfy the Hamilton–Jacobi–Bellman equation, which reduces to

$$\min_{u,v}(x^T(G+\rho^2H+\beta^2I)x+u^Tu+\rho^2v^Tv+V_x^T(A(p_o)x + BDu+\alpha(I-(BD)(BD)^+)v))=0$$

Since $u=Kx$, $v=Lx$ are the optimal controls, they must satisfy

$$x^T(G+\rho^2H+\beta^2I)x+x^TK^TKx+\rho^2x^TL^TLx + V_x^T(A(p_o)x+BDKx+\alpha(I-(BD)(BD)^+)Lx)=0 \quad (5.17)$$

$$2x^TK^T+V_x^TBD=0 \quad (5.18)$$

$$2\rho^2x^TL^T+V_x^T\alpha(I-(BD)(BD)^+)=0 \quad (5.19)$$

With the aid of the above three equations, we can show that $V(x)$ is a Lyapunov function for System (5.16). Clearly

$$V(x)>0 \qquad x\neq 0$$
$$V(x)=0 \qquad x=0$$

To show $\dot{V}(x)<0$ for all $x\neq 0$, we first use Equation (5.16)

$$\begin{aligned}\dot{V}(x) &= V_x^T\dot{x}\\ &=V_x^T(A(p)x+BD(Kx+E(p)Kx))\\ &=V_x^T(A(p_o)x+BDKx+\alpha(I-(BD)(BD)^+)Lx)\\ &\quad+V_x^T(A(p)-A(p_o))x-V_x^T\alpha(I-(BD)(BD)^+)Lx+V_x^TBDE(p)Kx\\ &=V_x^T(A(p_o)x+BDKx+\alpha(I-(BD)(BD)^+)Lx)+V_x^T(BD)(BD)^+\\ &\quad(A(p)-A(p_o))x+V_x^T(I-(BD)(BD)^+)(A(p)-A(p_o))x-V_x^T\alpha\\ &\quad(I-(BD)(BD)^+)Lx+V_x^T\,BDE(p)Kx\end{aligned}$$

By Equation (5.17)

$$\begin{aligned}&V_x^T(A(p_o)x+BDKx+\alpha(I-(BD)(BD)^+)Lx)\\ &=-x^T(G+\rho^2H+\beta^2I)x-x^TK^TKx-\rho^2x^TL^TLx\end{aligned}$$

By Equation (5.18)

$$V_x^T(BD)(BD)^+(A(p)-A(p_o))x=-2x^TK^T(BD)^+(A(p)-A(p_o))x$$
$$V_x^T(BD)E(p)Kx=-2x^TK^TE(p)Kx$$

By Equation (5.19)

$$V_x^{\mathrm{T}} \alpha (I - (BD)(BD)^{+}) Lx = -2\rho^2 x^{\mathrm{T}} L^{\mathrm{T}} Lx$$
$$V_x^{\mathrm{T}} (I - (BD)(BD)^{+})(A(p) - A(p_o))x = -2\alpha^{-1} \rho^2 x^{\mathrm{T}} L^{\mathrm{T}} (A(p) - A(p_o))x$$

Therefore

$$\begin{aligned}
\dot{V}(x) &= V_x^{\mathrm{T}}(A(p_o)x + BDKx + \alpha(I - (BD)(BD)^{+})Lx) + V_x^{\mathrm{T}}(BD)(BD)^{+} \\
&\quad (A(p) - A(p_o))x \\
&\quad + V_x^{\mathrm{T}}(I - (BD)(BD)^{+})(A(p) - A(p_o))x - V_x^{\mathrm{T}} \alpha(I - (BD)(BD)^{+})Lx \\
&\quad + V_x^{\mathrm{T}}\ BDE(p)Kx \\
&= -x^{\mathrm{T}}(G + \rho^2 H + \beta^2 I)x - x^{\mathrm{T}} K^{\mathrm{T}} Kx - \rho^2 x^{\mathrm{T}} L^{\mathrm{T}} Lx - 2x^{\mathrm{T}} K^{\mathrm{T}} (BD)^{+} \\
&\quad (A(p) - A(p_o))x \\
&\quad - 2\alpha^{-1}\rho^2 x^{\mathrm{T}} L^{\mathrm{T}}(A(p) - A(p_o))x + 2\rho^2 x^{\mathrm{T}} L^{\mathrm{T}} Lx - 2x^{\mathrm{T}} K^{\mathrm{T}} E(p)Kx
\end{aligned}$$

By Equation (5.15)

$$\begin{aligned}
&- x^{\mathrm{T}} K^{\mathrm{T}} Kx - 2x^{\mathrm{T}} K^{\mathrm{T}} (BD)^{+}(A(p) - A(p_o))x \\
&= - x^{\mathrm{T}}(K - (BD)^{+}(A(p) - A(p_o)))^{\mathrm{T}}(K - (BD)^{+}(A(p) - A(p_o)))x \\
&\quad + x^{\mathrm{T}}((BD)^{+}(A(p) - A(p_o)))^{\mathrm{T}}((BD)^{+}(A(p) - A(p_o)))x \\
&\le x^{\mathrm{T}}(A(p) - A(p_o))^{\mathrm{T}}(BD)^{+^{\mathrm{T}}}(BD)^{+}(A(p) - A(p_o))x \\
&\le x^{\mathrm{T}} Gx
\end{aligned}$$

By Equation (5.7)

$$\begin{aligned}
&- 2\alpha^{-1}\rho^2 x^{\mathrm{T}} L^{\mathrm{T}}(A(p) - A(p_o))x \\
&\le \rho^2 x^{\mathrm{T}} L^{\mathrm{T}} Lx + \rho^2 \alpha^{-2} x^{\mathrm{T}}(A(p) - A(p_o))^{\mathrm{T}}(A(p) - A(p_o))x \\
&\le \rho^2 x^{\mathrm{T}} L^{\mathrm{T}} Lx + \rho^2 x^{\mathrm{T}} Hx
\end{aligned}$$

Hence

$$\begin{aligned}
\dot{V}(x) &= - x^{\mathrm{T}}(G + \rho^2 H + \beta^2 I)x - x^{\mathrm{T}} K^{\mathrm{T}} Kx - \rho^2 x^{\mathrm{T}} L^{\mathrm{T}} Lx - 2x^{\mathrm{T}} K^{\mathrm{T}}(BD)^{+} \\
&\quad (A(p) - A(p_o))x \\
&\quad - 2\alpha^{-1}\rho^2 x^{\mathrm{T}} L^{\mathrm{T}}(A(p) - A(p_o))x + 2\rho^2 x^{\mathrm{T}} L^{\mathrm{T}} Lx - 2x^{\mathrm{T}} K^{\mathrm{T}} E(p)Kx \\
&\le - x^{\mathrm{T}}(G + \rho^2 H + \beta^2 I)x - \rho^2 x^{\mathrm{T}} L^{\mathrm{T}} Lx + x^{\mathrm{T}} Gx
\end{aligned}$$

$$
\begin{aligned}
&+\rho^2 x^{\mathrm{T}} L^{\mathrm{T}} L x+\rho^2 x^{\mathrm{T}} H x+2\rho^2 x^{\mathrm{T}} L^{\mathrm{T}} L x-2x^{\mathrm{T}} K^{\mathrm{T}} E(p) K x \\
&=-x^{\mathrm{T}}(\beta^2 I-2\rho^2 L^{\mathrm{T}} L)x-2x^{\mathrm{T}} K^{\mathrm{T}} E(p) K x \\
&\leq -x^{\mathrm{T}}(\beta^2 I-2\rho^2 L^{\mathrm{T}} L)x
\end{aligned}
$$

If the sufficient condition $\beta^2 I-2\rho^2 L^T L>0$ is satisfied, then

$$
\begin{aligned}
\dot{V}(x) &< 0 \qquad x \neq 0 \\
\dot{V}(x) &= 0 \qquad x = 0
\end{aligned}
$$

Therefore, by the Lyapunov stability theorem, System (5.16) is stable for all $p \in P$. In other words, $u = Kx$ is a solution to Robust Control Problem 5.9.

Q.E.D.

Example 5.6

Consider the following second-order system

$$
\begin{bmatrix} \dot{x}_1 \\ \dot{x}_2 \end{bmatrix} = \begin{bmatrix} p & 1+p \\ 1 & 0 \end{bmatrix}\begin{bmatrix} x_1 \\ x_2 \end{bmatrix} + \begin{bmatrix} 0 \\ q \end{bmatrix} u
$$

where $p \in [-2, \ \ 2]$ and $q \in [1, \ \ 10]$ are the uncertainties. We would like to design a robust control $u = Kx$ so that the closed-loop system is stable for all $p \in [-2, \ \ 2]$ and $q \in [1, \ \ 10]$.

To translate this problem into a LQR problem, let us pick $p_o = 0$ and check the controllability of $(A(p_o), B)$. As shown in Example 5.3, $(A(p_o), B)$ is controllable.

Since $q > 1 > 0$, we can take

$$
B = \begin{bmatrix} 0 \\ 1 \end{bmatrix}
$$

and $D = 1$. Then the LQR problem is the same as the LQR problem in Example 5.5. That is, for the auxiliary system

$$\begin{bmatrix} \dot{x}_1 \\ \dot{x}_2 \end{bmatrix} = \begin{bmatrix} 0 & 1 \\ 1 & 0 \end{bmatrix} \begin{bmatrix} x_1 \\ x_2 \end{bmatrix} + \begin{bmatrix} 0 \\ 1 \end{bmatrix} u + \begin{bmatrix} 0.05 & 0 \\ 0 & 0 \end{bmatrix} v$$

find a feedback control law $u = Kx$, $v = Lx$ that minimizes the cost functional

$$\int_0^\infty \left(x^{\mathrm{T}} \left(\begin{bmatrix} 0 & 0 \\ 0 & 0 \end{bmatrix} + \begin{bmatrix} 1600 & 1600 \\ 1600 & 1600 \end{bmatrix} + \begin{bmatrix} 100 & 0 \\ 0 & 100 \end{bmatrix} \right) x + u^{\mathrm{T}} u + v^{\mathrm{T}} v\right) dt$$

The solution is given by

$$u = Kx = \begin{bmatrix} -41.7185 & -42.1792 \end{bmatrix} x$$

$$v = Lx = \begin{bmatrix} -6.5577 & -2.0859 \\ 0 & 0 \end{bmatrix} x$$

For $u = Kx$ to be the solution to the robust control problem, we need to check the sufficient condition $\beta^2 I - 2\rho^2 L^{\mathrm{T}} L > 0$. Clearly

$$\beta^2 I - 2\rho^2 L^{\mathrm{T}} L = 100 \begin{bmatrix} 1 & 0 \\ 0 & 1 \end{bmatrix} - 2 \begin{bmatrix} -6.5577 & 0 \\ -2.0859 & 0 \end{bmatrix} \begin{bmatrix} -6.5577 & -2.0859 \\ 0 & 0 \end{bmatrix}$$

$$= \begin{bmatrix} 56.9969 & -13.6788 \\ -13.6788 & 95.6489 \end{bmatrix} > 0$$

Therefore, $u = Kx$ is the solution to the robust control problem.

To verify the results, we check the eigenvalues of the controlled system

$$\begin{bmatrix} \dot{x}_1 \\ \dot{x}_2 \end{bmatrix} = \begin{bmatrix} p & 1+p \\ 1 & 0 \end{bmatrix} \begin{bmatrix} x_1 \\ x_2 \end{bmatrix} + \begin{bmatrix} 0 \\ q \end{bmatrix} u$$

for different p and q as shown in Table 5.5. Clearly, the results show that the controlled system is robust.

Table 5.5 Eigenvalues for different values of p and q.

p	q	λ_1	λ_2
−2	1	−1.0109	−43.1683
−2	10	−1.0109	−422.7811
0	1	−0.9885	−41.1907
0	10	−0.9890	−420.8030
2	1	−0.9638	−39.2154
2	10	−0.9669	−418.8251

Case 3

Input uncertainty does not enter the system via B. So far, we have assumed that the input uncertainty $D(p)$ enters the system via the input matrix B. Now we will consider a more general case of the following system

$$\dot{x} = Ax + Bu + D(p)u$$

To focus on the problem, we assume that there is no uncertainty in A and B matrices. We also make the following assumptions.

Assumption 5.11

(A, B) is stabilizable.

Assumption 5.12

$D(p)$ is bounded.

Our goal is to solve the following robust control problem of stabilizing the system under uncertainty.

Robust Control Problem 5.11

Find a feedback control law $u = Kx$ such that the closed-loop system

$$\dot{x} = Ax + Bu + D(p)u = Ax + BKx + D(p)Kx$$

is asymptotically stable for all $p \in P$.

We would like to translate the above problem into the following LQR Problem.

LQR Problem 5.12

For the auxiliary system

$$\dot{x} = Ax + Bu + \alpha(I - BB^{+})v$$

find a feedback control law $u = Kx$, $v = Lx$ that minimizes the cost functional

$$\int_0^\infty (x^{\mathrm{T}}(M + \rho^2 N + \beta^2 I)x + u^{\mathrm{T}}u + \rho^2 v^T v)\mathrm{d}t,$$

where $\alpha \geq 0$, $\rho \geq 0$ and $\beta \geq 0$ are design parameters, and $M \geq 0$ and $N \geq 0$ are design parameter matrices.

The solution to LQR problem is given by first solving the Riccati equation

$$SA + A^{T}S + M + \rho^2 N + \beta^2 I - S(BB^{T} + \alpha^2\rho^{-2}(I - BB^{+})^2)S = 0$$

and then obtaining the control as

$$\begin{bmatrix} u \\ v \end{bmatrix} = \begin{bmatrix} -B^T S \\ -\alpha\rho^2(I - BB^{+})S \end{bmatrix} x = \begin{bmatrix} K \\ L \end{bmatrix} x$$

The following theorem states the relationship between Robust Control Problem 5.11 and LQR Problem 5.12.

Theorem 5.7

If one can choose α, β, ρ, M, and N such that the solution to LQR Problem 5.12 $u = Kx$, $v = Lx$ satisfies

$$\begin{aligned} &\beta^2 I - 2\rho^2 L^T L > 0 \\ &M - K^T D(p)^T B^{+^T} B^{+} D(p) K \geq 0 \qquad (5.20) \\ &N - \alpha^{-2} K^T D(p)^T D(p) K \geq 0 \end{aligned}$$

then $u = Kx$ is a solution to Robust Control Problem 5.11.

Proof

Since (A, B) is stabilizable and $M \geq K^T D(p)^T B^{+^T} B^{+} D(p) K > 0$ and $N \geq \alpha^{-2} K^T D(p)^T D(p) K > 0$, by Theorem 2.2, the solution to LQR Problem 5.12 exists. Denote the solution by $u = Kx$, $v = Lx$. We would like to prove that it is also a solution to the robust control problem; that is

$$\dot{x} = Ax + BKx + D(p)Kx \qquad (5.21)$$

is asymptotically stable for all $p \in P$.

To prove this, we define

$$V(x_o) = \min_{u,v} \int_0^\infty (x^T(M + \rho^2 N + \beta^2 I)x + u^T u + \rho^2 v^T v)dt$$

to be the minimum cost of the optimal control of the auxiliary system from some initial state x_o. We would like to show that $V(x)$ is a Lyapunov function for system (5.21). By definition, $V(x)$ must satisfy the Hamilton–Jacobi–Bellman equation, which reduces to

$$\min_{u,v}(x^T(M+\rho^2 N+\beta^2 I)x+u^T u+\rho^2 v^T v + V_x^T(Ax+Bu+\alpha(I-BB^+)v))=0$$

Since $u=Kx$, $v=Lx$ are the optimal controls, they must satisfy

$$x^T(M+\rho^2 N+\beta^2 I)x+x^T K^T Kx+\rho^2 x^T L^T Lx + V_x^T(Ax+BKx+\alpha(I-BB^+)Lx)=0 \quad (5.22)$$

$$2x^T K^T+V_x^T B=0 \quad (5.23)$$

$$2\rho^2 x^T L^T+V_x^T\alpha(I-BB^+)=0 \quad (5.24)$$

With the aid of the above three equations, we can show that $V(x)$ is a Lyapunov function for System (5.21). Clearly

$$V(x)>0 \qquad x\neq 0$$
$$V(x)=0, \qquad x=0$$

To show $\dot{V}(x)<0$ for all $x\neq 0$, we use Equations (5.21)–(5.24).

$$\begin{aligned}\dot{V}(x) &= V_x^T\dot{x}\\ &= V_x^T(Ax+BKx+D(p)Kx)\\ &= V_x^T(Ax+BKx+\alpha(I-BB^+)Lx)+V_x^T D(p)Kx-V_x^T\alpha(I-BB^+)Lx\\ &= V_x^T(Ax+BKx+\alpha(I-BB^+)Lx)+V_x^T BB^+D(p)Kx\\ &\quad +V_x^T(I-BB^+)D(p)Kx-V_x^T\alpha(I-BB^+)Lx\\ &= -x^T M-\rho^2 x^T N-\beta^2 x^T x-x^T K^T Kx-\rho^2 x^T L^T Lx\\ &\quad -2x^T K^T B^+D(p)Kx-2\alpha^{-1}\rho^2 x^T L^T D(p)Kx-2\rho^2 x^T L^T Lx\end{aligned}$$

By Equation (5.20)

$$\begin{aligned}&-x^T K^T Kx-2x^T K^T B^+D(p)Kx\\ &\leq x^T K^T D(p)^T B^{+^T} B^+D(p)Kx\\ &\leq x^T Mx\\ &\quad -2\alpha^{-1}\rho^2 x^T L^T D(p)Kx\\ &\leq \rho^2 x^T L^T Lx+\rho^2\alpha^{-2}x^T K^T D(p)^T D(p)Kx\\ &\leq \rho^2 x^T L^T Lx+\rho^2 x^T Nx\end{aligned}$$

Hence

$$\begin{aligned}\dot{V}(x) &= -x^{\mathrm{T}}M - \rho^2 x^{\mathrm{T}}Nx - \beta^2 x^{\mathrm{T}}x - x^{\mathrm{T}}K^{\mathrm{T}}Kx - \rho^2 x^{\mathrm{T}}L^{\mathrm{T}}Lx \\ &\quad -2x^{\mathrm{T}}K^{\mathrm{T}}B^{+}D(p)Kx - 2\alpha^{-1}\rho^2 x^{\mathrm{T}}L^{\mathrm{T}}D(p)Kx - 2\rho^2 x^{\mathrm{T}}L^{\mathrm{T}}Lx \\ &\le -x^{\mathrm{T}}M - \rho^2 x^{\mathrm{T}}Nx - \beta^2 x^{\mathrm{T}}x - \rho^2 x^{\mathrm{T}}L^{\mathrm{T}}Lx - 2\rho^2 x^{\mathrm{T}}L^{\mathrm{T}}Lx \\ &\quad +\rho^2 x^{\mathrm{T}}L^{\mathrm{T}}Lx + \rho^2 x^{\mathrm{T}}Nx + x^{\mathrm{T}}Mx \\ &= -x^{\mathrm{T}}(\beta^2 I - 2\rho^2 L^{\mathrm{T}}L)x\end{aligned}$$

In other words

$$\dot{V}(x) < 0 \qquad x \neq 0$$

$$\dot{V}(x) = 0 \qquad x = 0$$

Therefore, by the Lyapunov stability theorem , System (5.21) is stable for all $p \in P$. In other words, $u = Kx$ is a solution to Robust Control Problem 5.11.

Q.E.D.

5.5 NOTES AND REFERENCES

In this chapter, we have discussed the robust control design for linear systems. We first considered the case when the matching condition is satisfied. For the matched uncertainty, the robust control always exists as long as the system is stabilizable. The stabilizability condition is necessary because otherwise no control will exist. If we strengthen the condition from stabilizability to controllability, then we can not only make the system stable, but also place the poles to the left of any non-negative real number, which will strengthen the stability. For both cases, the solutions are obtained by solving some LQR problems.

The robust control problem is much more difficult if the matching condition is not satisfied. In fact, not too many results exist in the literature for unmatched uncertainties. We have partially solved the robust control problem for unmatched uncertainties in this chapter. Our approach is unconventional because we introduced an artificial control to handle the unmatched uncertainties. This control is used in solving the optimal control problem, but discarded in the robust control problem. When a sufficient condition is satisfied, the solution to the optimal control problem is a solution to the robust control problem.

The results in this chapter were first presented in references [104, 105, 107, 108]. The other works on related problems can be found in [17, 18, 33, 34, 49, 62, 99, 121, 124–126, 128].

5.6 PROBLEMS

5.1 Let us consider the following system

$$\dot{x} = \begin{bmatrix} 1 & 5 \\ -p & 1+p \end{bmatrix} x + \begin{bmatrix} 0 \\ 1 \end{bmatrix} u$$

where $p \in [-8, \ \ 2]$ is the uncertainty with $p_o = -1$. Find a robust feedback control by solving the corresponding LQR problem.

5.2 Prove that if $(A(p_o), B)$ is controllable, then $(A(p), B)$ is controllable, where $A(p) = A(p_o) + B\phi(p)$ for any matrix $\phi(p)$.

5.3 Consider the following robust control problem. For the system

$$\dot{x} = \begin{bmatrix} -3 & -1 & 0 \\ 0 & 0 & p \\ 2 & 1+p & -1 \end{bmatrix} x + \begin{bmatrix} 0 & 0 \\ 1 & 0 \\ 0 & 1 \end{bmatrix} u$$

design a state feedback to stabilize the system for all $p \in [-1, 3]$ with $p_o = 1$.

(a) Translate the robust control problem into an optimal control problem.
(b) Write the corresponding Riccati equation.
(c) Solve the problem using MATLAB.

5.4 For the system

$$\begin{bmatrix} \dot{x}_1 \\ \dot{x}_2 \end{bmatrix} = \begin{bmatrix} p & 1 \\ 1+p & 2 \end{bmatrix} \begin{bmatrix} x_1 \\ x_2 \end{bmatrix} + \begin{bmatrix} 1 \\ 1 \end{bmatrix} u$$

where $p \in [-10, \ \ 1]$ is the uncertainty, design a robust control $u = Kx$ so that the closed-loop system has all its poles on the left of -5 for all $p \in [-10, \ \ 1]$.

5.5 For the system in Problem 5.3, we would like to design a robust control $u = Kx$ so that the closed-loop system has all its poles on the left of -12 for all $p \in [-1, \ \ 3]$.

5.6 Prove that if (A, B) is controllable and D is an invertible matrix, then (A, BD) is also controllable.

5.7 Consider the following system

$$\begin{bmatrix} \dot{x}_1 \\ \dot{x}_2 \end{bmatrix} = \begin{bmatrix} p & 1+p \\ 1 & -2 \end{bmatrix} \begin{bmatrix} x_1 \\ x_2 \end{bmatrix} + \begin{bmatrix} 0 \\ 2 \end{bmatrix} u$$

where $p \in [-1, \ 2]$ is the uncertainty. Design a robust control $u = Kx$ so that the closed-loop system is stable for all $p \in [-1, \ 2]$.

5.8 For the following system

$$\dot{x} = \begin{bmatrix} 0 & 2 \\ 2+p & p \end{bmatrix} x + \begin{bmatrix} 0.5 \\ 0 \end{bmatrix} (1+q)u$$

where $p \in [-1, \ 2]$ and $q \in [0, \ 4]$, design a state feedback to stabilize the system for all uncertainties p and q. (Hint: start with $a = 0.04$, $\beta = 20$, $\rho = 1$.)

5.9 Consider the following system

$$\dot{x} = \begin{bmatrix} 1 & 1+p & 0 \\ 0 & 0 & 1-2p \\ 2-p & -3 & -8+p \end{bmatrix} x + \begin{bmatrix} 0 & 0 \\ 1 & 0 \\ 0 & 1 \end{bmatrix} (1+q)u$$

where $p \in [-1, \ 2]$ and $q \in [0, \ 4]$. The robust control problem is to design a state feedback to stabilize the system for all uncertainties p and q. Take $p_o = 1$ and $a = \beta = \rho = 1$.

(a) Translate the robust control problem into an optimal control problem.
(b) Write the corresponding Riccati equation.

5.10 In Robust Control Problem 5.5, assume that the system satisfies the matching condition. Show that by properly selecting the parameters α, ρ, β, LQR Problem 5.6 reduces to LQR Problem 5.2. Prove that the sufficient condition $\beta^2 I - 2\rho^2 L^T L > 0$ is satisfied.

6

Robust Control of Nonlinear Systems

In this chapter, we turn to nonlinear systems. It is well known that robust control design is more complex for nonlinear systems. In fact many approaches to robust control problems are applicable only to linear systems. However this is not the case for our optimal control approach. Conceptually, our approach applies equally to linear and nonlinear systems. The complexity is in terms of efficient computation. For linear systems, the optimal control problem becomes an LQR problem, whose solution always exists and can be easily obtained by solving an algebraic Riccati equation. However, for nonlinear systems, we may not be able to easily compute the solution to the optimal control problem as analytical solutions may not be available, forcing us to use numerical solutions.

We develop the theory for nonlinear systems in a manner similar to the theory for linear systems. We first study the case of matched uncertainty. We show that as long as the solution to the corresponding optimal control problem exists, it is a solution to the robust control problem. We then consider unmatched uncertainty. Again, we decompose the uncertainty into matched and unmatched components and introduce an augmented control for the unmatched uncertainty. A computable sufficient condition is also derived to ensure that the solution to the corresponding optimal control problem is a solution to the robust control problem. Finally, how to handle uncertainty in the input matrix will be discussed.

Robust Control Design: An Optimal Control Approach F. Lin

6.1 INTRODUCTION

In this chapter, we consider nonlinear systems of the form

$$\dot{x} = A(x) + B(x)u$$

where $A(x)$, $B(x)$ are nonlinear (matrix) functions of $x : A : R^n \rightarrow R^n$ and $B : R^m \rightarrow R^n$. Note that we assume that the control u enters the system linearly. This assumption is satisfied in most practical examples we investigated. Possible ways to relax this assumption is an open problem. Let us first look at the following example.

Example 6.1

Consider the mechanical system in Figure 6.1. y_1, y_2 are the displacements of masses M_1, M_2. The input to the system is the force f. K represents a spring and D_1, D_2, D_3 represent frictions. The force due to the spring is linear with respect to the corresponding displacement; that is, $K(y_1 - y_2)$. The forces due to the frictions are nonlinear functions of the displacements and velocities; they are denoted by $D_1(y_1, \dot{y}_1)$, $D_2(y_2, \dot{y}_2)$, $D_3(y_1 - y_2, \dot{y}_1 - \dot{y}_2)$, respectively.

The free body diagrams of two masses are shown in Figure 6.2. We consider only the forces in the horizontal direction. Applying Newton's second law, we obtain

$$M_1\ddot{y}_1 = f - K(y_1 - y_2) - D_3(y_1 - y_2, \dot{y}_1 - \dot{y}_2) - D_1(y_1, \dot{y}_1)$$
$$M_2\ddot{y}_2 = K(y_1 - y_2) + D_3(y_1 - y_2, \dot{y}_1 - \dot{y}_2) - D_2(y_1, \dot{y}_1)$$

From the above dynamic equations, we can obtain the state equations as follows. Define state variables as $x_1 = y_1$, $x_2 = \dot{y}_1$, $x_3 = y_2$, $x_4 = \dot{y}_2$, we have

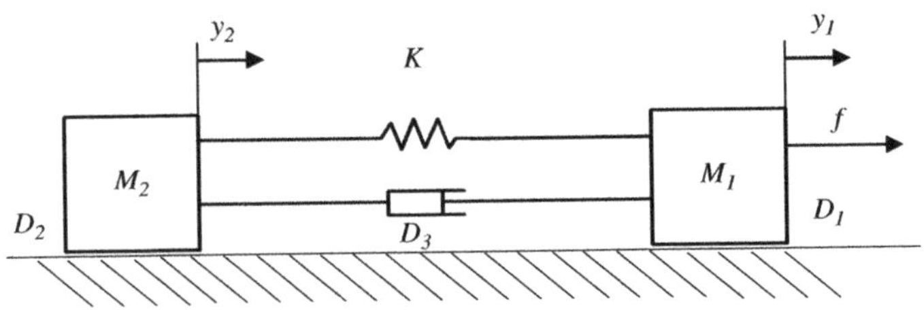

Figure 6.1 A mechanical system to illustrate various types of uncertainties.

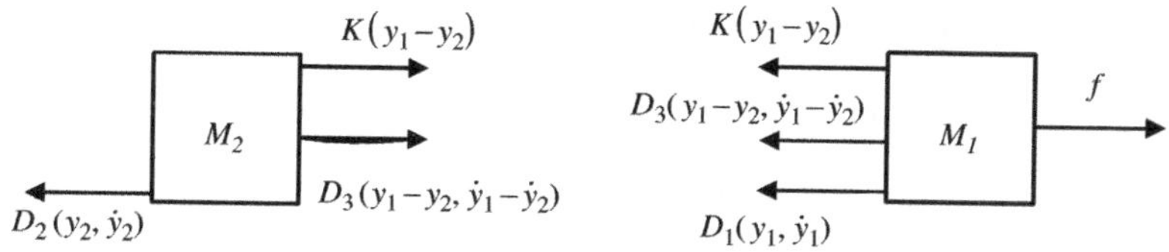

Figure 6.2 Free body diagrams of the mechanical system in Figure 6.1.

$$\begin{bmatrix} \dot{x}_1 \\ \dot{x}_2 \\ \dot{x}_3 \\ \dot{x}_4 \end{bmatrix} = \begin{bmatrix} x_2 \\ (Kx_3 - Kx_1 - D_3(x_1 - x_3, x_2 - x_4) - D_1(x_1, x_2))/M_1 \\ x_4 \\ (Kx_1 - Kx_3 + D_3(x_1 - x_3, x_2 - x_4) - D_2(x_3, x_4))/M_2 \end{bmatrix} + \begin{bmatrix} 0 \\ 1/M_1 \\ 0 \\ 0 \end{bmatrix} f$$

If $D_1(x_1, x_2)$ is uncertain, then the difference of A between the actual value $D_1(x_1, x_2)$ and the nominal value $D_{1o}(x_1, x_2)$ can be expressed as

$$A(D_1(x_1, x_2)) - A(D_{1o}(x_1, x_2)) = \begin{bmatrix} 0 \\ -(D_1(x_1, x_2) - D_{1o}(x_1, x_2))/M_1 \\ 0 \\ 0 \end{bmatrix} = \begin{bmatrix} 0 \\ 1/M_1 \\ 0 \\ 0 \end{bmatrix} (-\Delta D_1(x_1, x_2))$$

where $\Delta D_1(x_1, x_2) = D_1(x_1, x_2) - D_{1o}(x_1, x_2)$ is the deviation of $D_1(x_1, x_2)$ from its nominal value. Note that for $D_1(x_1, x_2)$, the uncertainty is in the range of B, that is, the matching condition is satisfied.

If $D_2(x_3, x_4)$ is uncertain, then the uncertainty in A is

$$A(D_2(x_3, x_4)) - A(D_{2o}(x_3, x_4)) = \begin{bmatrix} 0 \\ 0 \\ 0 \\ -(D_2(x_3, x_4) - D_{2o}(x_3, x_4))/M_2 \end{bmatrix}$$

In this case, the matching condition is not satisfied.

We will discuss both matched uncertainty and unmatched uncertainty. Similar to linear systems, the conditions for existence of robust control for systems with unmatched uncertainty are more difficult to meet. Therefore, let us first consider matched uncertainty.

6.2 MATCHED UNCERTAINTY

Consider the following nonlinear system

$$\dot{x} = A(x) + B(x)u + B(x)f(x)$$

where $B(x)f(x)$ models the uncertainty in the system dynamics. Since the uncertainty is in the range of $B(x)$, the matching condition is satisfied.

We make the following assumptions:

Assumption 6.1

$A(0) = 0$ and $f(0) = 0$ so that $x = 0$ is an equilibrium (it will be the only equilibrium if the robust control problem is solvable).

Assumption 6.2

The uncertainty $f(x)$ is bounded; that is, there exists a nonnegative function $f_{max}(x)$ such that

$$\|f(x)\| \leq f_{max}(x) \tag{6.1}$$

Our goal is to solve the following robust control problem of stabilizing the system under uncertainty.

Robust Control Problem 6.1

Find a feedback control law $u = u_o(x)$ such that the closed-loop system

$$\dot{x} = A(x) + B(x)u_o(x) + B(x)f(x)$$

is globally asymptotically stable for all uncertainties $f(x)$ satisfying $\|f(x)\| \leq f_{max}(x)$.

We will solve the above robust control problem indirectly by translating it into an optimal control problem.

Optimal Control Problem 6.2

For the nominal system

$$\dot{x} = A(x) + B(x)u$$

find a feedback control law $u = u_o(x)$ that minimizes the following cost functional

$$\int_0^\infty (f_{\max}(x)^2 + x^T x + u^T u)dt$$

The relation between the robust control problem and the optimal control problem is shown in the following theorem.

Theorem 6.1

If the solution to Optimal Control Problem 6.2 exists, then it is a solution to Robust Control Problem 6.1.

Proof

Let $u = u_o(x)$ be the solution to Optimal Control Problem 6.2. We would like to show that

$$\dot{x} = A(x) + B(x)u_o(x) + B(x)f(x) \tag{6.2}$$

is globally asymptotically stable for all uncertainties $f(x)$, satisfying $\|f(x)\| \leq f_{\max}(x)$.

To this end, we define

$$V(x_o) = \min_{u \in R^m} \int_0^\infty (f_{\max}(x)^2 + x^T x + u^T u)dt$$

to be the minimum cost of the optimal control of the nominal system from some initial state x_o. We would like to show that $V(x)$ is a Lyapunov function for system (6.2). By definition, $V(x)$ must satisfy the Hamilton–Jacobi–Bellman equation, which reduces to

$$\min_{u \in R^m} (f_{\max}(x)^2 + x^T x + u^T u + V_x^T(A(x) + B(x)u)) = 0$$

Since $u = u_o(x)$ is the optimal control, it must satisfy the above equation; that is

$$f_{\max}(x)^2 + x^T x + u_o(x)^T u_o(x) + V_x^T(A(x) + B(x)u_o(x)) = 0 \tag{6.3}$$

$$2u_o(x)^T + V_x^T B(x) = 0 \tag{6.4}$$

Using the above two equations, we can show that $V(x)$ is a Lyapunov function for System (6.2). Clearly

$$V(x) > 0 \qquad x \neq 0$$
$$V(x) = 0 \qquad x = 0$$

To show $\dot{V}(x) < 0$ for all $x \neq 0$, we use Equations (6.2)–(6.4)

$$\begin{aligned}
\dot{V}(x) &= V_x^{\mathrm{T}} \dot{x} \\
&= V_x^{\mathrm{T}}(A(x) + B(x)u_o(x) + B(x)f(x)) \\
&= V_x^{\mathrm{T}}(A(x) + B(x)u_o(x)) + V_x^{\mathrm{T}} B(x)f(x) \\
&= -f_{\max}(x)^2 - x^{\mathrm{T}}x - u_o(x)^{\mathrm{T}}u_o(x) + V_x^{\mathrm{T}} B(x)f(x) \\
&= -f_{\max}(x)^2 - x^{\mathrm{T}}x - u_o(x)^{\mathrm{T}}u_o(x) - 2u_o(x)^{\mathrm{T}}f(x) \\
&= -f_{\max}(x)^2 + f(x)^{\mathrm{T}}f(x) - x^{\mathrm{T}}x - u_o(x)^{\mathrm{T}}u_o(x) - 2u_o(x)^{\mathrm{T}}f(x) - f(x)^{\mathrm{T}}f(x) \\
&= -f_{\max}(x)^2 + f(x)^{\mathrm{T}}f(x) - x^{\mathrm{T}}x - (u_o(x) + f(x))^{\mathrm{T}}(u_o(x) + f(x)) \\
&\leq -f_{\max}(x)^2 + f(x)^{\mathrm{T}}f(x) - x^{\mathrm{T}}x
\end{aligned}$$

By Equation (6.1), $f(x)^{\mathrm{T}} f(x) \leq f_{\max}(x)^2$. Hence

$$\dot{V}(x) \leq -x^{\mathrm{T}}x$$

In other words

$$\dot{V}(x) < 0 \qquad x \neq 0$$
$$\dot{V}(x) = 0 \qquad x = 0$$

Thus, the conditions of the Lyapunov stability theorem are satisfied. Consequently, there exists a neighborhood of 0, $N = \{x : \|x\| < c\}$ for some $c > 0$ such that if $x(t)$ enters N, then

$$x(t) \to 0 \quad as \quad t \to \infty$$

But $x(t)$ cannot remain forever outside N. Otherwise,

$$\|x(t)\| \geq c$$

for all t > and

$$V(x(t)) - V(x(0)) = \int_0^t \dot{V}(x(\tau))\mathrm{d}\tau$$

$$\leq \int_0^t (-x^T x)d\tau$$
$$\leq -\int_0^t c^2 d\tau$$
$$\leq -c^2 t$$

Let $t \to \infty$, we have

$$V(x(t)) \leq V(x(0)) - c^2 t \to -\infty$$

which contradicts the fact that $V(x(t)) > 0$ for all $x(t)$. Therefore

$$x(t) \to 0 \quad as \quad t \to \infty$$

no matter where the trajectory begins. That is, system (6.2) is globally asymptotically stable for all admissible uncertainties. In other words, $u = u_o(x)$ is a solution to Robust Control Problem 6.1.

Q.E.D.

Example 6.2

Let us consider the following nonlinear system

$$\dot{x}_1 = x_2$$
$$\dot{x}_2 = p_1 x_1 \cos(p_2 \sqrt{x_2}) + u$$

where $p_1 \in [-0.2, 2]$ and $p_2 \in [-10, 100]$. The robust control problem is to find a control $u = u_o(x)$ so that the closed-loop system is stable for all p_1 and p_2.

To translate the robust control problem into an optimal control problem, we first rewrite the state equation as

$$\begin{bmatrix} \dot{x}_1 \\ \dot{x}_2 \end{bmatrix} = \begin{bmatrix} 0 & 1 \\ 0 & 0 \end{bmatrix} \begin{bmatrix} x_1 \\ x_2 \end{bmatrix} + \begin{bmatrix} 0 \\ 1 \end{bmatrix} u + \begin{bmatrix} 0 \\ 1 \end{bmatrix} p_1 x_1 \cos(p_2 \sqrt{x_2})$$

The uncertainty $f(x) = p_1 x_1 \cos(p_2 \sqrt{x_2})$ is bounded as follows.

$$\|f(x)\| = |p_1 x_1 \cos(p_2 \sqrt{x_2})| \leq 2|x_1| = f_{\max}(x)$$

Therefore, the optimal control problem is given as follows. For the nominal system

$$\begin{bmatrix} \dot{x}_1 \\ \dot{x}_2 \end{bmatrix} = \begin{bmatrix} 0 & 1 \\ 0 & 0 \end{bmatrix} \begin{bmatrix} x_1 \\ x_2 \end{bmatrix} + \begin{bmatrix} 0 \\ 1 \end{bmatrix} u$$

find a feedback control law $u = u_o(x)$ that minimizes the following cost functional

$$\int_0^\infty (f_{max}(x)^2 + x^T x + u^T u)dt = \int_0^\infty (5x_1^2 + x_2^2 + u^2)dt$$

This is an LQR problem with

$$A = \begin{bmatrix} 0 & 1 \\ 0 & 0 \end{bmatrix} \quad B = \begin{bmatrix} 0 \\ 1 \end{bmatrix}$$

$$Q = \begin{bmatrix} 5 & 0 \\ 0 & 1 \end{bmatrix} \quad R = 1$$

The solution to the LQR problem is given by

$$u = \begin{bmatrix} -2.2361 & -2.3393 \end{bmatrix} x$$

To test the robustness of the control, we use MATLAB to perform simulations of the closed-loop system under control with the initial conditions $x_1(0) = 10$ and $x_2(0) = -10$. We consider the following four cases.

Case A: $p_1 = -0.2$ $p_2 = -10$
Case B: $p_1 = -0.2$ $p_2 = 100$
Case C: $p_1 = 2$ $p_2 = -10$
Case D: $p_1 = 2$ $p_2 = 100$

The simulation results are shown in Figures 6.3–6.6. (Trajectory starting at 10 is for x_1.) Although the responses are different for these four cases, the difference is relatively small.

6.3 UNMATCHED UNCERTAINTY

Now we assume that uncertainty is not in the range of $B(x)$. Consider the following nonlinear system

$$\dot{x} = A(x) + B(x)u + C(x)f(x)$$

where $f(x)$models the uncertainty in the system dynamics and $C(x)$ can be any matrix. For example, if $C(x) = I$, then $C(x)f(x) = f(x)$. The reason for introducing $C(x)$ is to make the definition of uncertainty $f(x)$ more flexible. We make the following assumptions:

Assumption 6.3

$A(0) = 0$ and $f(0) = 0$ so that $x = 0$ is an equilibrium.

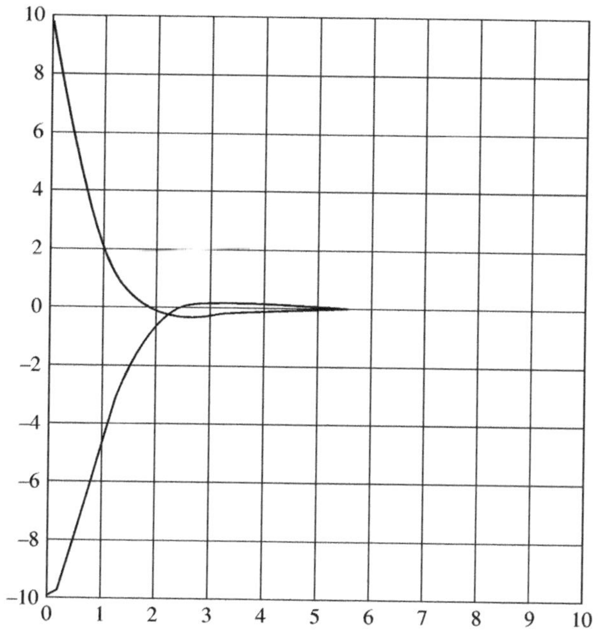

Figure 6.3 MATLAB simulation of the controlled system for Case A.

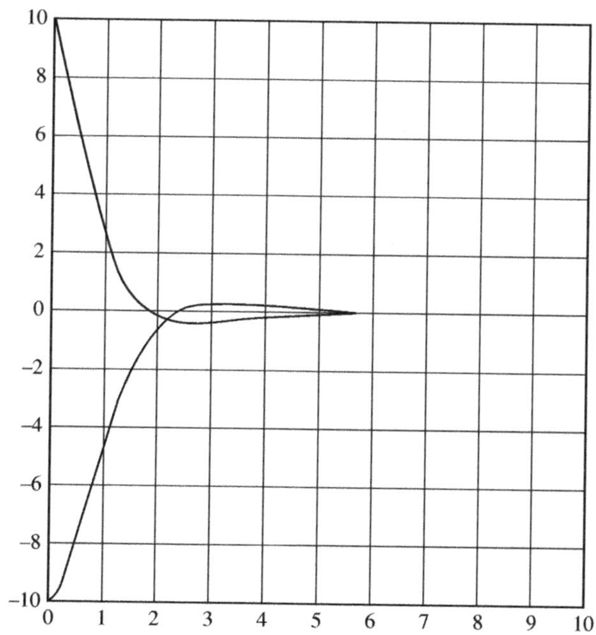

Figure 6.4 MATLAB simulation of the controlled system for Case B.

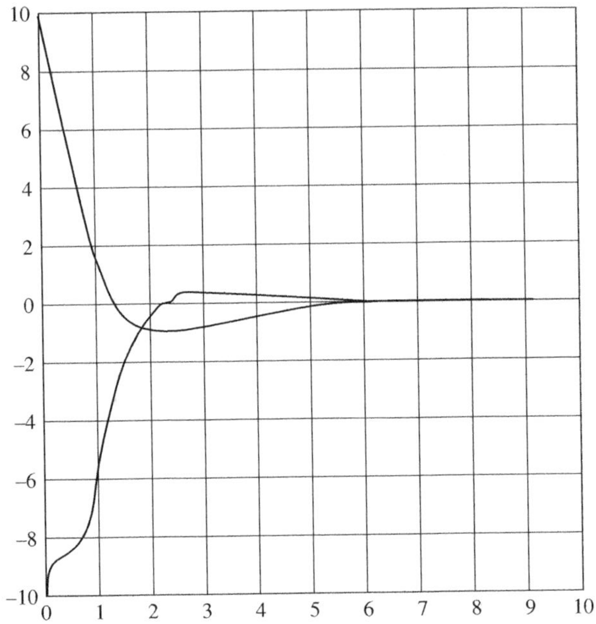

Figure 6.5 MATLAB simulation of the controlled system for Case C.

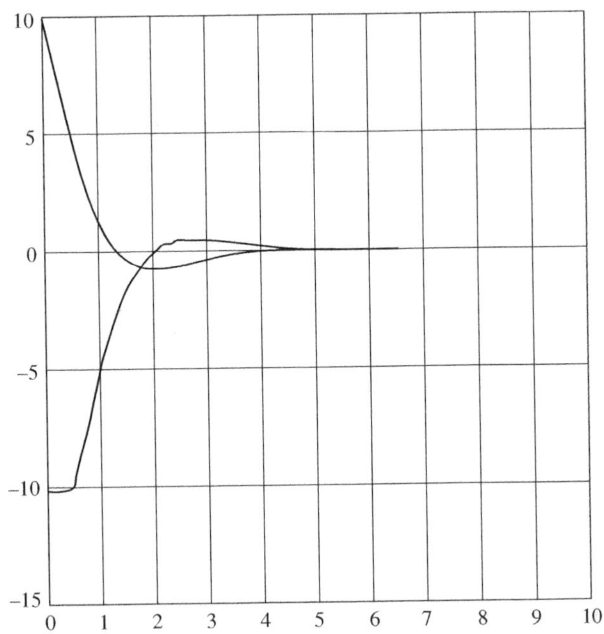

Figure 6.6 MATLAB simulation of the controlled system for Case D.

Assumption 6.4

The uncertainty $f(x)$ is bounded.

We would like to solve the following robust control problem.

Robust Control Problem 6.3

Find a feedback control law $u = u_o(x)$ such that the closed-loop system

$$\dot{x} = A(x) + B(x)u_o(x) + C(x)f(x)$$

is globally asymptotically stable for all uncertainties $f(x)$.

We will solve the above robust control problem indirectly by translating it into an optimal control problem.

Optimal Control Problem 6.4

For the auxiliary system

$$\dot{x} = A(x) + B(x)u + \alpha(I - B(x)B(x)^{+})C(x)v$$

find a feedback control law $(u_o(x),\ \ v_o(x))$ that minimizes the following cost functional

$$\int_0^\infty (f_{\max}(x)^2 + \rho^2 g_{\max}(x)^2 + \beta^2\|x\|^2 + \|u\|^2 + \rho^2\|v\|^2)dt,$$

where $\alpha \geq 0$, $\rho \geq 0$ and $\beta \geq 0$ are design parameters. $f_{\max}(x)$, $g_{\max}(x)$ are nonnegative functions such that

$$\|B(x)^{+}C(x)f(x)\| \leq f_{\max}(x) \tag{6.5}$$

$$\|\alpha^{-1}f(x)\| \leq g_{\max}(x) \tag{6.6}$$

The relation between the robust control problem and the optimal control problem is shown in the following theorem.

Theorem 6.2

If one can choose α, ρ and β such that the solution to Optimal Control Problem 6.4, denoted by $(u_o(x), v_o(x))$, exists and the following condition is satisfied

$$2\rho^2 \|v_o(x)\|^2 \leq \beta'^2 \|x\|^2 \qquad \forall x \in R^n$$

for some β' such that $|\beta'| < |\beta|$, then $u_o(x)$, the u-component of the solution to Optimal Control Problem 6.4, is a solution to Robust Control Problem 6.3.

Proof

Let $u_o(x)$, $v_o(x)$ be the solution to Optimal Control Problem 6.4. We would like to show that

$$\dot{x} = A(x) + B(x)u_o(x) + C(x)f(x) \tag{6.7}$$

is globally asymptotically stable for all uncertainties $f(x)$.

To prove this, we define

$$V(x_o) = \min_{u,v} \int_0^\infty (f_{\max}(x)^2 + \rho^2 g_{\max}(x)^2 + \beta^2 \|x\|^2 + \|u\|^2 + \rho^2 \|v\|^2)dt$$

to be the minimum cost of the optimal control of the auxiliary system from some initial state x_o. We would like to show that $V(x)$ is a Lyapunov function for system (6.7). By definition, $V(x)$ must satisfy the Hamilton–Jacobi–Bellman equation, which reduces to

$$\min_{u,v}(f_{\max}(x)^2 + \rho^2 g_{\max}(x)^2 + \beta^2 \|x\|^2 + \|u\|^2 + \rho^2 \|v\|^2 + V_x^{\mathrm{T}}(A(x) + B(x)u + \alpha(I - B(x)B(x)^+)C(x)v)) = 0$$

Since $u_o(x)$, $v_o(x)$ is the optimal control, it must satisfy the above equation; that is

$$f_{\max}(x)^2 + \rho^2 g_{\max}(x)^2 + \beta^2 \|x\|^2 + \|u_o(x)\|^2 + \rho^2 \|v_o(x)\|^2 + V_x^{\mathrm{T}}(A(x) + B(x)u_o(x) + \alpha(I - B(x)B(x)^+)C(x)v_o(x)) = 0 \tag{6.8}$$

$$2u_o(x)^{\mathrm{T}} + V_x^{\mathrm{T}} B(x) = 0 \tag{6.9}$$

$$2\rho^2 v_o(x)^{\mathrm{T}} + V_x^{\mathrm{T}} \alpha (I - B(x)B(x)^+)C(x) = 0 \tag{6.10}$$

With the aid of the above equations, we can show that $V(x)$ is a Lyapunov function for System (6.7). Clearly

$$V(x) > 0 \qquad x \neq 0$$
$$V(x) = 0 \qquad x = 0$$

To show $\dot{V}(x) < 0$ for all $x \neq 0$, we use Equations (6.7–6.10)

$$\begin{aligned}
\dot{V}(x) &= V_x^{\mathrm{T}}\dot{x} \\
&= V_x^{\mathrm{T}}(A(x) + B(x)u_o(x) + C(x)f(x)) \\
&= V_x^{\mathrm{T}}(A(x) + B(x)u_o(x) + \alpha(I - B(x)B(x)^{+})C(x)v_o(x)) \\
&\quad - V_x^{\mathrm{T}}\alpha(I - B(x)B(x)^{+})C(x)v_o(x) + V_x^{\mathrm{T}}C(x)f(x) \\
&= V_x^{\mathrm{T}}(A(x) + B(x)u_o(x) + \alpha(I - B(x)B(x)^{+})C(x)v_o(x)) \\
&\quad - V_x^{\mathrm{T}}\alpha(I - B(x)B(x)^{+})C(x)v_o(x) + V_x^{\mathrm{T}}B(x)B(x)^{+}C(x)f(x) \\
&\quad + V_x^{\mathrm{T}}(I - B(x)B(x)^{+})C(x)f(x) \\
&= -f_{\max}(x)^2 - \rho^2 g_{\max}(x)^2 - \beta^2\|x\|^2 - \|u_o(x)\|^2 - \rho^2\|v_o(x)\|^2 \\
&\quad + 2\rho^2 v_o(x)^{\mathrm{T}}v_o(x) - 2u_o(x)^{\mathrm{T}}B(x)^{+}C(x)f(x) - 2\alpha^{-1}\rho^2 v_o(x)^{\mathrm{T}}f(x) \\
&= -f_{\max}(x)^2 - \rho^2 g_{\max}(x)^2 - \beta^2\|x\|^2 - \|u_o(x)\|^2 + \rho^2\|v_o(x)\|^2 \\
&\quad - 2u_o(x)^{\mathrm{T}}B(x)^{+}C(x)f(x) - 2\alpha^{-1}\rho^2 v_o(x)^{\mathrm{T}}f(x)
\end{aligned}$$

On the other hand

$$\begin{aligned}
-\|u_o(x)\|^2 - 2u_o(x)^{\mathrm{T}}B(x)^{+}C(x)f(x) &\leq \|B(x)^{+}C(x)f(x)\|^2 \leq f_{\max}(x)^2 \\
-2\alpha^{-1}\rho^2 v_o(x)^{\mathrm{T}}f(x) &\leq \rho^2\|v_o(x)\|^2 + \rho^2\|\alpha^{-1}f(x)\|^2 \\
&\leq \rho^2\|v_o(x)\|^2 + \rho^2 g_{\max}(x)^2
\end{aligned}$$

Therefore, if the condition $2\rho^2\|v_o(x)\|^2 \leq \beta'^2\|x\|^2$, $\forall x \in R^n$ is satisfied

$$\begin{aligned}
\dot{V}(x) &\leq -\beta^2\|x\|^2 + 2\rho^2\|v_o(x)\|^2 \\
&= 2\rho^2\|v_o(x)\|^2 - \beta'^2\|x\|^2 - (\beta^2 - \beta'^2)\|x\|^2 \\
&\leq -(\beta^2 - \beta'^2)\|x\|^2
\end{aligned}$$

In other words,

$$\begin{aligned}
\dot{V}(x) &< 0 \qquad x \neq 0 \\
\dot{V}(x) &= 0 \qquad x = 0
\end{aligned}$$

Thus, the conditions of The Lyapunov stability theorem are satisfied. Consequently, there exists a neighbourhood of 0, $N = \{x : \|x\| < c\}$ for some $c > 0$ such that if $x(t)$ enters N, then

$$x(t) \to 0 \quad as \quad t \to \infty$$

But $x(t)$ cannot remain forever outside N. Otherwise

$$\|x(t)\| \geq c$$

for all t > 0, which implies

$$\begin{aligned} V(x(t)) - V(x(0)) &= \int_0^t \dot{V}(x(\tau))\mathrm{d}\tau \\ &\leq \int_0^t -(\beta^2 - \beta'^2)\|x\|^2 \mathrm{d}\tau \\ &\leq -\int_0^t (\beta^2 - \beta'^2)c^2 \mathrm{d}\tau \\ &\leq -(\beta^2 - \beta'^2)c^2 t \end{aligned}$$

Let $t \to \infty$, we have

$$V(x(t)) \leq V(x(0)) - (\beta^2 - \beta'^2)c^2 t \to -\infty$$

which contradicts the fact that $V(x(t)) > 0$ for all $x(t)$. Therefore

$$x(t) \to 0 \quad as \quad t \to \infty$$

no matter where the trajectory begins. This proves that $u_o(x)$ is a solution to Robust Control Problem 6.3.

Q.E.D.

Example 6.3

Let us consider the following nonlinear system

$$\begin{aligned} \dot{x}_1 &= x_2 + p_1 x_1 \cos\left(\frac{1}{x_2 + p_2}\right) + p_3 x_2 \sin(p_4 x_1 x_2) \\ \dot{x}_2 &= u \end{aligned}$$

where $p_1 \in [-0.2, 0.2]$, $p_2 \in [-10, 100]$, $p_3 \in [0, 0.2]$ and $p_4 \in [-100, 0]$. The robust control problem is to find a control $u = u_o(x)$ so that the closed-loop system is stable for all possible uncertainties.

To translate the robust control problem into an optimal control problem, we first rewrite the state equation as

$$\begin{bmatrix} \dot{x}_1 \\ \dot{x}_2 \end{bmatrix} = \begin{bmatrix} 0 & 1 \\ 0 & 0 \end{bmatrix} \begin{bmatrix} x_1 \\ x_2 \end{bmatrix} + \begin{bmatrix} 0 \\ 1 \end{bmatrix} u + \begin{bmatrix} 1 \\ 0 \end{bmatrix} f(x)$$

where

$$f(x) = p_1 x_1 \cos\left(\frac{1}{x_2 + p_2}\right) + p_3 x_2 \sin(p_4 x_1 x_2)$$

is the uncertainty. Let us take $a = 0.2$, $\beta = 1$, and $\rho = 1$. (This choice is obtained by trial and error.) $f_{\max}(x)$ and $g_{\max}(x)$ can then be calculated as follows.

$$\|B(x)^+ C(x) f(x)\| = \| \begin{bmatrix} 0 & 1 \end{bmatrix} \begin{bmatrix} 1 \\ 0 \end{bmatrix} f(x)\| = 0 = f_{\max}(x)$$

$$\|\alpha^{-1} f(x)\| = \|5(p_1 x_1 \cos\left(\frac{1}{x_2 + p_2}\right) + p_3 x_2 \sin(p_4 x_1 x_2))\| \le |x_1 + x_2| = g_{\max}(x)$$

Therefore

$$\begin{aligned} & f_{\max}(x)^2 + \rho^2 g_{\max}(x)^2 + \beta^2 \|x\|^2 + \|u\|^2 + \rho^2 \|v\|^2 \\ & = (x_1 + x_2)^2 + (x_1^2 + x_2^2) + u^2 + v^2 \end{aligned}$$

Hence, the corresponding optimal control problem is as follows. For the auxiliary system

$$\begin{bmatrix} \dot{x}_1 \\ \dot{x}_2 \end{bmatrix} = \begin{bmatrix} 0 & 1 \\ 0 & 0 \end{bmatrix} \begin{bmatrix} x_1 \\ x_2 \end{bmatrix} + \begin{bmatrix} 0 \\ 1 \end{bmatrix} u + \begin{bmatrix} 0.2 \\ 0 \end{bmatrix} v$$

find a feedback control law $(u_o(x), \quad v_o(x))$ that minimizes the following cost functional

$$\int_0^\infty (2x_1^2 + 2x_2^2 + 2x_1 x_2 + u^2 + v^2) dt$$

This is an LQR problem with

$$A = \begin{bmatrix} 0 & 1 \\ 0 & 0 \end{bmatrix} \quad B = \begin{bmatrix} 0 & 0.2 \\ 1 & 0 \end{bmatrix}$$

$$Q = \begin{bmatrix} 2 & 1 \\ 1 & 2 \end{bmatrix} \quad R = \begin{bmatrix} 1 & 0 \\ 0 & 1 \end{bmatrix}$$

The solution to the LQR problem is given by

$$\begin{aligned} u_o &= \begin{bmatrix} -1.3549 & -2.1532 \end{bmatrix} x \\ v_o &= \begin{bmatrix} -0.4054 & -0.2710 \end{bmatrix} x \end{aligned}$$

Let us check if the sufficient condition

$$2\rho^2 \|v_o(x)\|^2 \leq \beta^2 \|x\|^2 \qquad \forall x \in R^n$$

is satisfied. Clearly

$$\begin{aligned}
&\beta^2 \|x\|^2 - 2\rho^2 \|v_o(x)\|^2 \\
&= \|x\|^2 - 2\|v_o(x)\|^2 \\
&= x_1^2 + x_2^2 - 2(-0.4054x_1 - 0.2710x_2)^2 \\
&= x^T \begin{bmatrix} 0.6713 & -0.2197 \\ -0.2197 & 0.8531 \end{bmatrix} x \\
&\geq 0
\end{aligned}$$

Therefore

$$u_o = \begin{bmatrix} -1.3549 & -2.1532 \end{bmatrix} x$$

is a solution to the robust control problem.

To test the robustness of the control, we use MATLAB to perform simulations of the closed-loop system under control with the initial conditions $x_1(0) = 100$ and $x_2(0) = -50$. We consider the following four cases.

Case A: $p_1 = -0.2 \quad p_2 = -10 \quad p_3 = 0 \quad p_4 = -100$
Case B: $p_1 = 0.2 \quad p_2 = 100 \quad p_3 = 0.2 \quad p_4 = 0$
Case C: $p_1 = 0 \quad p_2 = 0 \quad p_3 = 0 \quad p_4 = 0$
Case D: $p_1 = -0.2 \quad p_2 = 100 \quad p_3 = -0.2 \quad p_4 = -100$

The simulation results are shown in Figures 6.7–6.10, respectively. (Trajectory starting at 100 is for x_1.) Results for Cases A, C, and D are similar. The closed-loop systems are all rather stable. However, the settling time for Case B is different from the settling times for other cases. We believe that this is because p_1 for Case B is significantly different from those for other cases.

6.4 UNCERTAINTY IN THE INPUT MATRIX

We now allow uncertainty in the input matrix. We consider three cases: (1) input uncertainty enters the system via $B(x)$ and uncertainty in $A(x)$ satisfies the matching condition; (2) input uncertainty enters the system via $B(x)$ and uncertainty in $A(x)$ does not satisfy the matching condition; (3) input uncertainty does not enter the system via $B(x)$.

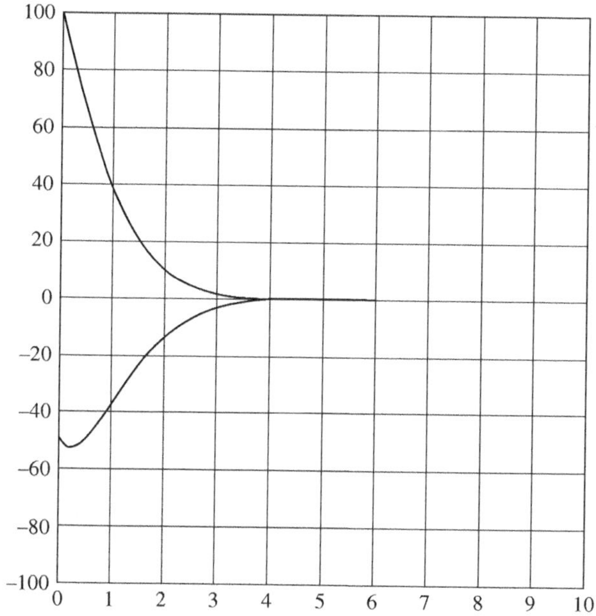

Figure 6.7 MATLAB simulation of the controlled system for Case A.

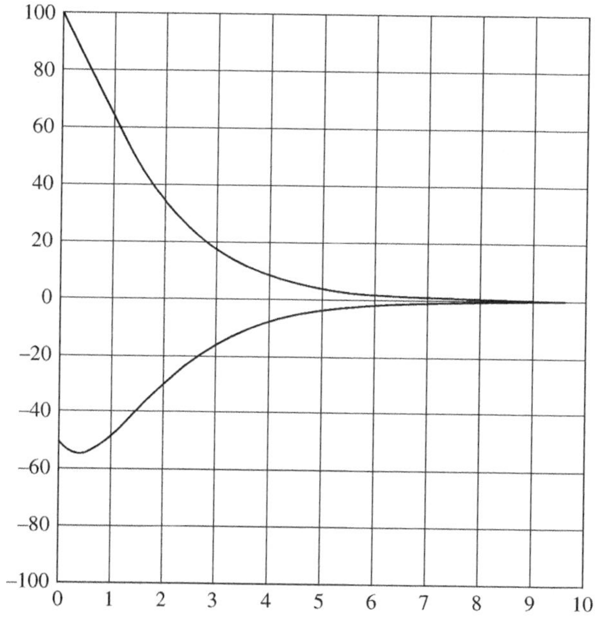

Figure 6.8 MATLAB simulation of the controlled system for Case B.

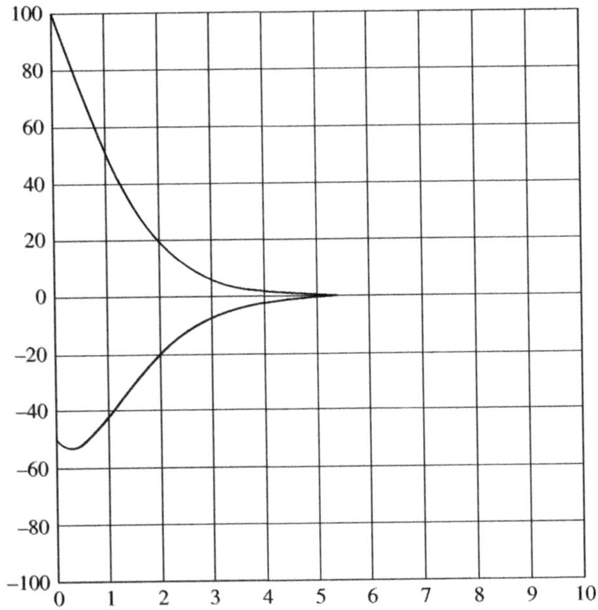

Figure 6.9 MATLAB simulation of the controlled system for Case C.

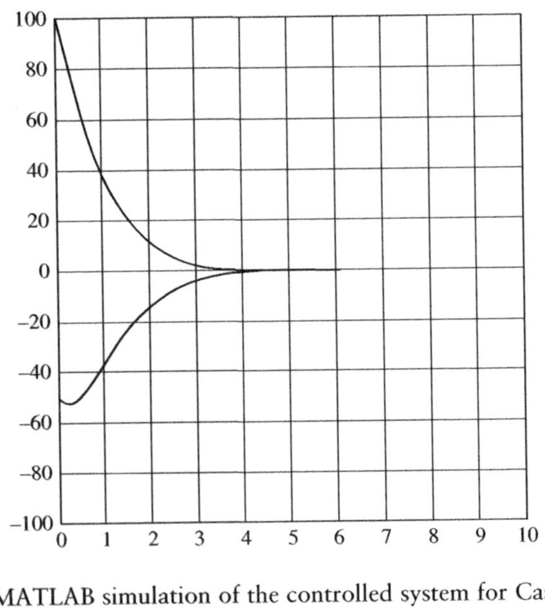

Figure 6.10 MATLAB simulation of the controlled system for Case D.

Case 1

We consider the following system

$$\dot{x} = A(x) + B(x)(u + h(x)u) + B(x)f(x)$$

where $h(x)$ is an $m \times m$ matrix representing the uncertainty in the input matrix. For Case 1, we make the following assumptions.

Assumption 6.5

$A(0) = 0$ and $f(0) = 0$ so that $x = 0$ is an equilibrium.

Assumption 6.6

The uncertainty $f(x)$ is bounded; that is, there exists a nonnegative function $f_{max}(x)$ such that

$$\|f(x)\| \leq f_{max}(x)$$

Assumption 6.7

$h(x)$ is a positive semi-definite matrix

$$h(x) \geq 0$$

We would like to solve the following robust control problem of stabilizing the above nonlinear system under uncertainty.

Robust Control Problem 6.5

Find a feedback control law $u = u_o(x)$ such that the closed-loop system

$$\dot{x} = A(x) + B(x)(u_o(x) + h(x)u_o(x)) + B(x)f(x)$$

is globally asymptotically stable for all uncertainties $f(x)$, $h(x)$ satisfying Assumptions 6.6 and 6.7.

We will solve the above robust control problem indirectly by translating it into an optimal control problem.

Optimal Control Problem 6.6

For the nominal system

$$\dot{x} = A(x) + B(x)u$$

find a feedback control law $u = u_o(x)$ that minimizes the following cost functional

$$\int_0^\infty (f_{max}(x)^2 + x^T x + u^T u)dt$$

We can solve the robust control problem by solving the optimal control problem as shown in the following theorem.

Theorem 6.3

If the solution to Optimal Control Problem 6.6 exists, then it is a solution to Robust Control Problem 6.5.

Proof

Let $u = u_o(x)$ be the solution to Optimal Control Problem 6.6. We would like to show that

$$\dot{x} = A(x) + B(x)(u_o(x) + h(x)u_o(x)) + B(x)f(x) \tag{6.11}$$

is globally asymptotically stable for all uncertainties $f(x)$, $h(x)$ satisfying Assumptions 6.6 and 6.7.

To this end, we define

$$V(x_o) = \min_{u \in R^m} \int_0^\infty (f_{max}(x)^2 + x^T x + u^T u)dt$$

to be the minimum cost of the optimal control of the nominal system from some initial state x_o. We would like to show that $V(x)$ is a Lyapunov function for system (6.11). By definition, $V(x)$ must satisfy the Hamilton–Jacobi–Bellman equation, which reduces to

$$\min_{u \in R^m} (f_{max}(x)^2 + x^T x + u^T u + V_x^T (A(x) + B(x)u)) = 0$$

where $V_x = \partial V / \partial x$. Since $u = u_o(x)$ is the optimal control, it must satisfy the above equation; that is

$$f_{max}(x)^2 + x^T x + u_o(x)^T u_o(x) + V_x^T (A(x) + B(x)u_o(x)) = 0$$

$$2u_o(x)^T + V_x^T B(x) = 0$$

Using the above two equations, we can show that $V(x)$ is a Lyapunov function for System (6.11). Clearly

$$V(x) > 0 \qquad x \neq 0$$
$$V(x) = 0 \qquad x = 0$$

To show $\dot{V}(x) < 0$ for all $x \neq 0$, we have

$$\begin{aligned}
\dot{V}(x) &= V_x^{\mathrm{T}} \dot{x} \\
&= V_x^{\mathrm{T}}(A(x) + B(x)(u_o(x) + h(x)u_o(x)) + B(x)f(x)) \\
&= V_x^{\mathrm{T}}(A(x) + B(x)u_o(x)) + V_x^{\mathrm{T}}B(x)f(x) + V_x^{\mathrm{T}}B(x)h(x)u_o(x) \\
&= -f_{\max}(x)^2 - x^{\mathrm{T}}x - u_o(x)^{\mathrm{T}}u_o(x) + V_x^{\mathrm{T}}B(x)f(x) + V_x^{\mathrm{T}}B(x)h(x)u_o(x) \\
&= -f_{\max}(x)^2 - x^{\mathrm{T}}x - u_o(x)^{\mathrm{T}}u_o(x) - 2u_o(x)^{\mathrm{T}}f(x) - 2u_o(x)^{\mathrm{T}}h(x)u_o(x) \\
&\leq -f_{\max}(x)^2 - x^{\mathrm{T}}x - u_o(x)^{\mathrm{T}}u_o(x) - 2u_o(x)^{\mathrm{T}}f(x) \\
&= -f_{\max}(x)^2 + f(x)^{\mathrm{T}}f(x) - x^{\mathrm{T}}x - u_o(x)^{\mathrm{T}}u_o(x) - 2u_o(x)^{\mathrm{T}}f(x) - f(x)^{\mathrm{T}}f(x) \\
&= -f_{\max}(x)^2 + f(x)^{\mathrm{T}}f(x) - x^{\mathrm{T}}x - (u_o(x) + f(x))^{\mathrm{T}}(u_o(x) + f(x)) \\
&\leq -f_{\max}(x)^2 + f(x)^{\mathrm{T}}f(x) - x^{\mathrm{T}}x
\end{aligned}$$

Since $f(x)^{\mathrm{T}}f(x) \leq f_{\max}(x)^2$

$$\dot{V}(x) \leq -x^T x$$

In other words

$$\dot{V}(x) < 0 \qquad x \neq 0$$
$$\dot{V}(x) = 0 \qquad x = 0$$

Thus, the conditions of The Lyapunov stability theorem are satisfied. Consequently, there exists a neighbourhood of 0, $N = \{x : \|x\| < c\}$ for some $c > 0$ such that if $x(t)$ enters N, then

$$x(t) \to 0 \qquad \text{as} \qquad t \to \infty$$

But $x(t)$ cannot remain forever outside N. Otherwise

$$\|x(t)\| \geq c$$

for all t > 0, which implies

$$\begin{aligned} V(x(t)) - V(x(0)) &= \int_0^t \dot{V}(x(\tau))d\tau \\ &\leq \int_0^t (-x^T x)d\tau \\ &\leq -\int_0^t c^2 d\tau \\ &\leq -c^2 t \end{aligned}$$

Let $t \to \infty$, we have

$$V(x(t)) \leq V(x(0)) - c^2 t \to -\infty$$

which contradicts the fact that $V(x(t)) > 0$ for all $x(t)$. Therefore

$$x(t) \to 0 \quad as \quad t \to \infty$$

no matter where the trajectory begins. That is, the system (6.11) is globally asymptotically stable for all admissible uncertainties. In other words, $u = u_o(x)$ is a solution to Robust Control Problem 6.5.

Q.E.D.

Example 6.4

Let us consider the following nonlinear system

$$\begin{aligned} \dot{x}_1 &= x_2 \\ \dot{x}_2 &= p_1 x_1 \cos(\sqrt{x_2}) + (1 + p_2 x_2^2)u \end{aligned}$$

where $p_1 \in [-0.2, 2]$ and $p_2 \in [0.1, 1]$. The robust control problem is to find a control $u = u_o(x)$ so that the closed-loop system is stable for all p_1 and p_2.

To translate the robust control problem into an optimal control problem, we first rewrite the state equation as

$$\begin{bmatrix} \dot{x}_1 \\ \dot{x}_2 \end{bmatrix} = \begin{bmatrix} 0 & 1 \\ 0 & 0 \end{bmatrix} \begin{bmatrix} x_1 \\ x_2 \end{bmatrix} + \begin{bmatrix} 0 \\ 1 \end{bmatrix} (1 + p_2 x_2^2)u + \begin{bmatrix} 0 \\ 1 \end{bmatrix} p_1 x_1 \cos(\sqrt{x_2})$$

The uncertainty $h(x) = p_2 x_2^2 \geq 0$ and $f(x) = p_1 x_1 \cos(\sqrt{x_2})$ is bounded as follows.

$$\|f(x)\| = |p_1 x_1 \cos(\sqrt{x_2})| \leq 2|x_1| = f_{max}(x)$$

Therefore, the optimal control problem is given as follows. For the nominal system

$$\begin{bmatrix} \dot{x}_1 \\ \dot{x}_2 \end{bmatrix} = \begin{bmatrix} 0 & 1 \\ 0 & 0 \end{bmatrix} \begin{bmatrix} x_1 \\ x_2 \end{bmatrix} + \begin{bmatrix} 0 \\ 1 \end{bmatrix} u,$$

find a feedback control law $u = u_o(x)$ that minimizes the following cost functional

$$\int_0^\infty (f_{max}(x)^2 + x^T x + u^T u)dt = \int_0^\infty (4x_1^2 + x_2^2 + u^2)dt.$$

This is an LQR problem with

$$A = \begin{bmatrix} 0 & 1 \\ 0 & 0 \end{bmatrix} \quad B = \begin{bmatrix} 0 \\ 1 \end{bmatrix}$$

$$Q = \begin{bmatrix} 5 & 0 \\ 0 & 1 \end{bmatrix} \quad R = 1$$

The solution to the LQR problem is given by

$$u = \begin{bmatrix} -2.2361 & -2.3393 \end{bmatrix} x$$

To test the robustness of the control, we use MATLAB to perform simulations of the closed-loop system under control with the initial conditions $x_1(0) = 10$ and $x_2(0) = -20$. We consider the following four cases.

Case A: $p_1 = -0.2 \quad p_2 = 0.1$
Case B: $p_1 = -0.2 \quad p_2 = 1$
Case C: $p_1 = 2 \quad p_2 = 0.1$
Case D: $p_1 = 2 \quad p_2 = 1$

The simulation results are shown in Figures 6.11–6.14, respectively. (Trajectory starting at 10 is for x_1.) We can see that the response in Case C is a little slower than the other cases, but the difference is small.

Case 2

We consider the following system

$$\dot{x} = A(x) + B(x)(u + h(x)u) + C(x)f(x),$$

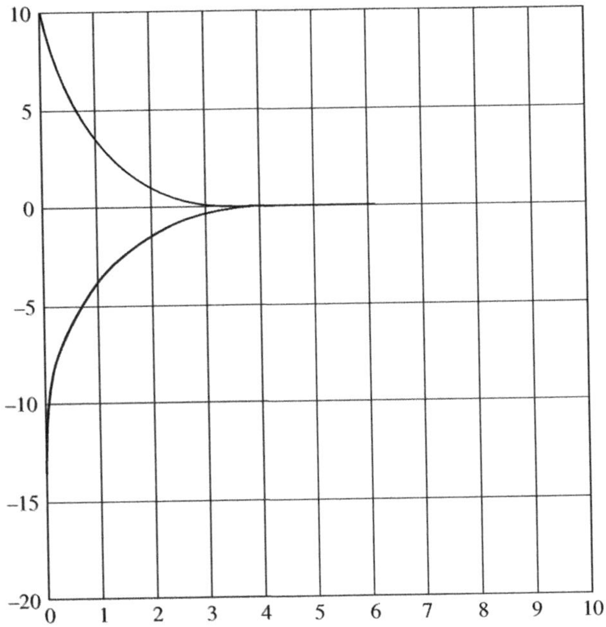

Figure 6.11 MATLAB simulation of the controlled system for Case A.

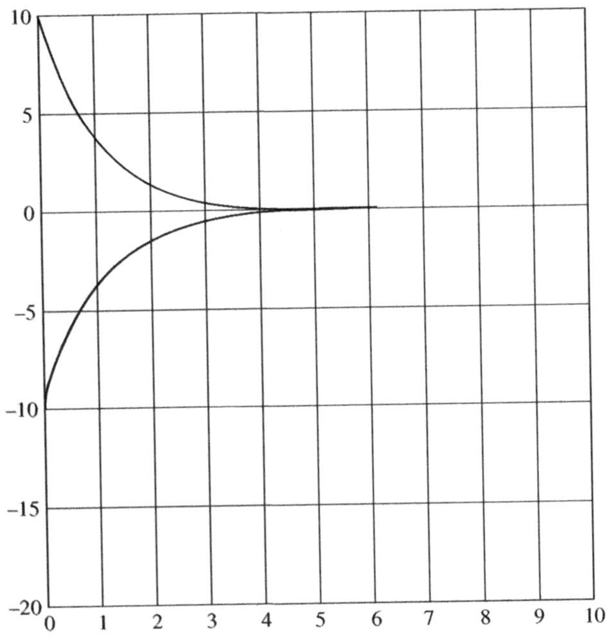

Figure 6.12 MATLAB simulation of the controlled system for Case B.

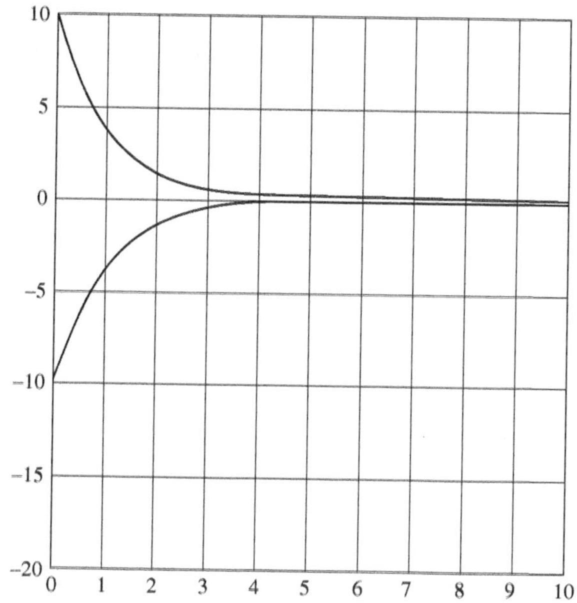

Figure 6.13 MATLAB simulation of the controlled system for Case C.

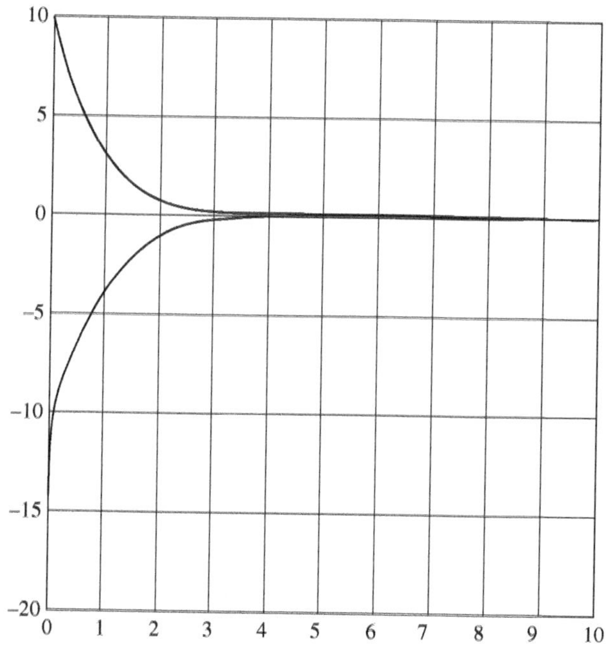

Figure 6.14 MATLAB simulation of the controlled system for Case D.

where $h(x)$ is an $m \times m$ matrix representing the uncertainty in the input matrix and the uncertainty $f(x)$ does not satisfy the matching condition. Hence we make the following assumptions.

Assumption 6.8

$A(0) = 0$ and $f(0) = 0$ so that $x = 0$ is an equilibrium.

Assumption 6.9

The uncertainty $f(x)$ is bounded.

Assumption 6.10

$h(x)$ is a positive semi-definite matrix: $h(x) \geq 0$.

We would like to solve the following robust control problem of stabilizing the above nonlinear system under uncertainty.

Robust Control Problem 6.7

Find a feedback control law $u = u_o(x)$ such that the closed-loop system

$$\dot{x} = A(x) + B(x)(u_o(x) + h(x)u_o(x)) + C(x)f(x)$$

is globally asymptotically stable for all uncertainties $f(x)$, $h(x)$ satisfying Assumptions 6.9 and 6.10.

We will solve the above robust control problem indirectly by translating it into an optimal control problem.

Optimal Control Problem 6.8

For the auxiliary system

$$\dot{x} = A(x) + B(x)u + \alpha(I - B(x)B(x)^{+})C(x)v$$

find a feedback control law $(u_o(x), v_o(x))$ that minimizes the following cost functional

$$\int_0^{\infty} (f_{\max}(x)^2 + \rho^2 g_{\max}(x)^2 + \beta^2 \|x\|^2 + \|u\|^2 + \rho^2 \|v\|^2)dt$$

where $\alpha \geq 0$, $\rho \geq 0$, and $\beta \geq 0$ are design parameters. $f_{max}(x)$, $g_{max}(x)$ are nonnegative functions such that

$$\|B(x)^{+}C(x)f(x)\| \leq f_{max}(x)$$
$$\|\alpha^{-1}f(x)\| \leq g_{max}(x)$$

We can solve the robust control problem by solving the optimal control problem as shown in the following theorem.

Theorem 6.4

If one can choose α, ρ and β such that the solution to Optimal Control Problem 6.8, denoted by $(u_o(x), v_o(x))$, exists and the following condition is satisfied

$$2\rho^2\|v_o(x)\|^2 \leq \beta'^2\|x\|^2 \qquad \forall x \in R^n$$

for some β' such that $|\beta'| < |\beta|$, then $u_o(x)$, the u-component of the solution to Optimal Control Problem 6.8, is a solution to Robust Control Problem 6.7.

Proof

Let $u_o(x)$, $v_o(x)$ be the solution to Optimal Control Problem 6.8. We would like to show that

$$\dot{x} = A(x) + B(x)(u_o(x) + h(x)u_o(x)) + C(x)f(x) \tag{6.12}$$

is globally asymptotically stable for all uncertainties $f(x)$.

To prove this, we define

$$V(x_o) = \min_{u,v}\int_0^{\infty}(f_{max}(x)^2 + \rho^2 g_{max}(x)^2 + \beta^2\|x\|^2 + \|u\|^2 + \rho^2\|v\|^2)dt$$

to be the minimum cost of the optimal control of the auxiliary system from some initial state x_o. We would like to show that $V(x)$ is a Lyapunov function for system (6.12). Since $V(x)$ is the minimal cost, it must satisfy the Hamilton–Jacobi–Bellman equation, which reduces to

$$\min_{u,v}(f_{max}(x)^2 + \rho^2 g_{max}(x)^2 + \beta^2\|x\|^2 + \|u\|^2 + \rho^2\|v\|^2$$
$$+ V_x^T\,(A(x) + B(x)u + \alpha(I - B(x)B(x)^{+})C(x)v)) = 0$$

Since $u_o(x)$, $v_o(x)$ is the optimal control, the following must be satisfied:

$$f_{max}(x)^2+\rho^2 g_{max}(x)^2+\beta^2\|x\|^2+\|u_o(x)\|^2+\rho^2\|v_o(x)\|^2$$
$$+V_x^{\mathrm{T}}\ (A(x)+B(x)u_o(x)+\alpha(I-B(x)B(x)^+)C(x)v_o(x))=0$$
$$2u_o(x)^{\mathrm{T}}+V_x^{\mathrm{T}}\ B(x)=0$$
$$2\rho^2 v_o(x)^{\mathrm{T}}+V_x^{\mathrm{T}}\ \alpha(I-B(x)B(x)^+)C(x)=0$$

From the above equations, we can show that $V(x)$ is a Lyapunov function for System (6.12). Clearly

$$V(x)>0 \qquad x\neq 0$$
$$V(x)=0 \qquad x=0$$

On the other hand, we have

$$\begin{aligned}
\dot{V}(x) &= V_x^{\mathrm{T}}\dot{x}\\
&= V_x^{\mathrm{T}}\ (A(x)+B(x)(u_o(x)+h(x)u_o(x))+C(x)f(x))\\
&= V_x^{\mathrm{T}}\ (A(x)+B(x)u_o(x))+C(x)f(x))+V_x^{\mathrm{T}}\ B(x)h(x)u_o(x)\\
&= V_x^{\mathrm{T}}\ (A(x)+B(x)u_o(x))+C(x)f(x))-2u_o(x)^{\mathrm{T}}h(x)u_o(x)\\
&\leq V_x^{\mathrm{T}}\ (A(x)+B(x)u_o(x))+C(x)f(x))\\
&= V_x^{\mathrm{T}}\ (A(x)+B(x)u_o(x)+\alpha(I-B(x)B(x)^+)C(x)v_o(x))\\
&\quad -V_x^{\mathrm{T}}\ \alpha(I-B(x)B(x)^+)C(x)v_o(x)+V_x^{\mathrm{T}}\ C(x)f(x)\\
&= V_x^{\mathrm{T}}\ (A(x)+B(x)u_o(x)+\alpha(I-B(x)B(x)^+)C(x)v_o(x))\\
&\quad +V_x^{\mathrm{T}}\ B(x)B(x)^+C(x)f(x)\\
&\quad -V_x^{\mathrm{T}}\ \alpha(I-B(x)B(x)^+)C(x)v_o(x)+V_x^{\mathrm{T}}\ (I-B(x)B(x)^+)C(x)f(x)\\
&= -f_{max}(x)^2-\rho^2 g_{max}(x)^2-\beta^2\|x\|^2-\|u_o(x)\|^2-\rho^2\|v_o(x)\|^2\\
&\quad +2\rho^2 v_o(x)^{\mathrm{T}}v_o(x)-2u_o(x)^{\mathrm{T}}B(x)^+C(x)f(x)-2\alpha^{-1}\rho^2 v_o(x)^{\mathrm{T}}C(x)f(x)\\
&= -f_{max}(x)^2-\rho^2 g_{max}(x)^2-\beta^2\|x\|^2-\|u_o(x)\|^2+\rho^2\|v_o(x)\|^2\\
&\quad -2u_o(x)^{\mathrm{T}}B(x)^+C(x)f(x)-2\alpha^{-1}\rho^2 v_o(x)^{\mathrm{T}}C(x)f(x).
\end{aligned}$$

But

$$-\|u_o(x)\|^2-2u_o(x)^{\mathrm{T}}B(x)^+C(x)f(x)\leq\|B(x)^+C(x)f(x)\|^2\leq f_{max}(x)^2$$
$$-2\alpha^{-1}\rho^2 v_o(x)^{\mathrm{T}}f(x)\leq\rho^2\|v_o(x)\|^2+\rho^2\|\alpha^{-1}f(x)\|^2$$
$$\leq\rho^2\|v_o(x)\|^2+\rho^2 g_{max}(x)^2$$

Therefore, by the condition $2\rho^2\|v_o(x)\|^2 \le \beta'^2\|x\|^2$, $\forall x \in R^n$,

$$\begin{aligned}\dot{V}(x) &\le -\beta^2\|x\|^2 + 2\rho^2\|v_o(x)\|^2 \\ &= 2\rho^2\|v_o(x)\|^2 - \beta'^2\|x\|^2 - (\beta^2 - \beta'^2)\|x\|^2 \\ &\le -(\beta^2 - \beta'^2)\|x\|^2\end{aligned}$$

In other words,

$$\begin{aligned}\dot{V}(x) &< 0 \qquad x \ne 0 \\ \dot{V}(x) &= 0 \qquad x = 0\end{aligned}$$

Thus, the conditions of the Lyapunov stability theorem are satisfied. Consequently, there exists a neighbourhood of 0, $N = \{x : \|x\| < c\}$ for some $c > 0$ such that if $x(t)$ enters N, then

$$x(t) \to 0 \quad as \quad t \to \infty$$

But $x(t)$ cannot remain forever outside N. Otherwise

$$\|x(t)\| \ge c$$

for all t > 0, which implies

$$\begin{aligned}V(x(t)) - V(x(0)) &= \int_0^t \dot{V}(x(\tau))\mathrm{d}\tau \\ &\le \int_0^t -(\beta^2 - \beta'^2)\|x\|^2 \mathrm{d}\tau \\ &\le -\int_0^t (\beta^2 - \beta'^2)c^2 \mathrm{d}\tau \\ &\le -(\beta^2 - \beta'^2)c^2 t\end{aligned}$$

Let $t \to \infty$, we have

$$V(x(t)) \le V(x(0)) - (\beta^2 - \beta'^2)c^2 t \to -\infty$$

which contradicts the fact that $V(x(t)) > 0$ for all $x(t)$. Therefore

$$x(t) \to 0 \quad as \quad t \to \infty$$

no matter where the trajectory begins. This proves that $u_o(x)$ is a solution to Robust Control Problem 6.7.

Q.E.D.

Example 6.5

Let us consider the following nonlinear system

$$\dot{x}_1 = x_2 + p_1 x_1 \cos(1/x_2) + p_3 x_2 \sin(p_4 x_1 x_2)$$
$$\dot{x}_2 = (1 + p_2 x_2^2) u$$

where $p_1 \in [-0.2, 0.2]$, $p_2 \in [0.2, 3]$, $p_3 \in [0, 0.2]$ and $p_4 \in [-100, 0]$. The robust control problem is to find a control $u = u_o(x)$ so that the closed-loop system is stable for all possible uncertainties.

To translate the robust control problem into an optimal control problem, we first rewrite the state equation as

$$\begin{bmatrix} \dot{x}_1 \\ \dot{x}_2 \end{bmatrix} = \begin{bmatrix} 0 & 1 \\ 0 & 0 \end{bmatrix} \begin{bmatrix} x_1 \\ x_2 \end{bmatrix} + \begin{bmatrix} 0 \\ 1 \end{bmatrix} (1 + p_2 x_2^2) u + \begin{bmatrix} 1 \\ 0 \end{bmatrix} f(x)$$

where $f(x) = p_1 x_1 \cos(1/x_2) + p_3 x_2 \sin(p_4 x_1 x_2)$ is the uncertainty. Let us take $a = 0.2$, $\beta = 1$, and $\rho = 1$. Then $f_{max}(x)$ and $g_{max}(x)$ can be calculated as follows.

$$\|B(x)^+ C(x) f(x)\| = \| \begin{bmatrix} 0 & 1 \end{bmatrix} \begin{bmatrix} 1 \\ 0 \end{bmatrix} f(x) \| = 0 = f_{max}(x)$$

$$\|\alpha^{-1} f(x)\| = \|5(p_1 x_1 \cos(1/x_2) + p_3 x_2 \sin(p_4 x_1 x_2))\| \leq |x_1 + x_2| = g_{max}(x)$$

Therefore,

$$f_{max}(x)^2 + \rho^2 g_{max}(x)^2 + \beta^2 \|x\|^2 + \|u\|^2 + \rho^2 \|v\|^2$$
$$= (x_1 + x_2)^2 + (x_1^2 + x_2^2) + u^2 + v^2.$$

Hence, the corresponding optimal control problem is as follows. For the auxiliary system

$$\begin{bmatrix} \dot{x}_1 \\ \dot{x}_2 \end{bmatrix} = \begin{bmatrix} 0 & 1 \\ 0 & 0 \end{bmatrix} \begin{bmatrix} x_1 \\ x_2 \end{bmatrix} + \begin{bmatrix} 0 \\ 1 \end{bmatrix} u + \begin{bmatrix} 0.2 \\ 0 \end{bmatrix} v$$

find a feedback control law $(u_o(x), v_o(x))$ that minimizes the following cost functional

$$\int_0^\infty (2x_1^2 + 2x_2^2 + 2x_1 x_2 + u^2 + v^2) dt$$

This is an LQR problem with

$$A = \begin{bmatrix} 0 & 1 \\ 0 & 0 \end{bmatrix} \quad B = \begin{bmatrix} 0 & 0.2 \\ 1 & 0 \end{bmatrix}$$

$$Q = \begin{bmatrix} 2 & 1 \\ 1 & 2 \end{bmatrix} \quad R = \begin{bmatrix} 1 & 0 \\ 0 & 1 \end{bmatrix}$$

The solution to the LQR problem is given by

$$u_o = \begin{bmatrix} -1.3549 & -2.1532 \end{bmatrix} x$$
$$v_o = \begin{bmatrix} -0.4054 & -0.2710 \end{bmatrix} x$$

Let us check if the sufficient condition

$$2\rho^2 \|v_o(x)\|^2 \leq \beta^2 \|x\|^2 \qquad \forall x \in R^n$$

is satisfied. Clearly

$$\begin{aligned} &\beta^2 \|x\|^2 - 2\rho^2 \|v_o(x)\|^2 \\ &= \|x\|^2 - 2\|v_o(x)\|^2 \\ &= x_1^2 + x_2^2 - 2(-0.4054x_1 - 0.2710x_2)^2 \\ &= x^{\mathrm{T}} \begin{bmatrix} 0.6713 & -0.2197 \\ -0.2197 & 0.8531 \end{bmatrix} x \\ &\geq 0 \end{aligned}$$

Therefore

$$u_o = \begin{bmatrix} -1.3549 & -2.1532 \end{bmatrix} x$$

is a solution to the robust control problem.

To test the robustness of the control, we use MATLAB to perform simulations of the closed-loop system under control with the initial conditions $x_1(0) = 10$ and $x_2(0) = 10$. We consider the following four cases.

Case A: $p_1 = -0.2$ $\quad p_2 = 0.2$ $\quad p_3 = 0$ $\quad p_4 = -100$
Case B: $P_1 = 0.2$ $\quad p_2 = 0.2$ $\quad p_3 = 0.2$ $\quad p_4 = 0$
Case C: $p_1 = 0$ $\quad p_2 = 3$ $\quad p_3 = 0$ $\quad p_4 = 0$
Case D: $p_1 = -0.2$ $\quad p_2 = 3$ $\quad p_3 = -0.2$ $\quad p_4 = -100$

The simulation results are shown in Figures 6.15–6.18, respectively. From the results we can see that the performance depends more on p_2. However, they are robust for all the parameters.

Case 3

We no longer assume that the input uncertainty enters the system via the input matrix $B(x)$; that is, we consider the following system

$$\dot{x} = A(x) + B(x)u + C(x)D(x)u,$$

where $D(x)$ is the (only) uncertainty. We make the following assumptions.

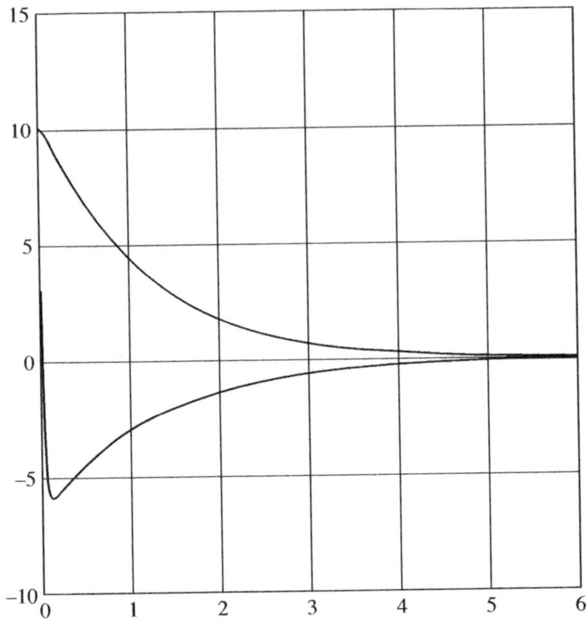

Figure 6.15 MATLAB simulation of the controlled system for Case A.

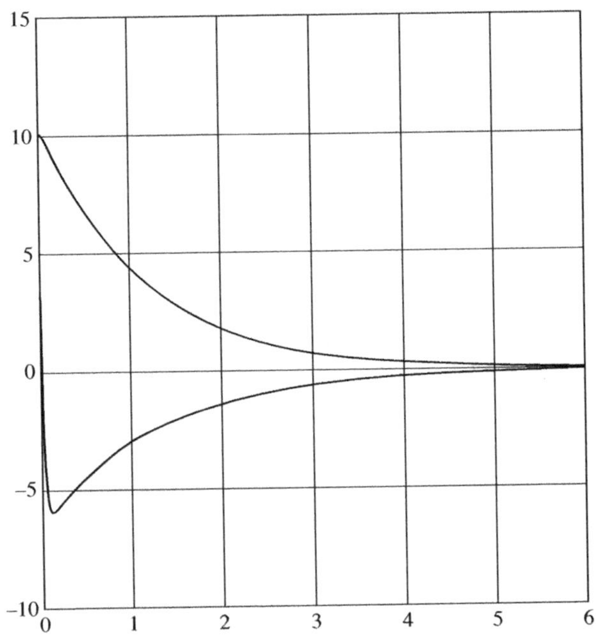

Figure 6.16 MATLAB simulation of the controlled system for Case B.

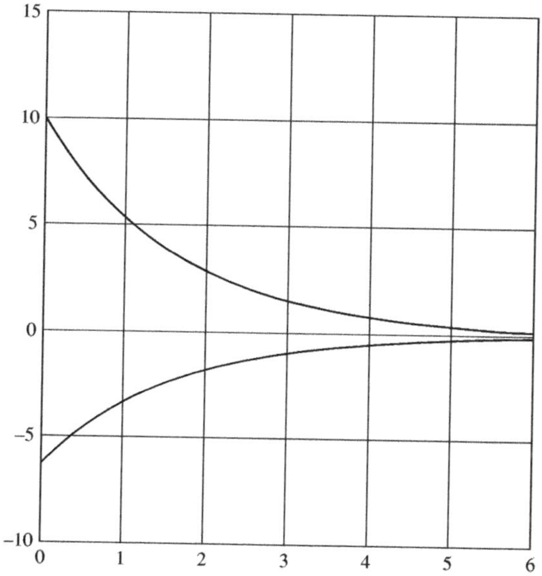

Figure 6.17 MATLAB simulation of the controlled system for Case C.

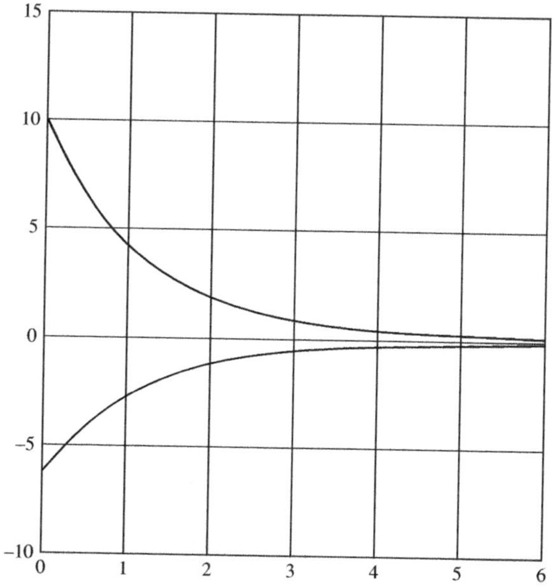

Figure 6.18 MATLAB simulation of the controlled system for Case D.

Assumption 6.11

$A(0) = 0$ so that $x = 0$ is an equilibrium.

Assumption 6.12

The uncertainty $D(x)$ is bounded: $\|D(x)\| \leq D_{max}(x)$ for some $D_{max}(x)$.

We would like to solve the following robust control problem of stabilizing the above nonlinear system under uncertainty.

Robust Control Problem 6.9

Find a feedback control law $u = u_o(x)$ such that the closed-loop system

$$\dot{x} = A(x) + B(x)u_o(x) + C(x)D(x)u_o(x)$$

is globally asymptotically stable for all uncertainties $D(x)$ satisfying $\|D(x)\| \leq D_{max}(x)$.

Since u will be a function of x, $u = u_o(x)$, we can view $f(x) = D(x)u_o(x)$ as the uncertainty and guess its bound:

$$\|B(x)^{+}C(x)D(x)u_o(x)\| \leq f_{max}(x)$$
$$\|\alpha^{-1}D(x)u_o(x)\| \leq g_{max}(x)$$

We can then solve the above robust control problem indirectly by translating it into an optimal control problem.

Optimal Control Problem 6.10

For the auxiliary system

$$\dot{x} = A(x) + B(x)u + \alpha(I - B(x)B(x)^{+})C(x)v$$

find a feedback control law $(u_o(x), v_o(x))$ that minimizes the following cost functional

$$\int_0^{\infty} (f_{max}(x)^2 + \rho^2 g_{max}(x)^2 + \beta^2\|x\|^2 + \|u\|^2 + \rho^2\|v\|^2)dt$$

where $\alpha \geq 0$, $\rho \geq 0$, and $\beta \geq 0$ are design parameters and $f_{max}(x)$, $g_{max}(x)$ are design functions.

The solution to the optimal control problem is a solution to the robust control problem if certain conditions are satisfied as shown in the following theorem.

Theorem 6.5

If one can choose design parameters α, ρ, β, and functions $f_{max}(x)$, $g_{max}(x)$ such that the solution to Optimal Control Problem 6.10, denoted by $(u_o(x), v_o(x))$, exists and the following conditions are satisfied

$$\|B(x)^+C(x)D(x)u_o(x)\| \leq f_{max}(x)$$
$$\|\alpha^{-1}D(x)u_o(x)\| \leq g_{max}(x)$$
$$2\rho^2\|v_o(x)\|^2 \leq \beta'^2\|x\|^2, \quad \forall x \in R^n$$

for some β' such that $|\beta'| < |\beta|$, then $u_o(x)$, the u-component of the solution to Optimal Control Problem 6.10, is a solution to Robust Control Problem 6.9.

Proof

Let $u_o(x)$, $v_o(x)$ be the solution to Optimal Control Problem 6.10. We would like to show that

$$\dot{x} = A(x) + B(x)(u_o(x) + h(x)u_o(x)) + C(x)D(x)u_o(x) \qquad (6.13)$$

is globally asymptotically stable for all uncertainties $D(x)$.

To prove this, we define

$$V(x_o) = \min_{u,v} \int_0^\infty (f_{max}(x)^2 + \rho^2 g_{max}(x)^2 + \beta^2\|x\|^2 + \|u\|^2 + \rho^2\|v\|^2)dt$$

to be the minimum cost of the optimal control of the auxiliary system from some initial state x_o. We would like to show that $V(x)$ is a Lyapunov function for system (6.13). Since $V(x)$ is the minimal cost, it must satisfy the Hamilton–Jacobi–Bellman equation, which reduces to

$$\min_{u,v}(f_{max}(x)^2 + \rho^2 g_{max}(x)^2 + \beta^2\|x\|^2 + \|u\|^2 + \rho^2\|v\|^2$$
$$+ V_x^T\ (A(x) + B(x)u + \alpha(I - B(x)B(x)^+)C(x)v)) = 0$$

Since $u_o(x)$, $v_o(x)$ is the optimal control, the following must be satisfied:

$$f_{\max}(x)^2+\rho^2 g_{\max}(x)^2+\beta^2\|x\|^2+\|u_o(x)\|^2+\rho^2\|v_o(x)\|^2$$
$$+V_x^{\mathrm{T}}\ (A(x)+B(x)u_o(x)+\alpha(I-B(x)B(x)^+)C(x)v_o(x))=0$$
$$2u_o(x)^{\mathrm{T}}+V_x^{\mathrm{T}}\ B(x)=0$$
$$2\rho^2 v_o(x)^{\mathrm{T}}+V_x^{\mathrm{T}}\ \alpha(I-B(x)B(x)^+)C(x)=0$$

From the above equations, we can show that $V(x)$ is a Lyapunov function for System (6.13). Clearly

$$V(x)>0 \qquad x\neq 0$$
$$V(x)=0 \qquad x=0$$

On the other hand, we have

$$\begin{aligned}
\dot V(x) &= V_x^{\mathrm{T}}\dot x\\
&= V_x^{\mathrm{T}}(A(x)+B(x)(u_o(x)+h(x)u_o(x))+C(x)D(x)u_o(x))\\
&= V_x^{\mathrm{T}}(A(x)+B(x)u_o(x))+C(x)D(x)u_o(x))+V_x^{\mathrm{T}}B(x)h(x)u_o(x)\\
&= V_x^{\mathrm{T}}(A(x)+B(x)u_o(x))+C(x)D(x)u_o(x))-2u_o(x)^{\mathrm{T}}h(x)u_o(x)\\
&\leq V_x^{\mathrm{T}}(A(x)+B(x)u_o(x))+C(x)D(x)u_o(x))\\
&= V_x^{\mathrm{T}}(A(x)+B(x)u_o(x)+\alpha(I-B(x)B(x)^+)C(x)v_o(x))\\
&\quad -V_x^{\mathrm{T}}\alpha(I-B(x)B(x)^+)C(x)v_o(x)+V_x^{\mathrm{T}}C(x)D(x)u_o(x)\\
&= V_x^{\mathrm{T}}(A(x)+B(x)u_o(x)+\alpha(I-B(x)B(x)^+)C(x)v_o(x))\\
&\quad +V_x^{\mathrm{T}}B(x)B(x)^+C(x)D(x)u_o(x)\\
&\quad -V_x^{\mathrm{T}}\alpha(I-B(x)B(x)^+)C(x)v_o(x)+V_x^{\mathrm{T}}(I-B(x)B(x)^+)C(x)D(x)u_o(x)\\
&= -f_{\max}(x)^2-\rho^2 g_{\max}(x)^2-\beta^2\|x\|^2-\|u_o(x)\|^2-\rho^2\|v_o(x)\|^2\\
&\quad +2\rho^2 v_o(x)^{\mathrm{T}}v_o(x)-2u_o(x)^{\mathrm{T}}B(x)^+C(x)D(x)u_o(x)\\
&\quad -2\alpha^{-1}\rho^2 v_o(x)^{\mathrm{T}}C(x)D(x)u_o(x)\\
&= -f_{\max}(x)^2-\rho^2 g_{\max}(x)^2-\beta^2\|x\|^2-\|u_o(x)\|^2+\rho^2\|v_o(x)\|^2\\
&\quad -2u_o(x)^{\mathrm{T}}B(x)^+C(x)D(x)u_o(x)-2\alpha^{-1}\rho^2 v_o(x)^{\mathrm{T}}C(x)D(x)u_o(x).
\end{aligned}$$

By conditions $\|B(x)^+C(x)D(x)u_o(x)\|\leq f_{\max}(x)$ and $\|\alpha^{-1}D(x)u_o(x)\|\leq g_{\max}(x)$

$$-\|u_o(x)\|^2-2u_o(x)^{\mathrm{T}}B(x)^+C(x)D(x)u_o(x)$$

$$
\begin{aligned}
&\leq \|B(x)^{+}C(x)D(x)u_{o}(x)\|^2 \\
&\leq f_{\max}(x)^2 \\
&\quad -2\alpha^{-1}\rho^2 v_o(x)^{\mathrm{T}}D(x)u_o(x) \\
&\leq \rho^2\|v_o(x)\|^2+\rho^2\|\alpha^{-1}D(x)u_o(x)\|^2 \\
&\leq \rho^2\|v_o(x)\|^2+\rho^2 g_{\max}(x)^2
\end{aligned}
$$

Furthermore, by condition $2\rho^2\|v_o(x)\|^2 \leq \beta'^2\|x\|^2, \quad \forall x \in R^n$

$$
\begin{aligned}
\dot{V}(x) &\leq -\beta^2\|x\|^2+2\rho^2\|v_o(x)\|^2 \\
&= 2\rho^2\|v_o(x)\|^2-\beta'^2\|x\|^2-(\beta^2-\beta'^2)\|x\|^2 \\
&\leq -(\beta^2-\beta'^2)\|x\|^2
\end{aligned}
$$

In other words

$$
\begin{aligned}
\dot{V}(x) &< 0 \qquad x \neq 0 \\
\dot{V}(x) &= 0 \qquad x = 0.
\end{aligned}
$$

Thus, the conditions of The Lyapunov stability theorem are satisfied. Consequently, there exists a neighbourhood of 0, $N=\{x : \|x\| < c\}$ for some $c > 0$ such that if $x(t)$ enters N, then

$$x(t) \to 0 \quad \text{as} \quad t \to \infty$$

But $x(t)$ cannot remain forever outside N. Otherwise

$$\|x(t)\| \geq c$$

for all t>0, which implies

$$
\begin{aligned}
V(x(t))-V(x(0)) &= \int_0^t \dot{V}(x(\tau))\mathrm{d}\tau \\
&\leq \int_0^t -(\beta^2-\beta'^2)\|x\|^2\mathrm{d}\tau \\
&\leq -\int_0^t (\beta^2-\beta'^2)c^2\mathrm{d}\tau \\
&\leq -(\beta^2-\beta'^2)c^2 t
\end{aligned}
$$

Let $t \to \infty$, we have

$$V(x(t)) \leq V(x(0))-(\beta^2-\beta'^2)c^2 t \to -\infty$$

which contradicts the fact that $V(x(t)) > 0$ for all $x(t)$. Therefore

$$x(t) \to 0 \quad \text{as} \quad t \to \infty$$

no matter where the trajectory begins. This proves that $u_o(x)$ is a solution to Robust Control Problem 6.9.

Q.E.D.

6.5 NOTES AND REFERENCES

This chapter deals with nonlinear systems. Although the idea is conceptually similar to that used in linear systems, the proofs of the results for nonlinear systems are much more complex. Furthermore, the computations of solutions are much more difficult. We have provided some examples to illustrate how to solve the optimal control problems, but there are many systems for which we do not know how to obtain solutions.

We first considered nonlinear systems satisfying the matching condition. Unlike linear systems, where the solution to robust control problems always exists, the solution for nonlinear systems may not exist if the corresponding optimal control problem cannot be solved. This is unfortunate, but it is the nature of nonlinear systems. This result was first published in reference [105].

We then investigated nonlinear systems with unmatched uncertainty. Here we require not only that the solution to the optimal control problem must exist, but also that the solution needs to satisfy an additional condition. Although this restricts our approach further, we do not consider the results as inadequate, because many other approaches cannot deal with unmatched uncertainties at all. Our result was published in reference [108]; other results on robust control of nonlinear systems can be found in [42, 77, 134]

We have also discussed robust control problems for systems with uncertainty in the input matrix. This introduces additional difficulty to the problem. Nevertheless, our approach provides a unique way to handle this most difficult robust control problem.

6.6 PROBLEMS

6.1 Consider the following nonlinear system

$$\dot{x} = -x^2 \cos(2x) + (x^2 + 4)u + 2p_1 x \sin(x + p_2)$$

where $p_1 \in [-1, 1]$, $p_2 \in [10, 100]$ are the uncertainties. Find a feedback control law $u = u_o(x)$, if possible, such that the closed-loop system is globally asymptotically stable for all uncertainties.

6.2 For the following system

$$\dot{x}_1 = u + p_1 x_1 \sin(x_1) - p_2 x_2 \sin(x_2)$$
$$\dot{x}_2 = x_1$$

where $p_1 \in [-1, 1]$, $p_2 \in [0, 3]$, design a feedback control $u = u_o(x)$ so that the closed-loop system is globally asymptotically stable for all uncertainties.

6.3 Consider the following nonlinear system

$$\dot{x}_1 = x_1 + x_2$$
$$\dot{x}_2 = p_1 x_1 \cos\left(p_2 \sqrt{x_1 + x_2^2}\right) + u$$

where $p_1 \in [-1, 1]$, $p_2 \in [10, 100]$ are the uncertainties. Find a feedback control law $u = u_o(x)$, if possible, such that the closed-loop system is globally asymptotically stable for all uncertainties.

6.4 For the following system

$$\dot{x}_1 = u + p_1 x_2 \sin(x_1 + p_2)$$
$$\dot{x}_2 = x_3$$
$$x_3 = u + p_3 x_1 \cos(x_2^2)$$

where $p_1 \in [-0.1, 1]$, $p_2 \in [-50, 70]$, $p_3 \in [0, 3]$, design a feedback control $u = u_o(x)$ so that the closed-loop system is globally asymptotically stable for all uncertainties.

6.5 Use MATLAB to simulate the closed-loop system obtained in Problem 6.4.

6.6 Let us consider the following nonlinear system

$$\dot{x}_1 = -x_1 + 3x_2 + p_1 x_1 \cos(x_2 + p_2) + p_3 x_2 \sin(p_4 x_1 x_2)$$
$$\dot{x}_2 = u$$

where $p_1 \in [-0.2, \ 0.2], p_2 \in [-10, \ 20], p_3 \in [0, \ 0.2]$ and $p_4 \in [-20, \ 0]$. Find a robust control $u = u_o(x)$ so that the closed-loop system is stable for all possible uncertainties.

6.7 Use MATLAB to simulate the closed-loop system obtained in Problem 6.6.

6.8 Consider the following nonlinear system

$$\dot{x}_1 = x_2$$
$$\dot{x}_2 = -2x_1 + p_1 x_1 \cos(3x_2) + (1 + p_2 x_2^2)u$$

where $p_1 \in [-0.7, \quad 2]$ and $p_2 \in [0.1, \quad 5]$. Find a robust control $u = u_o(x)$ so that the closed-loop system is stable for all possible uncertainties.

6.9 Use MATLAB to simulate the closed-loop system obtained in Problem 6.8.

6.10 Let us consider the following nonlinear system

$$\dot{x}_1 = -3x_1 + x_2 + p_1 x_1 \cos(x_1 x_2) + p_3 x_2 \sin(p_4 x_1)$$
$$\dot{x}_2 = 2x_1 + (1 + p_2 x_2^2)u$$

where $p_1 \in [-0.2, \quad 0.2]$, $p_2 \in [0.2, \quad 3]$, $p_3 \in [0, \quad 0.2]$ and $p_4 \in [-100, \quad 0]$. The robust control problem is to find a control $u = u_o(x)$ so that the closed-loop system is stable for all possible uncertainties.

6.11 Use MATLAB to simulate the closed-loop system obtained in Problem 6.10.

7

Kharitonov Approach

In Chapters 5 and 6, we presented an optimal control approach to robust control. We will show three applications of this approach in Chapters 9, 10 and 11. In this chapter and the next one, we will discuss two other main approaches to robust control that have been studied extensively in the literature. They are parametric approach, sometimes called the Kharitonov approach and the H_∞/H_2 approach.

This method was originated in the seminal paper [89] published by Kharitonov in 1978, which shows how to check the stability of a set of polynomials with uncertain parameters. Since then the initial results have been extended in several directions. We will focus on the main results of the Kharitonov approach, provide a complete proof, apply it to feedback control, and compare it with the optimal control approach discussed earlier.

7.1 INTRODUCTION

As the name implies, parametric approach studies the robust stability of a system when its parameters are uncertain. In other words, we would like to check if a system is stable when its parameters are uncertain and vary over an interval. Two questions are of importance in this regard: (1) given a set of intervals where parameters very, can we check if the system is stable for all possible parameters?; (2) if so, how many values of the parameters do we need to check?

Robust Control Design: An Optimal Control Approach F. Lin

For example, consider a system with the following characteristic polynomial

$$\varphi(s,p) = p_0 + p_1 s + p_2 s^2 + p_3 s^3$$

where $p_i \in [p_i^-, p_i^+]$, $i = 0, 1, 2, 3$ are coefficients whose values are uncertain, but we know their lower and upper bounds. To check if the system is stable for all possible parameters, a naïve approach is to check a set of polynomials

$$\Psi = \{p_0 + p_1 s + p_2 s^2 + p_3 s^3 : p_i \in [p_i^-, p_i^+], i = 0, 1, 2, 3\}$$

Since this set is infinite, it is not possible to check its elements one by one. In fact, we really do not need to do so. By the Routh–Hurwitz criterion, we know that a third-order system is stable if and only if the coefficients of its characteristic polynomial satisfy the following condition:

$$p_0 > 0, p_1 > 0, p_2 > 0, p_3 > 0, p_1 p_2 > p_0 p_3$$

Hence, the entire set of polynomials Φ is stable if and only if

$$p_0^- > 0, p_1^- > 0, p_2^- > 0, p_3^- > 0, p_1^- p_2^- > p_0^+ p_3^+$$

In other words, to check if all polynomials in Φ are stable we only need to check if the following two polynomials are stable:

$$\varphi_1 = p_0^- + p_1^- s + p_2^- s^2 + p_3^- s^3$$
$$\varphi_2 = p_0^+ + p_1^- s + p_2^- s^2 + p_3^+ s^3$$

From this example, we see that in order to determine the stability of an infinite set Φ, only some finite numbers of polynomials need to be checked. In the example, only two polynomials need to be checked, but this is a simple case. In general, we want to know how many polynomials need to be checked. Also, in this example, the two polynomials to be checked correspond to two corners of the polytope representing the region of possible parameter values in the parameter space. In general, do we always check the corners? How many corners do we need to check? Note that the number of corners increases exponentially as the number of parameters increases. The Kharitonov theorem answers the above question elegantly: no matter how many parameters are involved, we only need to check four specific corners.

7.2 PRELIMINARY THEOREMS

In this section, we will discuss two preliminary results that will be used to prove the main results of the Kharitonov approach. These two preliminary results are the Boundary Crossing theorem and Interlacing theorem. We will first present the Boundary Crossing theorem. Let

$$\varphi(s,p) = a_0(p) + a_1(p)s + \cdots + a_{n-1}(p)s^{n-1} + s^n$$

be a family of polynomials, where the coefficients $a_i(p)$, $i = 0, 1, \ldots, n-1$, are continuous functions of uncertain parameter p over a fixed interval $p \in P = [p^-, p^+]$.

Assume that the complex plane C is divided into three disjoint parts:

$$C = S \cup \partial S \cup U$$

where S is an open set which is interpreted as the 'stable region', ∂S is the boundary of S, and U is the complement of $S \cup \partial S$ which is also an open set and is interpreted as the 'unstable region'. For linear time-invariant systems with continuous time, S is the open left half of the s-plane, ∂S is the imaginary axis, and U is open right-half of the s-plane as shown in Figure 7.1.

We now present the following Boundary Crossing theorem.

Theorem 7.1

If $\varphi(s, p^-)$ has all its roots in S and $\varphi(s, p^+)$ has at least one root in U, then there exists at least one $p \in [p^-, p^+]$ such that $\varphi(s, p)$ has all its roots in $S \cup \partial S$ and at least one root in ∂S. Similarly, if $\varphi(s, p^+)$ has all its roots in S and $\varphi(s, p^-)$ has at least one root in U, then there exists at least one $p \in [p^-, p^+]$ such that $\varphi(s, p)$ has all its roots in $S \cup \partial S$ and at least one root in ∂S.

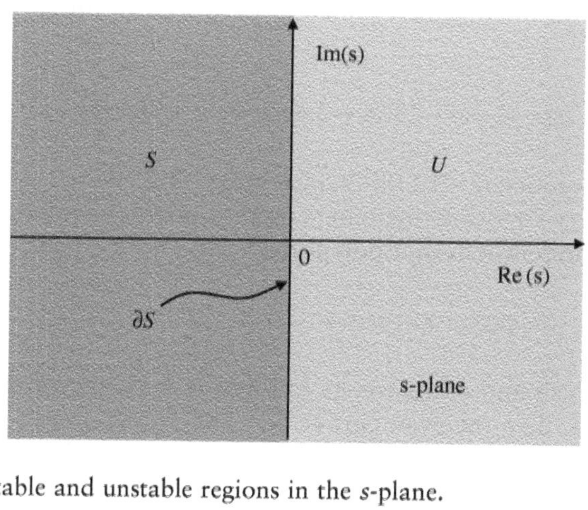

Figure 7.1 Stable and unstable regions in the s-plane.

Proof

The result follows from the assumption that $a_i(p), i = 0, 1, \ldots, n-1$, are continuous functions of p and the fact that roots of a polynomial are continuous functions of its coefficients.

Q.E.D.

Example 7.1

Let us consider the following polynomial:

$$\varphi(s, p) = (40 - p^2) + (1 + p)s + (9 + 3p)s^2 + s^3$$

where $p \in [p^-, p^+] = [1, 2]$. Let S be the open left half of the s-plane. For $p = p^- = 1$,

$$\varphi(s, p^-) = 39 + 2s + 12s^2 + s^3$$

has the roots at $-12.1011, 0.0505 + j1.7945, 0.0505 - j1.7945$ (which corresponds to an unstable system). For $p = p^+ = 2$

$$\varphi(s, p^-) = 36 + 3s + 15s^2 + s^3$$

has the roots at $-14.9603, -0.0198 + j1.5511j, -0.0198 - j1.5511$ (which corresponds to a stable system).

By the Boundary Crossing theorem, there exists at least one $p \in [p^-, p^+]$ such that $\varphi(s, p)$ has all its roots in the closed left half of the s-plane and at least one root on the imaginary axis. Indeed, we can find $p = 1.6623$ such that

$$\varphi(s, p) = 37.2368 + 2.6623s + 13.9869s^2 + s^3$$

and $\varphi(s, p)$ has roots at $-13.9869, j1.6316, -j1.6316$.

The Boundary Crossing theorem is used to prove the Interlacing theorem that we now present. Given a polynomial

$$\varphi(s) = a_0 + a_1 s + \cdots + a_{n-1} s^{n-1} + a_n s^n$$

we define its even and odd parts as follows. If the order of the polynomial $n = 2m$ is even, then the even part is

$$\varphi^e(s) = a_0 + a_2 s^2 + \cdots + a_{n-2} s^{n-2} + a_n s^n$$

and the odd part without s is

$$\varphi^{o}(s) = a_1 + a_3 s^2 + \cdots + a_{n-3} s^{n-4} + a_{n-1} s^{n-2}$$

On the other hand, if the order of the polynomial $n = 2m+1$ is odd, then the even part is

$$\varphi^{e}(s) = a_0 + a_2 s^2 + \cdots + a_{n-3} s^{n-3} + a_{n-1} s^{n-1}$$

and the odd part without s is

$$\varphi^{o}(s) = a_1 + a_3 s^2 + \cdots + a_{n-2} s^{n-3} + a_n s^{n-1}$$

The intuition behind the definitions of $\varphi^{e}(s)$ and $\varphi^{o}(s)$ can be seen as follows.

Let $s = j\omega$, then

$$\varphi(j\omega) = a_0 + a_1(j\omega) + a_2(j\omega)^2 + \cdots + a_{n-1}(j\omega)^{n-1} + a_n(j\omega)^n$$

The real part of $\varphi(j\omega)$ can be expressed as follows. For $n = 2m$

$$Re(\varphi(j\omega)) = a_0 + a_2(j\omega)^2 + \cdots + a_{n-2}(j\omega)^{n-2} + a_n(j\omega)^n$$

For $n = 2m+1$

$$Re(\varphi(j\omega)) = a_0 + a_2(j\omega)^2 + \cdots + a_{n-3}(j\omega)^{n-3} + a_{n-1}(j\omega)^{n-1}$$

In both cases,

$$Re(\varphi(j\omega)) = \varphi^{e}(j\omega)$$

Similarly, the imaginary part of $\varphi(j\omega)$ can be expressed as follows. For $n = 2m$

$$j\mathrm{Im}(\varphi(j\omega)) = j\omega(a_1 + a_3(j\omega)^2 + \cdots + a_{n-3}(j\omega)^{n-4} + a_{n-1}(j\omega)^{n-2})$$

For $n = 2m+1$

$$j\mathrm{Im}(\varphi(j\omega)) = j\omega(a_1 + a_3(j\omega)^2 + \cdots + a_{n-2}(j\omega)^{n-3} + a_n(j\omega)^{n-1})$$

In both cases,

$$j\mathrm{Im}(\varphi(j\omega)) = j\omega\varphi^{o}(j\omega)$$

or

$$\mathrm{Im}(\varphi(j\omega)) = \omega\varphi^{o}(j\omega)$$

Therefore, $\varphi^{e}(s)$ and $\varphi^{o}(s)$ are related to the real and imaginary parts of $\varphi(j\omega)$. We say that $\varphi(s) = a_0 + a_1 s + \cdots + a_{n-1}s^{n-1} + a_n s^n$ satisfies the interlacing property if

1. a_n and a_{n-1} have the same sign
2. all roots of $\varphi^e(j\omega)$ and $\varphi^o(j\omega)$ are real and distinct
3. if $n = 2m$ is even, then $\varphi^e(j\omega)$ has m positive roots $\omega_1^e < \omega_2^e < \cdots < \omega_m^e$, $\varphi^o(j\omega)$ has $m-1$ positive roots $\omega_1^o < \omega_2^o < \cdots < \omega_{m-1}^o$, and they must interlace in the following manner: $0 < \omega_1^e < \omega_1^o < \omega_2^e < \cdots < \omega_{m-1}^o < \omega_m^e$
4. if $n = 2m+1$ is odd, then $\varphi^e(j\omega)$ has m positive roots $\omega_1^e < \omega_2^e < \cdots < \omega_m^e$, $\varphi^o(j\omega)$ has m positive roots $\omega_1^o < \omega_2^o < \cdots < \omega_m^o$, and they must interlace in the following manner: $0 < \omega_1^e < \omega_1^o < \omega_2^e < \cdots < \omega_{m-1}^o < \omega_m^e < \omega_m^o$

The conditions for the interlacing property can be illustrated in Figures 7.2 and 7.3. Figure 7.2 shows the $\varphi(j\omega)$ curve in the $\varphi(j\omega)$-plane when ω goes from 0 to $+\infty$. The curve intersects with the imaginary axis and the real axis at $\omega = \omega_1, \omega_2, \omega_3, \ldots$, etc. Figure 7.3 shows the curves of $\varphi^e(j\omega)$ and $\varphi^o(j\omega)$ as ω goes from 0 to $+\infty$. Two curves interlace.

The following Interlacing theorem relates the interlacing property with stability.

Theorem 7.2

A polynomial $\varphi(s) = a_0 + a_1 s + \cdots + a_{n-1} s^{n-1} + a_n s^n$ has all its roots in the open left half of the complex plane if and only if it satisfies the interlacing property.

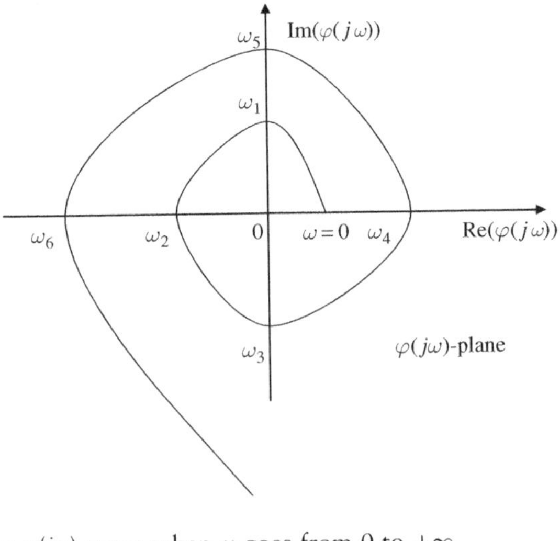

Figure 7.2 The $\varphi(j\omega)$ curve when ω goes from 0 to $+\infty$.

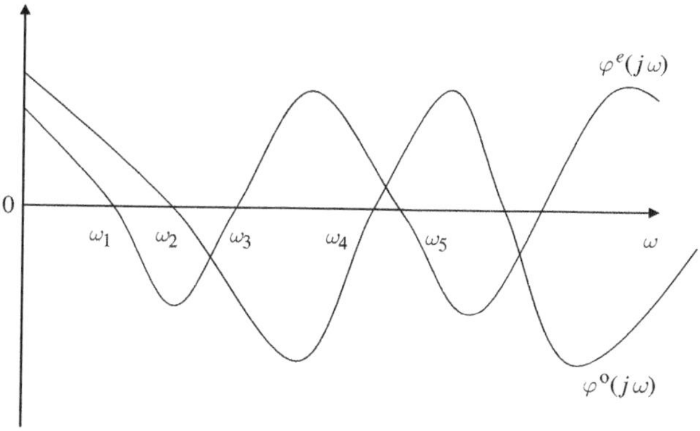

Figure 7.3 Interlacing of $\varphi^e(j\omega)$ and $\varphi^o(j\omega)$.

Before proving Theorem 7.2, let us first recall (Lemma 3.1) that if a polynomial $\varphi(s) = a_0 + a_1 s + \cdots + a_{n-1}s^{n-1} + a_n s^n$ has all its roots in the open left half of the complex plane, then all its coefficients a_i, $i = 0, 1, 2, \ldots, n$ must have the same sign.

For convenience, we shall assume, without loss of generality, that all coefficients a_i, $i = 0, 1, 2, \cdots, n$ satisfy $a_i \geq 0$. (Actually, if a polynomial $\varphi(s) = a_0 + a_1 s + \cdots + a_{n-1}s^{n-1} + a_n s^n$ has all its roots in the open left half of the complex plane, then all its coefficients must be nonzero: $a_i \neq 0$.)

Let us now consider $\varphi(j\omega)$. When ω changes from $-\infty$ to $+\infty$ along the imaginary axis, the phase of $\varphi(j\omega)$, denoted by $\angle\varphi(j\omega)$, also changes. The amount of change can be determined using the following lemma.

Lemma 7.1

For a polynomial $\varphi(s) = a_0 + a_1 s + \cdots + a_{n-1}s^{n-1} + a_n s^n$ having all its roots in the open left half of the complex plane, the phase of $\varphi(j\omega)$ is a continuous and strictly increasing function of ω on $(-\infty, +\infty)$. Furthermore, the amount of phase change in $\angle\varphi(j\omega)$ from $-\infty$ to $+\infty$ is given by

$$\angle\varphi(+j\omega) - \angle\varphi(-j\omega) = n\pi$$

or the amount of phase change in $\angle\varphi(j\omega)$ from 0 to $+\infty$ is given by

$$\angle\varphi(+j\omega) - \angle\varphi(j0) = \frac{n\pi}{2}$$

Proof

Denote the roots of $\varphi(s)$ by $r_1, r_2, \ldots, r_n$, where

$$r_i = c_i + jd_i \qquad \text{with } c_i < 0 \text{ for } i = 1, 2, \ldots, n$$

Write the polynomial $\varphi(s)$ as

$$\varphi(s) = a_n(s - r_1)(s - r_2)\ldots(s - r_n)$$

Here $a_n > 0$. The phase of $\varphi(s)$ is then given by

$$\angle\varphi(j\omega) = \angle(j\omega - r_1) + \angle(j\omega - r_2) + \cdots + \angle(j\omega - r_n)$$

Since $c_i < 0$, none of the roots are in the imaginary axis and all the roots are in the open left half of the complex plane. Therefore, for all $i = 1, 2, \ldots, n$, $\angle(j\omega - r_i) = \angle(j\omega - c_i - jd_i)$ is a continuous and strictly increasing function of ω on $(-\infty, +\infty)$.

For $\omega < d_i$, $\angle(j\omega - c_i - jd_i)$ is negative. In particular, when $\omega = -\infty$, $\angle(-j\infty - c_i - jd_i) = -\pi/2$. For $\omega > d_i$, $\angle(j\omega - c_i - jd_i)$ is positive. In particular, when $\omega = +\infty$, $\angle(+j\infty - c_i - jd_i) = \pi/2$. Hence, the amount of phase change in $\angle(j\omega - c_i - jd_i)$ from $-\infty$ to $+\infty$ is given by

$$\angle(+j\infty - c_i - jd_i) - \angle(-j\infty - c_i - jd_i) = \frac{\pi}{2} - (-\frac{\pi}{2}) = \pi$$

The total amount of phase change in $\angle\varphi(j\omega)$ from $-\infty$ to $+\infty$ is

$$\angle\varphi(+j\omega) - \angle\varphi(-j\omega) = \sum_{i=1}^{n}(\angle(+j\infty - c_i - jd_i) - \angle(-j\infty - c_i - jd_i)) = n\pi$$

Since the roots of $\varphi(s)$ are symmetric with respect to the real axis, it is not difficult to see that the amount of phase change in $\angle\varphi(j\omega)$ from $-\infty$ to 0 is the same as that from 0 to $+\infty$. In other words

$$\angle\varphi(+j\omega) - \angle\varphi(j0) = \frac{n\pi}{2}$$

Q.E.D.

Let us now study the trajectory of $\varphi(j\omega)$ when ω varies from $-\infty$ to $+\infty$. The trajectory is symmetric with respect to the real axis because $\varphi(-j\omega)$ and $\varphi(j\omega)$ are complex conjugates. Hence, we only need to consider the trajectory of $\varphi(j\omega)$ when ω varies from 0 to $+\infty$. Since $\varphi(j\omega) = a_0 > 0$, the trajectory starts at the positive real axis. Since $\angle\varphi(j\omega)$ is a strictly increasing function of ω, the trajectory of $\varphi(j\omega)$ circles around the origin counter-clockwise. It will cross the imaginary axis and the real axis alternately. Let

$\omega_1, \omega_2, \omega_3, \ldots$, with $\omega_1 < \omega_2 < \omega_3 < \ldots$, be the frequencies at which the trajectory of $\varphi(j\omega)$ crosses the imaginary axis and the real axis. Clearly, $\omega_1, \omega_3, \ldots$ are the frequencies at which the trajectory crosses the imaginary axis. In other words

$$Re(\varphi(j\omega_i)) = \varphi^e(j\omega_i) = 0, i = 1, 3, \ldots$$

Similarly, $\omega_2, \omega_4, \ldots$ are the frequencies at which the trajectory crosses the real axis:

$$\mathrm{Im}(\varphi(j\omega_j)) = \omega_j \varphi^o(j\omega_j) = 0, j = 2, 4, \ldots$$

If $n = 2m$ is even, then the amount of phase change in $\angle\varphi(j\omega)$ from 0 to $+\infty$ is $\angle\varphi(+j\omega) - \angle\varphi(j0) = n\pi/2 = m\pi$. Therefore, the trajectory of $\varphi(j\omega)$ will cross the imaginary axis m times corresponding to the m positive root of $\varphi^e(j\omega)$: $\omega_1^e < \omega_2^e < \cdots < \omega_m^e$. By the definition of $\omega_1, \omega_2, \omega_3, \ldots$, it is clear that $\omega_1 = \omega_1^e, \omega_3 = \omega_2^e, \ldots \omega_{2m-1} = \omega_m^e$. Similarly, the trajectory of $\varphi(j\omega)$ will cross the real axis $m-1$ times corresponding to the $m-1$ positive root of $\varphi^o(j\omega)$: $\omega_1^o < \omega_2^o < \cdots < \omega_{m-1}^o$. In other words, $\omega_2 = \omega_1^o, \omega_4 = \omega_2^o, \ldots \omega_{2m-2} = \omega_{m-1}^o$.

If $n = 2m+1$ is odd, then the amount of phase change in $\angle\varphi(j\omega)$ from 0 to $+\infty$ is $\angle\varphi(+j\omega) - \angle\varphi(j0) = n\pi/2 = m\pi + \pi/2$. Therefore, the trajectory of $\varphi(j\omega)$ will cross the imaginary axis m times, corresponding to the m positive roots of $\varphi^e(j\omega)$: $\omega_1^e < \omega_2^e < \cdots < \omega_m^e$. We have $\omega_1 = \omega_1^e, \omega_3 = \omega_2^e, \ldots \omega_{2m-1} = \omega_m^e$. Also, the trajectory of $\varphi(j\omega)$ will cross the real axis m times, corresponding to the m positive root of $\varphi^o(j\omega)$: $\omega_1^o < \omega_2^o < \cdots < \omega_m^o$ and $\omega_2 = \omega_1^o, \omega_4 = \omega_2^o, \ldots \omega_{2m} = \omega_m^o$.

Lemma 7.2

For a polynomial $\varphi(s) = a_0 + a_1 s + \cdots + a_{n-1}s^{n-1} + a_n s^n$ having all its roots in the open left half of the complex plane, roots of $\varphi^e(j\omega)$ and $\varphi^o(j\omega)$ satisfy the following properties. If $n = 2m$ is even, then $0 < \omega_1^e < \omega_1^o < \omega_2^e < \cdots < \omega_{m-1}^o < \omega_m^e$. If $n = 2m+1$ is odd, then $0 < \omega_1^e < \omega_1^o < \omega_2^e < \cdots < \omega_{m-1}^o < \omega_m^e < \omega_m^o$

Proof

Based on the above discussion, if $n = 2m$ is even, then

$$0 < \omega_1 < \omega_2 < \omega_3 < \cdots < \omega_{2m-1} \Rightarrow 0 < \omega_1^e < \omega_1^o < \omega_2^e < \cdots < \omega_{m-1}^o < \omega_m^e$$

If $n = 2m+1$ is odd, then

$$0 < \omega_1 < \omega_2 < \omega_3 < \cdots < \omega_{2m} \quad \Rightarrow \quad 0 < \omega_1^{\mathrm{e}} < \omega_1^{\mathrm{o}} < \omega_2^{\mathrm{e}} < \cdots < \omega_{m-1}^{\mathrm{o}} < \omega_m^{\mathrm{e}} < \omega_m^{\mathrm{o}}$$

Q.E.D.

With the above preparation, we can now prove Theorem 7.2.

Proof of Theorem 7.2

(ONLY IF) We first prove the only if part of Theorem 7.2. We assume that $\varphi(s) = a_0 + a_1 s + \cdots + a_{n-1}s^{n-1} + a_n s^n$ has all its roots in the open left half of the complex plane and we show that $\varphi(s)$ satisfies the interlacing property as follows.

1. a_n and a_{n-1} have the same sign; this is by Lemma 3.1
2. all roots of $\varphi^{\mathrm{e}}(j\omega)$ and $\varphi^{\mathrm{o}}(j\omega)$ are real and distinct; this is because the roots correspond to the frequencies at which the trajectory of $\varphi(j\omega)$ crosses the imaginary axis and the real axis
3. if $n = 2m$ is even, then $\varphi^{\mathrm{e}}(j\omega)$ has m positive roots $\omega_1^{\mathrm{e}} < \omega_2^{\mathrm{e}} < \cdots < \omega_m^{\mathrm{e}}$, $\varphi^{\mathrm{o}}(j\omega)$ has $m-1$ positive roots $\omega_1^{\mathrm{o}} < \omega_2^{\mathrm{o}} < \cdots < \omega_{m-1}^{\mathrm{o}}$, and they must interlace in the following manner: $0 < \omega_1^{\mathrm{e}} < \omega_1^{\mathrm{o}} < \omega_2^{\mathrm{e}} < \cdots < \omega_{m-1}^{\mathrm{o}} < \omega_m^{\mathrm{e}}$; this is by Lemma 7.2
4. if $n = 2m+1$ is odd, then $\varphi^{e}(j\omega)$ has m positive roots $\omega_1^{\mathrm{e}} < \omega_2^{\mathrm{e}} < \cdots < \omega_m^{\mathrm{e}}$, $\varphi^{\mathrm{o}}(j\omega)$ has m positive roots $\omega_1^{\mathrm{o}} < \omega_2^{\mathrm{o}} < \cdots < \omega_m^{\mathrm{o}}$, and they must interlace in the following manner: $0 < \omega_1^{\mathrm{e}} < \omega_1^{\mathrm{o}} < \omega_2^{e} < \cdots < \omega_{m-1}^{o} < \omega_m^{e} < \omega_m^{\mathrm{o}}$; this is by Lemma 7.2

(IF) We now prove the 'if' part of Theorem 7.2. We assume that $\varphi(s)$ satisfies the interlacing property, we want to show that $\varphi(s)$ has all its roots in the open left half of the complex plane.

Let us consider the case when $n = 2m$ is even. The other case of $n = 2m+1$ is similar.

Without loss of generality, let us assume that both a_n and a_{n-1} are positive. Since $\varphi^{\mathrm{e}}(j\omega)$ has m positive roots $\omega_1^{\mathrm{e}} < \omega_2^{\mathrm{e}} < \cdots < \omega_m^{\mathrm{e}}$, we can write

$$\varphi^{\mathrm{e}}(j\omega) = a_{2m}(\omega^2 - \omega_1^{\mathrm{e}^2})(\omega^2 - \omega_2^{\mathrm{e}^2}) \ldots (\omega^2 - \omega_m^{\mathrm{e}^2})$$

Similarly

$$\varphi^{\mathrm{o}}(j\omega) = a_{2m-1}(\omega^2 - \omega_1^{\mathrm{o}^2})(\omega^2 - \omega_2^{\mathrm{o}^2}) \ldots (\omega^2 - \omega_{m-1}^{\mathrm{o}\,2})$$

Now, let

$$\psi(s) = b_0 + b_1 s + \cdots + b_{n-1} s^{n-1} + b_n s^n$$

be a polynomial with all its roots in the open left half of the complex plane. By the proof of the only if part, $\psi(s)$ satisfies the interlacing property. Let $\omega_1'^{e} < \omega_2'^{e} < \cdots < \omega_m'^{e}$ be the positive roots of $\psi^e(j\omega)$ and $\omega_1'^{o} < \omega_2'^{o} < \cdots < \omega_{m-1}'^{o}$ be the positive roots of $\psi^o(j\omega)$. Then we can write

$$\psi^e(j\omega) = b_{2m}(\omega^2 - \omega_1'^{e2})(\omega^2 - \omega_2'^{e2}) \ldots (\omega^2 - \omega_m'^{e2})$$
$$\psi^o(j\omega) = b_{2m-1}(\omega^2 - \omega_1'^{o2})(\omega^2 - \omega_2'^{o2}) \ldots (\omega^2 - \omega_{m-1}'^{o2})$$

By the definitions of $\varphi^e(j\omega)$, $\varphi^o(j\omega)$, $\psi^e(j\omega)$, and $\psi^o(j\omega)$, we have

$$\varphi(j\omega) = \varphi^e(j\omega) + j\omega\varphi^o(j\omega)$$
$$\psi(j\omega) = \psi^e(j\omega) + j\omega\psi^o(j\omega)$$

Let us define the following polynomial with a parameter $\lambda \in [0, 1]$:

$$\begin{aligned}\phi(j\omega, \lambda) =& (\lambda a_{2m} + (1-\lambda) b_{2m})(\omega^2 - (\lambda\omega_1^{e2} + (1-\lambda)\omega_1'^{e2}))\\ &\times \cdots \times (\omega^2 - (\lambda\omega_m^{e2} + (1-\lambda)\omega_m'^{e2}))\\ &+ j\omega(\lambda a_{2m-1} + (1-\lambda) b_{2m-1})(\omega^2 - (\lambda\omega_1^{o2} + (1-\lambda)\omega_1'^{o2}))\\ &\times \cdots \times (\omega^2 - (\lambda\omega_{m-1}^{o}{}^2 + (1-\lambda)\omega_{m-1}'^{o}{}^2))\end{aligned}$$

Clearly, if $\lambda = 0$, then $\phi(s, 0) = \psi(s)$, and if $\lambda = 1$, then $\phi(s, 1) = \varphi(s)$.

Polynomial $\psi(s)$ has all its roots in the open left half of the complex plane. We prove that $\varphi(s)$ also has all its roots in the open left half of the complex plane by contradiction. Suppose $\varphi(s)$ also has at least one root outside the open left half of the complex plane, then by Theorem 7.1, there exists at least one $\lambda \in [0, 1]$ such that $\phi(s, \lambda)$ has all its roots in the closed left half of the complex plane and at least one root on the imaginary axis. Denote the root on the imaginary axis by $s = j\omega_o$, that is, $\phi(j\omega_o, \lambda) = 0$. Since both the real part and the imaginary part of $\phi(j\omega_o, \lambda)$ must be zero, there must exist $i \in \{1, 2, \ldots, m\}$ and $j \in \{1, 2, \ldots, m-1\}$ such that the following two conditions are both satisfied.

$$\omega_o^2 = \lambda\omega_i^{e2} + (1-\lambda)\omega_i'^{e2} \qquad \text{and} \qquad \omega_o^2 = \lambda\omega_j^{o2} + (1-\lambda)\omega_j'^{o2}$$

But this is impossible because by the interlacing property, we have either: (1) $\omega_i^e < \omega_j^o$ and $\omega_i'^e < \omega_j'^o$, or (2) $\omega_i^e > \omega_j^o$ and $\omega_i'^e > \omega_j'^o$. In the first case, we conclude

$$\lambda\omega_i^{e2} + (1-\lambda)\omega_i'^{e2} < \lambda\omega_j^{o2} + (1-\lambda)\omega_j'^{o2}$$

In the second case, we have

$$\lambda\omega_i^{e^2} + (1-\lambda)\omega_i'^{e^2} > \lambda\omega_j^{o^2} + (1-\lambda)\omega_j'^{o^2}$$

Therefore, in either case, we cannot have both $\omega_o^2 = \lambda\omega_i^{e^2} + (1-\lambda)\omega_i'^{e^2}$ and $\omega_o^2 = \lambda\omega_j^{o^2} + (1-\lambda)\omega_j'^{o^2}$ satisfied.

Q.E.D.

Example 7.2

Let us consider the following polynomial

$$\varphi(s) = 1 + 2s + 4s^2 + 2s^3 + s^4$$

Its even and odd parts can be obtained respectively as

$$\varphi^e(s) = 1 + 4s^2 + s^4$$
$$\varphi^o(s) = 2 + 2s^2$$

The roots of $\varphi^e(j\omega) = 1 - 4\omega^2 + \omega^4$ are -1.9319, -0.5176, 0.5176, and 1.9319. Therefore, $\omega_1^e = 0.5176$ and $\omega_2^e = 1.9319$.

The roots of $\varphi^o(j\omega) = 2 - 2\omega^2$ are -1 and 1. Therefore, $\omega_1^o = 1$. Clearly, the interlacing condition is satisfied. In particular, $0 < \omega_1^e < \omega_1^o < \omega_2^e$.

Let us now check the roots of $\varphi(s) = 1 + 2s + 4s^2 + 2s^3 + s^4$, which are $-0.7429 + j1.5291$, $-0.7429 - j1.5291$, $-0.2571 + j0.5291$, $-0.2571 - j0.5291$ They are all in the open left half of the complex plane.

Example 7.3

Let us consider the polynomial

$$\varphi(s) = 11.1216 + 19.1866s + 95.7330s^2 + 67.3653s^3 + 66.0427s^4 + 42.2195s^5$$

Its even and odd parts are as follows.

$$\varphi^e(s) = 11.1216 + 95.7330s^2 + 66.0427s^4$$
$$\varphi^o(s) = 19.1866 + 67.3653s^2 + 42.2195s^4$$

The roots of $\varphi^e(j\omega) = 11.1216 - 95.7330\omega^2 + 66.0427\omega^4$ are

$$-1.1499, -0.3569, 0.3569, 1.1499$$

that is, $\omega_1^e = 0.3569$ and $\omega_2^e = 1.1499$.

The roots of $\varphi^o(s) = 19.1866 + 67.3653s^2 + 42.2195s^4$ are

$$-1.1066, -0.6092, 0.6092, 1.1066$$

that is, $\omega_1^o = 0.6092$ and $\omega_2^o = 1.1066$. Because $0 < \omega_1^e < \omega_1^o < \omega_2^e < \omega_2^o$ is not true, the interlacing condition is not satisfied. The roots of $\varphi(s) = 11.1216 + 19.1866s + 95.7330s^2 + 67.3653s^3 + 66.0427s^4 + 42.2195s^5$ can be calculated as follows.

$$-1.4518, 0.0182 + j1.1391, 0.0182 - j1.1391,$$
$$-0.0744 + j0.3664, -0.0744 - j0.3664$$

Clearly, two of the roots are not in the open left half of the complex plane.

In the rest of the chapter, we say that a polynomial is stable if all its roots are in the open left half of the complex plane. We will consider stability of a set of polynomials. Using the previous notation, we can write a polynomial as

$$\varphi(s) = \varphi^e(s) + s\varphi^o(s)$$

We present the following two theorems from the Interlacing theorem, which will be used to prove the Kharitonov theorem.

Theorem 7.3

Consider two stable polynomials of the same degree. They have the same even part, but different odd parts.

$$\varphi_1(s) = \varphi^e(s) + s\varphi_1^o(s)$$
$$\varphi_2(s) = \varphi^e(s) + s\varphi_2^o(s)$$

Assume the odd parts satisfy

$$\varphi_1^o(j\omega) \le \varphi_2^o(j\omega) \qquad \text{for all} \qquad \omega \in [0, \infty)$$

Then for any polynomial $\varphi(s) = \varphi^e(s) + s\varphi^o(s)$ satisfying

$$\varphi_1^o(j\omega) \le \varphi^o(j\omega) \le \varphi_2^o(j\omega) \qquad \text{for all} \qquad \omega \in [0, \infty)$$

$\varphi(s)$ is stable.

Proof

Since $\varphi_1(s)$ and $\varphi_2(s)$ are stable, they satisfy the interlacing property. To prove $\varphi(s)$ is stable, we show that $\varphi(s)$ satisfies the interlacing property as follows. Denote

$$\varphi^{\rm o}(s) = a_1 + a_3 s^2 + \cdots + a_{n-2} s^{n-3} + a_n s^{n-1}$$
$$\varphi_1^{\rm o}(s) = a_1^- + a_3^- s^2 + \cdots + a_{n-1}^- s^{n-3} + a_n^- s^{n-1}$$
$$\varphi_2^{\rm o}(s) = a_1^+ + a_3^+ s^2 + \cdots + a_{n-1}^+ s^{n-3} + a_n^+ s^{n-1}$$

The condition $\varphi_1^{\rm o}(j\omega) \le \varphi^{\rm o}(j\omega) \le \varphi_2^{\rm o}(j\omega)$ for all $\omega \in [0, \infty)$ implies the following.

A. The coefficients a_1^-, a_1^+, and a_1 have the same sign, because otherwise $\varphi_1^{\rm o}(j\omega) \le \varphi^{\rm o}(j\omega) \le \varphi_2^{\rm o}(j\omega)$ will be violated for $\omega = 0$. Let us assume they are all positive.
B. The polynomials $\varphi_1(s)$, $\varphi_2(s)$, and $\varphi(s)$ have the same degree, because otherwise $\varphi_1^{\rm o}(j\omega) \le \varphi^{\rm o}(j\omega) \le \varphi_2^{\rm o}(j\omega)$ will be violated for $\omega = \infty$.
C. Denote the positive real roots of $\varphi^{\rm o}(j\omega)$, $\varphi_1^{\rm o}(j\omega)$, and $\varphi_2^{\rm o}(j\omega)$ as $\omega_1, \omega_2, \ldots \omega_k$, $\omega_1^-, \omega_2^-, \ldots \omega_k^-$, and $\omega_1^+, \omega_2^+, \ldots \omega_k^+$, respectively. Then for all $i = 1, 2, \ldots, k$, the roots satisfy $\omega_i^- \le \omega_i \le \omega_i^+$. This is because the curve of $\varphi^{\rm o}(j\omega)$ vs $\omega \in [0, \infty)$ is bounded by the curves of $\varphi_1^{\rm o}(j\omega)$ and $\varphi_2^{\rm o}(j\omega)$.

From the above results, we can check the conditions for the interlacing property.

1. a_n and a_{n-1} have the same sign. This is because of condition A and the assumption that $\varphi_1(s)$ and $\varphi_2(s)$ satisfy the interlacing property.
2. All roots of $\varphi^{\rm e}(j\omega)$ and $\varphi^{\rm o}(j\omega)$ are real and distinct. This is because of conditions B and C and the assumption that $\varphi_1(s)$ and $\varphi_2(s)$ satisfy the interlacing property.
3. If $n = 2m$ is even, then $\varphi^{\rm e}(j\omega)$ has m positive roots $\omega_1^{\rm e} < \omega_2^{\rm e} < \cdots < \omega_m^{\rm e}$, $\varphi^{\rm o}(j\omega)$ has $m-1$ positive roots $\omega_1^{\rm o} < \omega_2^{\rm o} < \cdots < \omega_{m-1}^{\rm o}$, and they must interlace in the following manner: $0 < \omega_1^{\rm e} < \omega_1^{\rm o} < \omega_2^{\rm e} < \cdots < \omega_{m-1}^{\rm o} < \omega_m^{\rm e}$. This is because of condition C and the assumption that $\varphi_1(s)$ and $\varphi_2(s)$ satisfy the interlacing property.
4. If $n = 2m+1$ is odd, then $\varphi^{\rm e}(j\omega)$ has m positive roots $\omega_1^{\rm e} < \omega_2^{\rm e} < \cdots < \omega_m^{\rm e}$, $\varphi^{\rm o}(j\omega)$ has m positive roots $\omega_1^{\rm o} < \omega_2^{\rm o} < \cdots < \omega_m^{\rm o}$, and they must interlace in the following manner: $0 < \omega_1^{\rm e} < \omega_1^{\rm o} < \omega_2^{\rm e} < \cdots < \omega_{m-1}^{\rm o} < \omega_m^{\rm e} < \omega_m^{\rm o}$. This is because of condition C and the assumption that $\varphi_1(s)$ and $\varphi_2(s)$ satisfy the interlacing property.

Finally, by the Interlacing theorem, polynomial $\varphi(s)$ is stable.

Q.E.D.

Theorem 7.4

Consider two stable polynomials of the same degree. They have the same odd parts, but different even parts.

$$\varphi_1(s) = \varphi_1^e(s) + s\varphi^o(s)$$
$$\varphi_2(s) = \varphi_2^e(s) + s\varphi^o(s)$$

Assume the even parts satisfy

$$\varphi_1^e(j\omega) \leq \varphi_2^e(j\omega) \qquad \text{for all} \qquad \omega \in [0, \infty)$$

Then for any polynomial $\varphi(s) = \varphi^e(s) + s\varphi^o(s)$ satisfying

$$\varphi_1^e(j\omega) \leq \varphi^e(j\omega) \leq \varphi_2^e(j\omega) \text{ for all} \omega \in [0, \infty)$$

$\varphi(s)$ is stable.

Proof

The proof is similar to that of Theorem 7.3.

Q.E.D.

The Theorems proven in this section will be used in the next section to prove our main results of this chapter.

7.3 KHARITONOV THEOREM

In this section, we consider the key question of this chapter. Given a set of polynomials, how to determine if all these polynomials are stable? We call such a problem a robust stability problem. Clearly, we cannot check all polynomials because the set is often infinite. So we need to find a smart way to check only a finite number of polynomials to determine the stability of an infinite set. This goal is obviously not always achievable. The Kharitonov result shows that for a set of 'interval' polynomials, the robust stability problem can be solved by checking only four polynomials. Let us now see how this is done. Given a set of polynomials

$$\varphi(s, p) = p_0 + p_1 s + \cdots + p_{n-1}s^{n-1} + p_n s^n$$

where $p_i \in [p_i^-, p_i^+]$, $i = 0, 1, \ldots, n-1, n$ are coefficients whose values are uncertain. We would like to know if all the polynomials in the set are stable, that is, if the set is robustly stable. In other words, let

$$p = [p_0 \cdots p_n]$$

be the vector of uncertain coefficients, and

$$P = [p_0^-, p_0^+] \times \cdots \times [p_n^-, p_n^+]$$

be the set of possible values of p.

Define the set of admissible polynomials as

$$\Psi(s, p) = \{\varphi(s, p) : p \in P\}$$

Obviously, this set is infinite. To check if, for all $\varphi(s, p) \in \Psi(s, p)$, $\varphi(s, p)$ is stable, we will check the following four Kharitonov polynomials.

$$K_1(s) = p_0^- + p_1^- s + p_2^+ s^2 + p_3^+ s^3 + p_4^- s^4 + p_5^- s^5 + \cdots$$
$$K_2(s) = p_0^- + p_1^+ s + p_2^+ s^2 + p_3^- s^3 + p_4^- s^4 + p_5^+ s^5 + \cdots$$
$$K_3(s) = p_0^+ + p_1^- s + p_2^- s^2 + p_3^+ s^3 + p_4^+ s^4 + p_5^- s^5 + \cdots$$
$$K_4(s) = p_0^+ + p_1^+ s + p_2^- s^2 + p_3^- s^3 + p_4^+ s^4 + p_5^+ s^5 + \cdots$$

The Kharitonov theorem states that the stability of the above four polynomials is necessary and sufficient for the stability of all polynomials in the infinite set of $\Psi(s, p)$. This result is rather surprising because, intuitively, we expect that we need to check at least the set of all the extreme polynomials, which consists of 2^{n+1} polynomials. But in fact, four are enough, as shown in Figure 7.4. Let us state this result formally.

Theorem 7.5

The set of polynomials $\Psi(s, p)$ has the property that every polynomial in the set is stable if and only if the four Kharitonov polynomials are stable.

Proof

For the set of polynomials

$$\varphi(s, p) = p_0 + p_1 s + \cdots + p_{n-1} s^{n-1} + p_n s^n$$

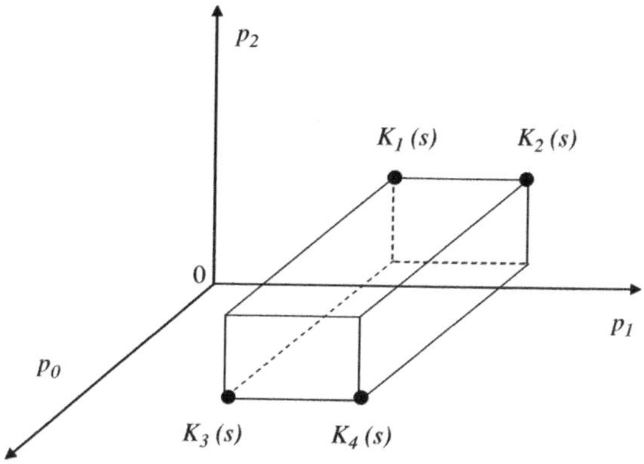

Figure 7.4 Kharitonov polynomials in parameter space.

where $p_i \in [p_i^-, p_i^+]$, $i = 0, 1, \ldots, n-1, n$, define the minimum and maximum even parts as

$$\varphi_{\min}^{e}(s) = p_0^- + p_2^+ s^2 + p_4^- s^4 + p_6^+ s^6 + \cdots$$
$$\varphi_{\max}^{e}(s) = p_0^+ + p_2^- s^2 + p_4^+ s^4 + p_6^- s^6 + \cdots$$

and the minimum and maximum odd parts without s as

$$\varphi_{\min}^{o}(s) = p_1^- + p_3^+ s^2 + p_5^- s^4 + p_7^+ s^6 + \cdots$$
$$\varphi_{\max}^{o}(s) = p_1^+ + p_3^- s^2 + p_5^+ s^4 + p_7^- s^6 + \cdots$$

Any polynomial $\varphi^e(s) + s\varphi^o(s) \in \Psi(s, p)$ is bounded by the minimum and maximum parts as follows. For all $\omega \in [0, \infty)$

$$\varphi_{\min}^{e}(j\omega) \le \varphi^{e}(j\omega) \le \varphi_{\max}^{e}(j\omega)$$
$$\varphi_{\min}^{o}(j\omega) \le \varphi^{o}(j\omega) \le \varphi_{\max}^{o}(j\omega)$$

On the other hand, by the definitions of the Kharitonov polynomials, it is clear that

$$K_1(s) = \varphi_{\min}^{e}(s) + s\varphi_{\min}^{o}(s)$$
$$K_2(s) = \varphi_{\min}^{e}(s) + s\varphi_{\max}^{o}(s)$$
$$K_3(s) = \varphi_{\max}^{e}(s) + s\varphi_{\min}^{o}(s)$$
$$K_4(s) = \varphi_{\max}^{e}(s) + s\varphi_{\max}^{o}(s).$$

With these notations, we can now prove Theorem 7.5.

(ONLY IF) The proof is obvious because all Kharitonov polynomials belong to $\Psi(s, p)$.

(IF) Let us assume that four Kharitonov polynomials are stable. We want to show that any $\varphi(s) = \varphi^e(s) + s\varphi^o(s) \in \Psi(s, p)$ is stable.

Since $\varphi^o_{min}(j\omega) \le \varphi^o(j\omega) \le \varphi^o_{max}(j\omega)$, for all $\omega \in [0, \infty)$, by Theorem 7.3, the condition that $K_1(s) = \varphi^e_{min}(s) + s\varphi^o_{min}(s)$ and $K_2(s) = \varphi^e_{min}(s) + s\varphi^o_{max}(s)$ are stable implies that polynomial $\varphi^e_{min}(s) + s\varphi^o(s)$ is stable. Similarly, the condition that $K_3(s) = \varphi^e_{max}(s) + s\varphi^o_{min}(s)$ and $K_4(s) = \varphi^e_{max}(s) + s\varphi^o_{max}(s)$ are stable implies that $\varphi^e_{max}(s) + s\varphi^o(s)$ is stable.

We also have $\varphi^e_{min}(j\omega) \le \varphi^e(j\omega) \le \varphi^e_{max}(j\omega)$, for all $\omega \in [0, \infty)$. Therefore, by Theorem 7.4, the condition that $\varphi^e_{min}(s) + s\varphi^o(s)$ and $\varphi^e_{max}(s) + s\varphi^o(s)$ are stable implies that $\varphi(s) = \varphi^e(s) + s\varphi^o(s)$ is stable.

Q.E.D.

Let us now consider two examples of using the Kharitonov Theorem to determine the robust stability of a set of polynomials.

Example 7.4

Let us consider the set of polynomials $\Psi(s, p) = \{\varphi(s, p) : p \in P\}$

$$\varphi(s, p) = p_0 + p_1 s + p_2 s^2 + p_3 s^3 + p_4 s^4 + p_5 s^5 + p_6 s^6$$

where $p_0 \in [1, 3], p_1 \in [9, 13], p_2 \in [2, 4], p_3 \in [11, 14], p_4 \in [10, 12], p_5 \in [7, 10]$, and $p_6 \in [1, 1]$. To determine the robust stability of the set, let us construct the four Kharitonov polynomials

$$K_1(s) = 1 + 9s + 4s^2 + 14s^3 + 10s^4 + 7s^5 + s^6$$
$$K_2(s) = 1 + 13s + 4s^2 + 11s^3 + 10s^4 + 10s^5 + s^6$$
$$K_3(s) = 3 + 9s + 2s^2 + 14s^3 + 12s^4 + 7s^5 + s^6$$
$$K_4(s) = 3 + 13s + 2s^2 + 11s^3 + 12s^4 + 10s^5 + s^6$$

We can check that these four polynomials are not stable. Therefore, the set of polynomials is not robustly stable.

Example 7.5

Let us consider the set of fourth-degree polynomials

$$\varphi(s, p) = p_0 + p_1 s + p_2 s^2 + p_3 s^3 + p_4 s^4$$

where $p_0 \in [93, 98], p_1 \in [760, 761], p_2 \in [727, 733], p_3 \in [975, 980]$, and $p_4 \in [501, 508]$. To determine the stability of the set, let us construct the four Kharitonov polynomials

$$K_1(s) = 93 + 760s + 733s^2 + 980s^3 + 501s^4$$

$$K_2(s) = 93 + 761s + 733s^2 + 975s^3 + 501s^4$$

$$K_3(s) = 98 + 760s + 727s^2 + 980s^3 + 508s^4$$

$$K_4(s) = 98 + 761s + 727s^2 + 975s^3 + 508s^4$$

We can check that all these four polynomials are stable. Therefore, the set of polynomials is robustly stable.

7.4 CONTROL DESIGN USING KHARITONOV THEOREM

In this section, we discuss robust control design using the Kharitonov theorem. As we have shown in the previous section, the Kharitonov theorem is a very nice tool for robust stability analysis. However, it is not a convenient tool for robust control design. The optimal control approach to the robust control problem, as discussed in the previous two chapters, is inherently a design tool in the sense that it will design a controller that can robustly stabilize the system. The Kharitonov theorem is inherently an analysis tool in the sense that given a (closed-loop) system; it will analyse and verify if the system is robustly stable. With this difference in mind, let us discuss the best way to design a robust controller using the Kharitonov theorem.

Suppose the system to be controlled is given by the transfer function

$$G(s, p) = \frac{b_0 + b_1 s + \cdots + b_{n-1} s^{n-1}}{p_0 + p_1 s + \cdots + p_{n-1} s^{n-1} + s^n}$$

We can realize this system in state space representation in the following controllable canonical form.

$$\dot{x} = \begin{bmatrix} 0 & 1 & \ldots & 0 \\ \ldots & & & \\ 0 & 0 & \ldots & 1 \\ -p_0 & -p_1 & \ldots & -p_{n-1} \end{bmatrix} x + \begin{bmatrix} 0 \\ \ldots \\ 0 \\ 1 \end{bmatrix} u$$

$$y = \begin{bmatrix} b_0 & b_1 & \cdots & b_{n-1} \end{bmatrix} x$$

The characteristic equation of the open-loop system is given by

$$\varphi(s, p) = p_0 + p_1 s + \cdots + p_{n-1} s^{n-1} + s^n$$

with $p_i \in [p_i^-, p_i^+]$, $i = 0, 1, \ldots, n-1$. If the open-loop system is not robustly stable, then we want use state feedback

$$u = -Lx = -\begin{bmatrix} l_0 & l_1 & \ldots & l_{n-1} \end{bmatrix} x$$

so that the closed-loop system is robustly stable for all $p \in P$. Clearly the characteristic equation of the closed-loop system is given by

$$\varphi(s, p+l) = (p_0 + l_0) + (p_1 + l_1)s + \cdots + (p_{n-1} + l_{n-1})s^{n-1} + s^n$$

The question is then whether $\varphi(s, p+l)$ is stable for all $p \in P$. To check this, we can calculate the four Kharitonov polynomials as follows

$$K_1(s) = (l_0 + p_0^-) + (l_1 + p_1^-)s + (l_2 + p_2^+)s^2 + (l_3 + p_3^+)s^3 + (l_4 + p_4^-)s^4 + \cdots$$
$$K_1(s) = (l_0 + p_0^-) + (l_1 + p_1^+)s + (l_2 + p_2^+)s^2 + (l_3 + p_3^-)s^3 + (l_4 + p_4^-)s^4 + \cdots$$
$$K_1(s) = (l_0 + p_0^+) + (l_1 + p_1^-)s + (l_2 + p_2^-)s^2 + (l_3 + p_3^+)s^3 + (l_4 + p_4^+)s^4 + \cdots$$
$$K_1(s) = (l_0 + p_0^+) + (l_1 + p_1^+)s + (l_2 + p_2^-)s^2 + (l_3 + p_3^-)s^3 + (l_4 + p_4^+)s^4 + \cdots$$

If we can find $L = \begin{bmatrix} l_0 & l_1 & \ldots & l_{n-1} \end{bmatrix}$ such that the above Kharitonov polynomials are all stable, then by the Kharitonov Theorem, the closed-loop system is robustly stable.

Such a feedback map L always exists. Essentially, we need to take a very large L so that parameters p are relatively small and the uncertainty in p does not matter as far as stability is concerned. A large L means using high gain control which is undesirable in some applications. Therefore, ideally we would like to have an L large enough to ensure robust stability, but not too large. However, there is no simple and systematic way to design L just large enough for robust stability. Some *ad hoc* method must be used and the following example illustrates this.

Example 7.6

Let us consider an open-loop system with the following transfer function.

$$G(s, p) = \frac{5}{p_0 + p_1 s + p_2 s^2 + p_3 s^3 + p_4 s^4 + p_5 s^5 + s^6}$$

where $p_0 \in [1, 3]$, $p_1 \in [9, 13]$, $p_2 \in [2, 4]$, $p_3 \in [11, 14]$, $p_4 \in [10, 12]$, and $p_5 \in [7, 10]$. We can realize this system in the controllable canonical form:

$$\dot{x} = \begin{bmatrix} 0 & 1 & 0 & 0 & 0 & 0 \\ 0 & 0 & 1 & 0 & 0 & 0 \\ 0 & 0 & 0 & 1 & 0 & 0 \\ 0 & 0 & 0 & 0 & 1 & 0 \\ 0 & 0 & 0 & 0 & 0 & 1 \\ -p_0 & -p_1 & -p_2 & -p_3 & -p_4 & -p_5 \end{bmatrix} x + \begin{bmatrix} 0 \\ 0 \\ 0 \\ 0 \\ 0 \\ 1 \end{bmatrix} u$$

$$y = \begin{bmatrix} 5 & 0 & 0 & 0 & 0 & 0 \end{bmatrix} x$$

As shown in Example 7.4, the open-loop system is not robustly stable. To find a feedback map L, let us suppose that we would like to place all the poles of the closed-loop system to be near -100; that is,

$$(s+100)^6 = 10^{12} + 6\times 10^{10}s + 15\times 10^8 s^2 + 2\times 10^7 s^3 + 15\times 10^4 s^4 + 6 \times 10^2 s^5 + s^6$$

So, let the feedback map be

$$L = \begin{bmatrix} 10^{12} & 6\times 10^{10} & 15\times 10^8 & 2\times 10^7 & 15\times 10^4 & 6\times 10^2 \end{bmatrix}$$

Then, using the feedback control $u = -Lx$, the characteristic equation of the closed-loop system is given by

$$(10^{12}+p_0) + (6\times 10^{10}+p_1)s + (15\times 10^8+p_2)s^2 + (2\times 10^7+p_3)s^3 + (15\times 10^4+p_4)s^4 + (6\times 10^2+p_5)s^5 + s^6 = 0$$

Using the Kharitonov theorem, we can check that the closed-loop system is indeed robustly stable.

The procedure in the above example is an *ad hoc* approach. A systematic approach to robust design of feedback control using the Kharitonov theorem is hard to obtain in general. However, the same problem can be solved using the optimal control approach discussed earlier. Let us illustrate this by considering the following general linear system

$$\dot{x} = \begin{bmatrix} 0 & 1 & \cdots & 0 \\ \cdots & & & \\ 0 & 0 & \cdots & 1 \\ -p_0 & -p_1 & \cdots & -p_{n-1} \end{bmatrix} x + \begin{bmatrix} 0 \\ \cdots \\ 0 \\ 1 \end{bmatrix} u$$

with parameter uncertainty described by $p_i \in [p_i^-, p_i^+]$, $i = 0, 1, \ldots, n-1$. Let the nominal value of the parameters be $[p_0^- \ \ p_1^- \ \ \cdots \ \ p_{n-1}^-]$. (We can also take the nominal value to be $[p_0^+ \; p_1^+ \; \ldots \; p_{n-1}^+]$ or any value in between.) Then the nominal system is given by

$$\dot{x} = \begin{bmatrix} 0 & 1 & \ldots & 0 \\ \ldots & & & \\ 0 & 0 & \ldots & 1 \\ -p_0^- & -p_1^- & \ldots & -p_{n-1}^- \end{bmatrix} x + \begin{bmatrix} 0 \\ \ldots \\ 0 \\ 1 \end{bmatrix} u$$
$$= A_0 x + Bu$$

The uncertainty can be written as

$$\begin{bmatrix} 0 & 1 & \ldots & 0 \\ \ldots & & & \\ 0 & 0 & \ldots & 1 \\ -p_0 & -p_1 & \ldots & -p_{n-1} \end{bmatrix} - \begin{bmatrix} 0 & 1 & \ldots & 0 \\ \ldots & & & \\ 0 & 0 & \ldots & 1 \\ -p_0^- & -p_1^- & \ldots & -p_{n-1}^- \end{bmatrix}$$
$$= \begin{bmatrix} 0 & 0 & \ldots & 0 \\ \ldots & & & \\ 0 & 0 & \ldots & 0 \\ p_0^- - p_0 & p_1^- - p_1 & \ldots & p_{n-1}^- - p_{n-1} \end{bmatrix} = \begin{bmatrix} 0 \\ \ldots \\ 0 \\ 1 \end{bmatrix}$$
$$[p_0^- - p_0 \quad p_1^- - p_1 \quad \ldots \quad p_{n-1}^- - p_{n-1}].$$

Hence, the uncertainty satisfies the matching condition. The uncertainty is given by

$$\phi = [p_0^- - p_0 \quad p_1^- - p_1 \quad \ldots \quad p_{n-1}^- \quad - p_{n-1}].$$

which is bounded by

$$\phi^{\mathrm{T}}\phi \le \begin{bmatrix} (p_0^+ - p_0^-)(p_0^+ - p_0^-) & (p_0^+ - p_0^-)(p_1^+ - p_1^-) & \ldots & (p_0^+ - p_0^-)(p_{n-1}^+ - p_{n-1}^-) \\ (p_1^+ - p_1^-)(p_0^+ - p_0^-) & (p_1^+ - p_1^-)(p_1^+ - p_1^-) & \ldots & (p_1^+ - p_1^-)(p_{n-1}^+ - p_{n-1}^-) \\ \ldots & & & \\ (p_{n-1}^+ - p_{n-1}^-)(p_0^+ - p_0^-) & (p_{n-1}^+ - p_{n-1}^-)(p_1^+ - p_1^-) & \ldots & (p_{n-1}^+ - p_{n-1}^-)(p_{n-1}^+ - p_{n-1}^-) \end{bmatrix}$$
$$= F$$

Therefore, to design a robust feedback control, all we need is to solve the following optimal control problem.

For the nominal system

$$\dot{x} = A_0 x + Bu$$

find a feedback control law $u = -Lx$ that minimizes the cost functional

$$\int_0^\infty (x^{\mathrm{T}} F x + x^{\mathrm{T}} x + u^{\mathrm{T}} u) dt$$

Let us use the same system as in Example 7.7 to illustrate the approach.

Example 7.7

Let us consider the system

$$\dot{x} = \begin{bmatrix} 0 & 1 & 0 & 0 & 0 & 0 \\ 0 & 0 & 1 & 0 & 0 & 0 \\ 0 & 0 & 0 & 1 & 0 & 0 \\ 0 & 0 & 0 & 0 & 1 & 0 \\ 0 & 0 & 0 & 0 & 0 & 1 \\ -p_0 & -p_1 & -p_2 & -p_3 & -p_4 & -p_5 \end{bmatrix} x + \begin{bmatrix} 0 \\ 0 \\ 0 \\ 0 \\ 0 \\ 1 \end{bmatrix} u$$

where $p_0 \in [1, 3], p_1 \in [9, 13], p_2 \in [2, 4], p_3 \in [11, 14], p_4 \in [10, 12]$, and $p_5 \in [7, 10]$. To design a feedback control, we solve the following optimal control problem. The nominal system is given by

$$\dot{x} = \begin{bmatrix} 0 & 1 & 0 & 0 & 0 & 0 \\ 0 & 0 & 1 & 0 & 0 & 0 \\ 0 & 0 & 0 & 1 & 0 & 0 \\ 0 & 0 & 0 & 0 & 1 & 0 \\ 0 & 0 & 0 & 0 & 0 & 1 \\ -1 & -9 & -2 & -11 & -10 & -7 \end{bmatrix} x + \begin{bmatrix} 0 \\ 0 \\ 0 \\ 0 \\ 0 \\ 1 \end{bmatrix} u$$

The bound on the uncertainty can be found as

$$F = \begin{bmatrix} 2 \\ 4 \\ 2 \\ 3 \\ 2 \\ 3 \end{bmatrix} \begin{bmatrix} 2 & 4 & 2 & 3 & 2 & 3 \end{bmatrix} = \begin{bmatrix} 4 & 8 & 4 & 6 & 4 & 6 \\ 8 & 16 & 8 & 12 & 8 & 12 \\ 4 & 8 & 4 & 6 & 4 & 6 \\ 6 & 12 & 6 & 9 & 6 & 9 \\ 4 & 8 & 4 & 6 & 8 & 6 \\ 6 & 12 & 6 & 9 & 6 & 9 \end{bmatrix}$$

The cost functional is

$$\int_0^\infty (x^T F x + x^T x + u^T u) dt$$

This is a LQR problem with $Q = F + I$ and $R = 1$. Using MATLAB, we can find the following feedback map

$$L = [1.4495 \quad 5.7302 \quad 24.7358 \quad 23.3119 \quad 12.6652 \quad 2.1832]$$

With the feedback control $u = -Lx$, the characteristic equation of the closed-loop system is given by

$$\begin{aligned} &(1.4495 + p_0) + (5.7302 + p_1)s + (24.7358 + p_2)s^2 + (23.3119 + p_3)s^3 \\ &+ (12.6652 + p_4)s^4 + (2.1832 + p_5)s^5 + s^6 = 0 \end{aligned}$$

This closed-loop system is robustly stable.

7.5 NOTES AND REFERENCES

The Kharitonov Theorem is an important and elegant result. It gives a surprisingly simple solution to a seeming complex problem. Its derivation is based on some elementary properties of the complex polynomials. The Kharitonov approach is most suitable for analysis problems, where we are given a system and need to check its robust stability with respect to the parameter uncertainty. It is less suitable for design problems, where a feedback control must be designed. For design problems, the optimal control approach becomes preferred. The material covered in this chapter can be found in many papers and books, including references [19, 55, 56, 63, 68, 78, 86, 89, 103, 114, 119, 120, 129, 142, 150, 168].

7.6 PROBLEMS

7.1 Consider the following polynomial:

$$\varphi(s,p) = (60 + 2p^2) + (10 + 2p)s + (9 - p)s^2 + s^3$$

where $p \in [0, 5]$. Let S be the open left half of the s-plane. Find the roots of $\varphi(s,p)$ for $p = 0$ and $p = 5$. States the result from the Boundary Crossing theorem.

7.2 Consider a stable polynomial

$$\varphi(s) = a_0 + a_1 s + \cdots + a_{n-1}s^{n-1} + a_n s^n$$

Write it as $\varphi(s) = \varphi^e(s) + s\varphi^o(s)$, where

$$\varphi^e(s) = a_0 + a_2 s^2 + .a_4 s^4 + a_6 s^6 + \cdots,$$

$$\varphi^o(s) = a_1 + a_3 s^2 + a_5 s^4 + a_7 s^6 + \cdots$$

Prove the following polynomial is also stable.

$$\varphi^e(s) + \frac{d\varphi^e(s)}{ds} = a_0 + a_2 s^2 + 2a_2 s + .a_4 s^4 + 4a_4 s^3 + a_6 s^6 + 6a_6 s^5 + \cdots$$

7.3 Assume that a polynomial

$$\varphi(s) = a_0 + a_1 s + \cdots + a_{n-1}s^{n-1} + a_n s^n$$

satisfies the following conditions: (1) a_n and a_{n-1} have the opposite signs; (2) all roots of $\varphi^e(j\omega)$ and $\varphi^o(j\omega)$ are real and distinct; (3) if $n = 2m$ is even, then $\varphi^e(j\omega)$ has m positive roots $\omega_1^e < \omega_2^e < \cdots < \omega_m^e$, $\varphi^o(j\omega)$

has $m-1$ positive roots $\omega_1^{\rm o} < \omega_2^{\rm o} < \cdots < \omega_{m-1}^{\rm o}$, and they must interlace in the following manner: $0 < \omega_1^{\rm e} < \omega_1^{\rm o} < \omega_2^{\rm e} < \cdots < \omega_{m-1}^{\rm o} < \omega_m^{\rm e}$; (4) if $n = 2m+1$ is odd, then $\varphi^{\rm e}(j\omega)$ has m positive roots $\omega_1^{\rm e} < \omega_2^{\rm e} < \cdots < \omega_m^{\rm e}$, $\varphi^{\rm o}(j\omega)$ has m positive roots $\omega_1^{\rm o} < \omega_2^{\rm o} < \cdots < \omega_m^{\rm o}$, and they must interlace in the following manner: $0 < \omega_1^{\rm e} < \omega_1^{\rm o} < \omega_2^{\rm e} < \cdots < \omega_{m-1}^{\rm o} < \omega_m^{\rm e} < \omega_m^{\rm o}$. Prove that $\varphi(s)$ has all its roots in the open right half of the complex plane.

7.4 Use the Interlacing theorem to check if the following polynomials are stable.

(a) $\varphi_1(s) = 4+2s+5s^2+s^3+8s^4+6s^5+3s^6+5s^7+2s^7$
(b) $\varphi_2(s) = 3+7s+2s^2+4s^3+s^4+5s^5+9s^6+8s^7$
(c) $\varphi_3(s) = 6+8s+2s^2+7s^3+s^4+9s^5+7s^6$

7.5 Determine if the following systems are robustly stable for all admissible uncertainties using the Kharitonov theorem.

(a) $\varphi(s,p) = p_0+p_1s+p_2s^2+p_3s^3+p_4s^4$, where $p_0 \in [1,3]$, $p_1 \in [7,13]$, $p_2 \in [2,5]$, $p_3 \in [4,14]$, and $p_4 \in [3,12]$
(b) $\varphi(s,p) = p_0+p_1s+p_2s^2+p_3s^3+p_4s^4+p_5s^5$, where $p_0 \in [1,5]$, $p_1 \in [4,7]$, $p_2 \in [3,8]$, $p_3 \in [5,11]$, $p_4 \in [80,12]$, and $p_5 \in [7,12]$
(c) $\varphi(s,p) = p_0+p_1s+p_2s^2+p_3s^3+p_4s^4+p_5s^5+p_6s^6$, where $p_0 \in [720,725]$, $p_1 \in [1448,1458]$, $p_2 \in [1213,1227]$, $p_3 \in [535,542]$, $p_4 \in [131,142]$, $p_5 \in [17,19]$, and $p_6 \in [1,1]$

7.6 In the closed-loop system shown in Figure 7.5, the coefficients are uncertain with the following bounds: $a_0 \in [1,3]$, $a_1 \in [0,2]$, $a_2 \in [-1,3]$, $a_3 \in [2,4]$, $b_0 \in [4,8]$, $b_1 \in [0.5,1.5]$, $b_2 \in [1,4]$, $b_3 \in [7,9]$, and $b_4 \in [1,1.5]$. Determine if the system is robustly stable for all admissible uncertainties using the Kharitonov theorem.

7.7 A third-order system has the following characteristic equation

$$a_0+a_1s+a_2s^2+a_3s^3 = 0$$

The nominal values of the coefficients are : $a_0 = 3$, $a_1 = 5$, $a_2 = 7$, and $a_3 = 4$. Find the largest equal intervals around the nominal values

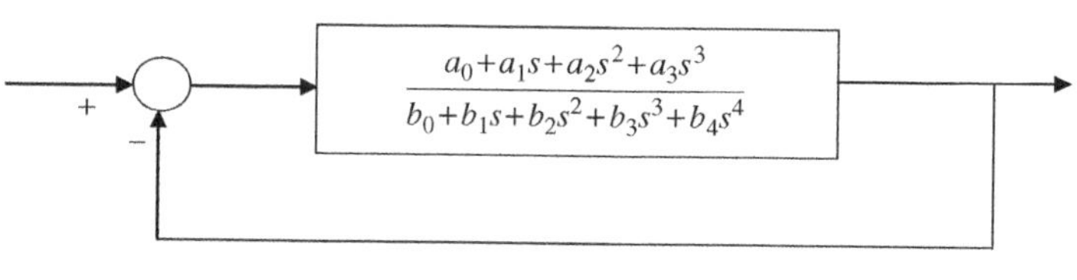

Figure 7.5 Closed-loop system for Problem 7.6.

in which the coefficients can vary while the system remains robustly stable.

7.8 Given the set of polynomials

$$\varphi(s,p) = p_0 + p_1 s + \cdots + p_{n-1} s^{n-1} + p_n s^n$$

where $p_i \in [p_i^-, p_i^+]$, $i = 0, 1, \ldots, n-1, n$. Assume that the four Kharitonov polynomials

$$K_1(s) = p_0^- + p_1^- s + p_2^+ s^2 + p_3^+ s^3 + p_4^- s^4 + p_5^- s^5 + \cdots$$
$$K_2(s) = p_0^- + p_1^+ s + p_2^+ s^2 + p_3^- s^3 + p_4^- s^4 + p_5^+ s^5 + \cdots$$
$$K_3(s) = p_0^+ + p_1^- s + p_2^- s^2 + p_3^+ s^3 + p_4^+ s^4 + p_5^- s^5 + \cdots$$
$$K_4(s) = p_0^+ + p_1^+ s + p_2^- s^2 + p_3^- s^3 + p_4^+ s^4 + p_5^+ s^5 + \cdots$$

all have the following property: Their roots are all in the open right half of the complex plane. Prove that any polynomial in the set also has its roots all in the open right half of the complex plane.

7.9 Consider an open loop system with the following transfer function

$$G(s,p) = \frac{10}{p_0 + p_1 s + p_2 s^2 + p_3 s^3 + p_4 s^4 + s^5}$$

where $p_0 \in [1, 2]$, $p_1 \in [7, 9]$, $p_2 \in [2, 4]$, $p_3 \in [6, 8]$, and $p_4 \in [4, 7]$.

(a) realize the transfer function using controllable canonical form;
(b) check the robust stability of the open-loop system using the Kharitonov theorem;
(c) design a state feedback control that robustly stabilizes the closed-loop system;
(d) translate the robust control problem into an optimal control problem. Design a robust control by solving the optimal control problem.

7.10 Use MATLAB to simulate the two closed-loop systems obtained in Problem 7.9.

8

H_∞ and H_2 Control

In this chapter we discuss another approach to robust control: the H_∞/H_2 approach. This approach uses the H_∞/H_2 norm. We first present some useful prerequisities pertaining to function spaces and their norms. We follow this with pertinent calculation schemes for the H_2 norm and the H_∞ norm. The goal of the H_∞/H_2 approach is to minimize the H_∞/H_2 norm of a transfer function. We show how this can be achieved by synthesizing a suitable controller.

8.1 INTRODUCTION

To motivate the H_∞/H_2 approach to robust control, let us recall the control problems discussed in Chapter 5. The system with uncertainty is modelled as

$$\dot{x} = A(p_o)x + Bu + B\phi(p)x$$

where $p \in P$ is an uncertain parameter vector and $p_o \in P$ is a nominal value of p. The uncertainty is described by $\phi(p)$. To solve the robust control problem, we translate it into the following LQR problem: for the nominal system

$$\dot{x} = A(p_o)x + Bu$$

Robust Control Design: An Optimal Control Approach F. Lin

find a feedback control law $u = Kx$ that minimizes the cost functional

$$\int_0^\infty (x^T Fx + x^T x + u^T u)dt$$

where F is an upper bound on the uncertainty $\phi(p)^T \phi(p)$; that is, $\phi(p)^T \phi(p) \leq F$.

Let us now reformulate the optimal control problem as the following equivalent problem. For a linear time-invariant system

$$\begin{aligned} \dot{x} &= A(p_o)x + w + Bu \\ z &= \begin{bmatrix} (F+I)^{1/2} \\ 0 \end{bmatrix} x + \begin{bmatrix} 0 \\ I \end{bmatrix} u \\ y &= x, \end{aligned}$$

find a feedback control law $u = Ky$ that minimizes

$$\int_0^\infty g(t)^2 dt$$

where $g(t)$ is the impulse response of the controlled system from input w to output z.

We will show that

$$\sqrt{\int_0^\infty g(t)^2 dt}$$

is the H_2 norm of the transfer function from w to z. In other words, the LRQ problem can be viewed as the problem of minimizing the H_2 norm.

8.2 FUNCTION SPACE

Let us start this section by recalling the definition of inner products. The inner product of vectors

$$x = \begin{bmatrix} x_1 \\ \cdots \\ x_n \end{bmatrix} \quad \text{and} \quad y = \begin{bmatrix} y_1 \\ \cdots \\ y_n \end{bmatrix}$$

in a Euclidean space C^n is defined as

$$\langle x, y \rangle = \overline{x}^T y = \sum_{i=1}^n \overline{x}_i y_i$$

where $\overline{x}$ denotes the conjugate of x. The concept of inner product can be extended to infinite vector space V over C^nas follows. An inner product on V is a complex valued function

$$\langle .,. \rangle : V \times V \rightarrow C$$

satisfying the following conditions for all $x, y, z \in V$ and $\alpha, \beta \in C$

1. $\langle x, \alpha y + \beta z \rangle = \alpha \langle x, y \rangle + \beta \langle x, z \rangle$ (linearity)
2. $\langle x, y \rangle = \overline{\langle y, x \rangle}$ (conjugate symmetry)
3. $\langle x, x \rangle > 0$, if $x \neq 0$ (non-negativity)

Note that $\langle x, x \rangle$ is real because by symmetry condition (2): $\langle x, x \rangle = \overline{\langle x, x \rangle}$. Two vectors x, y in the inner product space V are orthogonal, denoted by $x \perp y$, if $\langle x, y \rangle = 0$.

A vector space V having an inner product is called an inner product space. The inner product reduces the following norm

$$\|x\| = \sqrt{\langle x, x \rangle}$$

The inner product space has the following properties.

1. $|\langle x, y \rangle| \leq \|x\| \times \|y\|$ (Cauchy–Schwarz inequality)
2. $\|x+y\|^2 + \|x-y\|^2 = 2\|x\|^2 + 2\|y\|^2$ (parallelogram law)
3. $\|x+y\|^2 = \|x\|^2 + \|y\|^2$ if $x \perp y$

Recall that a metric space M is said to be complete (or Cauchy) if every Cauchy sequence (sequence whose elements become close as the sequence progresses) of points in M has a limit that is also in M. Intuitively, a metric space M is complete if it does not have any holes. For instance, the rational numbers are not complete, because, for example, $\sqrt{2}$ is 'missing' even though we can construct a Cauchy sequence of rational numbers that converge to $\sqrt{2}$.

A Banach space is a complete vector space with a norm. A Hilbert space is a complete inner product space with norm induced by its inner product. Clearly, a Hilbert space is a Banach space, but a Banach space may not be a Hilbert space.

With the above basic concepts and definitions, let us now discuss various space of complex valued functions of time t.

Consider the set of all complex valued functions f over the interval $[a, b]$, $f : [a, b] \rightarrow C$, whose absolute value raised to the p-th power has a finite Lebesgue integral; that is, the p-norm defined below exists.

$$\|f\|_p = \left(\int_a^b |f(t)|^p \mathrm{d}t \right)^{1/p} < \infty$$

Clearly, this space is an infinite dimensional Banach space and is denoted by $L_p[a, b]$. In particular, $L_\infty[a, b]$ is the set of functions bounded almost everywhere on $[a, b]$, whose norm is given by

$$\|f\|_\infty = \lim_{p\to\infty} \|f\|_p = \inf\{B \geq 0 : |f(t)| \leq B \text{ for all most all } t \in [a, b]\}.$$

Also, for $L_2[a, b]$, we can define the following inner product

$$\langle f, g\rangle = \int_a^b \overline{f(t)} g(t)\mathrm{d}t$$

From the definition of the integral, it is obvious that the three requirements (linearity, conjugate symmetry, and nonnegativity) of an inner product are satisfied.

Therefore, $L_2[a, b]$ is an infinite dimensional Hilbert space. We are often interested in the Hilbert space $L_2[0, \infty)$, which consists of a set of bounded functions of time t. The norm of $L_2[0, \infty)$ is given by

$$\|f\|_2 = \sqrt{\int_0^\infty \overline{f(t)} f(t)\mathrm{d}t}$$

Next consider complex valued functions of complex frequency $s \in C$. Let $D \subseteq C$ be an open set and $s_o \in D$ be a point in D. Let $F: D \to C$ be a complex valued function defined on D. F is said to be analytic at s_o if it is differentiable at s_o and some neighborhood of s_o. F is said to be analytic at D if it is analytic in any point in D. Analytic functions have the following properties.

1. If F is analytic at s_o, then its derivative of any order exists and is continuous at s_o.
2. If F is analytic at s_o, then it has a power series representation at s_o.
3. If F has a power series representation at s_o, then it is analytic at s_o.

For example, any real stable transfer function is analytic in the right half of the complex plane.

We use $L_\infty(jR)$ to denote the Banach space of all complex valued functions $F: C \to C$ that are bounded on the imaginary axis jR with its norm given by

$$\|F\|_\infty = \sup_{\omega\in R} |F(j\omega)| \tag{8.1}$$

We use H_∞ to denote the subspace of $L_\infty(jR)$ where functions F are analytic and bounded in the open right half of the complex plane. For functions in H_∞, it can be shown that

$$\sup_{\mathrm{Re}(s)\geq 0} |F(s)| = \sup_{\omega\in R} |F(j\omega)| = \|F\|_\infty \tag{8.2}$$

We use $L_2(jR)$ to denote the Hilbert space of all complex valued functions $F: C \rightarrow C$ such that the following integral is bounded

$$\int_{-\infty}^{\infty} \overline{F(j\omega)} F(j\omega) d\omega < \infty \tag{8.3}$$

The inner product of $L_2(jR)$ is defined as follows. For $F, G \in L_2(jR)$

$$\langle F, G \rangle = \frac{1}{2\pi} \int_{-\infty}^{\infty} \overline{F(j\omega)} G(j\omega) d\omega \tag{8.4}$$

Clearly the three requirements (linearity, conjugate symmetry, and nonnegativity) of inner product are satisfied.

We use H_2 to denote the subspace of $L_2(jR)$ where functions F are analytic and bounded in the open right half of the complex plane. For a function $F \in H_2$, its norm $\|F\|_2$ can be written as

$$\|F\|_2^2 = \frac{1}{2\pi} \int_{-\infty}^{\infty} \overline{F(j\omega)} F(j\omega) d\omega = \sup_{\sigma > 0} \left\{ \frac{1}{2\pi} \int_{-\infty}^{\infty} \overline{F(\sigma + j\omega)} F(\sigma + j\omega) d\omega \right\} \tag{8.5}$$

Let F be the Laplace transform of f: $\mathsf{L}\,[f(t)] = F(s)$. Then we have the following relationship between time-domain function space and frequency-domain function space: $f(t) \in L_2[0, \infty) \Leftrightarrow F(s) \in H_2$.

In the above definitions of function spaces, we assume that functions are scalar valued. We can extend these definitions to matrix valued functions as follows.

For $L_\infty(jR)$, we substitute Equation (8.1) by

$$\|F\|_\infty = \sup_{\omega \in R} \overline{\sigma}(F(j\omega))$$

where $\overline{\sigma}(F(j\omega)) = \max \sqrt{\lambda(\overline{F(j\omega)}^T F(j\omega))}$ is the largest singular value of $F(j\omega)$ with $\lambda(.)$ denoting the eigenvalues.

For H_∞, we substitute Equation (8.2) by

$$\sup_{Re(s) \geq 0} \overline{\sigma}(F(s)) = \sup_{\omega \in R} \overline{\sigma}(F(j\omega)) = \|F\|_\infty$$

For $L_2(jR)$, we substitute Equation (8.3) by

$$\int_{-\infty}^{\infty} trace(\overline{F(j\omega)}^T F(j\omega)) d\omega < \infty$$

and substitute Equation (8.4) by

$$\langle F, G \rangle = \frac{1}{2\pi} \int_{-\infty}^{\infty} trace(\overline{F(j\omega)}^T G(j\omega)) d\omega$$

For H_2, we substitute Equation (8.5) by

$$\begin{aligned}\|F\|_2^2 &= \frac{1}{2\pi}\int_{-\infty}^{\infty} trace(\overline{F(j\omega)}^{\mathrm{T}} F(j\omega))d\omega \\ &= \sup_{\sigma>0}\left\{\frac{1}{2\pi}\int_{-\infty}^{\infty} trace(\overline{F(\sigma+j\omega)}^{\mathrm{T}} F(\sigma+j\omega))d\omega\right\}\end{aligned}$$

8.3 COMPUTATION OF H_2 AND H_∞ NORMS

Let us first discuss how to compute the H_2 norm. As defined in Section 8.2, for a function $G \in H_2$, its H_2 norm is given by

$$\|G\|_2 = \sqrt{\frac{1}{2\pi}\int_{-\infty}^{\infty} trace(\overline{G(j\omega)}^{\mathrm{T}} G(j\omega))d\omega} \tag{8.6}$$

Let $G(s)$ be the Laplace transform of $g(t)$; that is, $\mathrm{L}[g(t)] = G(s)$. Then H_2 norm of $G(s)$ can also be written as

$$\|G\|_2 = \|g\|_2 = \sqrt{\int_{-\infty}^{\infty} trace(\overline{g(t)}^{\mathrm{T}} g(t))dt} \tag{8.7}$$

Since

$$\begin{aligned}trace(\overline{G(j\omega)}^{\mathrm{T}} G(j\omega)) &= trace(G(j\omega)\overline{G(j\omega)}^{\mathrm{T}}) \\ trace(\overline{g(t)}^{\mathrm{T}} g(t)) &= trace(g(t)\overline{g(t)}^{\mathrm{T}})\end{aligned}$$

Equations (8.6) and (8.7) can also be written as

$$\|G\|_2 = \sqrt{\frac{1}{2\pi}\int_{-\infty}^{\infty} trace(G(j\omega)\overline{G(j\omega)}^{\mathrm{T}})d\omega} = \sqrt{\int_{-\infty}^{\infty} trace(g(t)\overline{g(t)}^{\mathrm{T}})dt}$$

In control problems that we are interested in, $G(s)$ is a strictly proper real rational stable transfer function. We say that

$$\begin{aligned}\dot{x} &= Ax + Bu \\ y &= Cx + Du\end{aligned}$$

is a realization of $G(s)$ if $G(s) = C(sI - A)^{-1}B + D$. We sometime denote a realization of $G(s)$ as

$$G(s) = \begin{pmatrix} A & B \\ C & D \end{pmatrix}$$

To compute $\|G\|_2$, we assume $D = 0$. Note that the inverse Laplace transform $g(t)$ of $G(s)$, which is also the impulse response, is given by

$$g(t) = \mathsf{L}^{-1}[G(s)] = \mathsf{L}^{-1}\left[C(sI - A)^{-1}B\right] = \begin{cases} Ce^{At}B & t \geq 0 \\ 0 & t < 0 \end{cases}$$

Therefore

$$\begin{aligned} \|G\|_2 &= \sqrt{\int_{-\infty}^{\infty} trace(\overline{g(t)}^{\mathrm{T}} g(t))\mathrm{d}t} \\ &= \sqrt{\int_0^{\infty} trace((Ce^{At}B)^{\mathrm{T}} Ce^{At}B)\mathrm{d}t} \\ &= \sqrt{trace(B^{\mathrm{T}}(\int_0^{\infty} e^{A^{\mathrm{T}}t}C^{\mathrm{T}}Ce^{At}\mathrm{d}t)B)} \end{aligned}$$

Define

$$S = \int_0^{\infty} e^{A^{\mathrm{T}}t}C^{\mathrm{T}}Ce^{At}\mathrm{d}t$$

which is called observability Gramian of (A, C). Then

$$\|G\|_2 = \sqrt{trace(B^{\mathrm{T}}SB)}$$

To compute S, let us proceed as follows. Define

$$S(\theta) = \int_0^{\theta} e^{A^{\mathrm{T}}t}C^{\mathrm{T}}Ce^{At}\mathrm{d}t$$

Clearly, $S = \lim_{\theta\to\infty} S(\theta)$. On the other hand, let $t = \theta - \tau$, we have

$$\begin{aligned} S(\theta) &= \int_0^{\theta} e^{A^{\mathrm{T}}t}C^{\mathrm{T}}Ce^{At}\mathrm{d}t \\ &= \int_{\theta}^{0} e^{A^{\mathrm{T}}(\theta-\tau)}C^{\mathrm{T}}Ce^{A(\theta-\tau)}(-\mathrm{d}\tau) \\ &= \int_0^{\theta} e^{A^{\mathrm{T}}(\theta-\tau)}C^{\mathrm{T}}Ce^{A(\theta-\tau)}\mathrm{d}\tau \end{aligned}$$

Therefore

$$\begin{aligned} \frac{\mathrm{d}S(\theta)}{\mathrm{d}\theta} &= \frac{\mathrm{d}}{\mathrm{d}\theta}\int_0^{\theta} e^{A^{\mathrm{T}}(\theta-\tau)}C^{\mathrm{T}}Ce^{A(\theta-\tau)}\mathrm{d}\tau \\ &= e^{A^{\mathrm{T}}(\theta-\tau)}C^{\mathrm{T}}Ce^{A(\theta-\tau)}\Big|_{\tau=\theta} \\ &\quad + \int_0^{\theta} A^{\mathrm{T}}e^{A^{\mathrm{T}}(\theta-\tau)}C^{\mathrm{T}}Ce^{A(\theta-\tau)}\mathrm{d}\tau + \int_0^{\theta} e^{A^{\mathrm{T}}(\theta-\tau)}C^{\mathrm{T}}Ce^{A(\theta-\tau)}A\mathrm{d}\tau \\ &= C^{\mathrm{T}}C + A^{\mathrm{T}}S(\theta) + S(\theta)A \end{aligned}$$

As $\theta \rightarrow \infty$

$$\frac{\mathrm{d}S(\theta)}{\mathrm{d}\theta} \rightarrow \frac{\mathrm{d}S}{\mathrm{d}\theta} = 0$$

Hence, S satisfies the equation

$$SA + A^{\mathrm{T}}S + C^{\mathrm{T}}C = 0 \tag{8.8}$$

Comparing Equation (8.8) with the algebraic Riccati equation

$$SA + A^{\mathrm{T}}S + Q - SBR^{-1}B^{\mathrm{T}}S = 0$$

we conclude that we can solve Equation (8.8) by solving the algebraic Riccati equation with A, $B = 0$, $Q = C^{\mathrm{T}}C$, $R = I$.

Similarly, we can also compute $\|G\|_2$ as follows

$$\begin{aligned}
\|G\|_2 &= \sqrt{\int_{-\infty}^{\infty} trace(g(t)\overline{g(t)}^{\mathrm{T}})\mathrm{d}t} \\
&= \sqrt{\int_{0}^{\infty} trace(Ce^{At}B(Ce^{At}B)^{\mathrm{T}})\mathrm{d}t} \\
&= \sqrt{trace(C(\int_{0}^{\infty} e^{A^{\mathrm{T}}t}BB^{\mathrm{T}}e^{At}dt)C^{\mathrm{T}})}
\end{aligned}$$

Define

$$U = \int_0^\infty e^{A^{\mathrm{T}}t}BB^{\mathrm{T}}e^{At}\mathrm{d}t$$

which is called controllability Gramian of (A, B). U can be computed by solving

$$AU + UA^{\mathrm{T}} + BB^{\mathrm{T}} = 0 \tag{8.8}$$

After computing U, the H_2 norm of G can be obtained as

$$\|G\|_2 = \sqrt{trace(CUC^{\mathrm{T}})}$$

Example 8.1

Consider the following system

$$\dot{x} = \begin{bmatrix} 0 & 1 & 0 \\ 0 & 0 & 1 \\ -2 & -5 & -1 \end{bmatrix} x + \begin{bmatrix} 0 \\ 0 \\ 1 \end{bmatrix} u$$

$$y = x$$

Its transfer function is given by

$$G(s) = \begin{bmatrix} \dfrac{1}{s^3+s^2+5s+2} \\ \dfrac{s}{s^3+s^2+5s+2} \\ \dfrac{s^2}{s^3+s^2+5s+2} \end{bmatrix}$$

To compute $\|G\|_2$, let us first solve

$$SA + A^{\mathrm{T}}S + C^{\mathrm{T}}C = 0$$

The solution is

$$S = \begin{bmatrix} 2.4167 & 2.4167 & 0.2500 \\ 2.4167 & 5.7500 & 0.5833 \\ 0.2500 & 0.5833 & 1.0833 \end{bmatrix}$$

Hence

$$\|G\|_2 = \sqrt{trace(B^{\mathrm{T}}SB)} = 1.0408$$

Next, we discuss how to compute the H_∞ norm. Unfortunately, this is much more difficult than computing the H_2 norm. However, it is relatively easy to check, for any $\gamma > 0$, whether $\|G\|_\infty < \gamma$ is true or not as shown in the following theorem.

Theorem 8.1

Let $G(s) = C(sI - A)^{-1}B + D$. Then $\|G\|_\infty < \gamma$ for some $\gamma > 0$ if and only if $\overline{\sigma}(D) < \gamma$ and the following matrix has no eigenvalues on the imaginary axis:

$$\begin{bmatrix} A + BV^{-1}D^{\mathrm{T}}C & BV^{-1}B^{\mathrm{T}} \\ -C^{\mathrm{T}}(I + DV^{-1}D^{\mathrm{T}})C & -(A + BV^{-1}D^{\mathrm{T}}C)^{\mathrm{T}} \end{bmatrix}$$

where $V = \gamma^2 I - D^{\mathrm{T}}D$.

Proof

Define

$$U(s) = \gamma^2 I - G(-s)^{\mathrm{T}}G(s)$$

Then

$$\begin{aligned}
&\|G\|_\infty < \gamma \\
&\Leftrightarrow \sup_{\omega\in R} \overline{\sigma}(G(j\omega)) < \gamma \\
&\Leftrightarrow (\forall\omega \in R)\overline{\sigma}(G(j\omega)) < \gamma \\
&\Leftrightarrow (\forall\omega \in R) \max \sqrt{\lambda(\overline{G(j\omega)}^{\mathrm{T}} G(j\omega))} < \gamma \\
&\Leftrightarrow (\forall\omega \in R) \max \lambda(\overline{G(j\omega)}^{\mathrm{T}} G(j\omega)) < \gamma^2 \\
&\Leftrightarrow (\forall\omega \in R) \max \lambda(G(-j\omega)^{\mathrm{T}} G(j\omega)) < \gamma^2 \\
&\Leftrightarrow (\forall\omega \in R)\gamma^2 I - G(-j\omega)^{\mathrm{T}} G(j\omega) > 0 \\
&\Leftrightarrow (\forall\omega \in R) U(j\omega) > 0 \\
&\Leftrightarrow U(j\infty) > 0 \wedge (\forall\omega \in R) U(j\omega) \text{ is nonsigular} \\
&\Leftrightarrow \gamma^2 I - D^{\mathrm{T}} D > 0 \wedge U(s) \text{ has no zeros on the imaginary axis} \\
&\Leftrightarrow \overline{\sigma}(D) < \gamma \wedge U(s)^{-1} \text{ has no poles on the imaginary axis}
\end{aligned}$$

It can be shown that $U(s)^{-1}$ has the following realization:

$$x = \begin{bmatrix} A+BV^{-1}D^{\mathrm{T}}C & BV^{-1}B^{\mathrm{T}} \\ -C^{\mathrm{T}}(I+DV^{-1}D^{\mathrm{T}})C & -(A+BV^{-1}D^{\mathrm{T}}C)^{T} \end{bmatrix} x + \begin{bmatrix} BV^{-1} \\ -C^{\mathrm{T}}DV^{-1} \end{bmatrix} u$$
$$y = \begin{bmatrix} V^{-1}D^{\mathrm{T}}C & V^{-1}B^{\mathrm{T}} \end{bmatrix} x + V^{-1}u$$

Therefore, $U(s)^{-1}$ has no poles on the imaginary axis if and only if the following matrix has no eigenvalues on the imaginary axis:

$$\begin{bmatrix} A+BV^{-1}D^{\mathrm{T}}C & BV^{-1}B^{\mathrm{T}} \\ -C^{\mathrm{T}}(I+DV^{-1}D^{\mathrm{T}})C & -(A+BV^{-1}D^{\mathrm{T}}C)^{\mathrm{T}} \end{bmatrix}$$

Q.E.D.

Example 8.2

This example illustrates that

$$x = \begin{bmatrix} A+BV^{-1}D^{\mathrm{T}}C & BV^{-1}B^{\mathrm{T}} \\ -C^{\mathrm{T}}(I+DV^{-1}D^{\mathrm{T}})C & -(A+BV^{-1}D^{T}C)^{T} \end{bmatrix} x + \begin{bmatrix} BV^{-1} \\ -C^{T}DV^{-1} \end{bmatrix} u$$
$$y = \begin{bmatrix} V^{-1}D^{\mathrm{T}}C & V^{-1}B^{\mathrm{T}} \end{bmatrix} x + V^{-1}u$$

is a realization of $U(s)^{-1}$. Let

$$A = \begin{bmatrix} -1 & 0 & 0 \\ 0 & -2 & 0 \\ 0 & 0 & -3 \end{bmatrix} \qquad B = \begin{bmatrix} 1 \\ 2 \\ 3 \end{bmatrix}$$

$$C = \begin{bmatrix} -1 & -2 & -3 \end{bmatrix} \qquad D = 1$$

Then

$$G(s) = \frac{s^3 - 8s^2 - 37s - 30}{s^3 + 6s^2 + 11s + 6}$$

Let $\gamma = 10$. Then

$$\begin{aligned}
U(s) &= \gamma^2 I - G(-s)^{\mathrm{T}} G(s) \\
&= 10^2 - \left(\frac{-s^3 - 8s^2 + 37s - 30}{-s^3 + 6s^2 - 11s + 6}\right)\left(\frac{s^3 - 8s^2 - 37s - 30}{s^3 + 6s^2 + 11s + 6}\right) \\
&= 100 - \left(\frac{s^3 + 8s^2 - 37s + 30}{s^3 - 6s^2 + 11s - 6}\right)\left(\frac{s^3 - 8s^2 - 37s - 30}{s^3 + 6s^2 + 11s + 6}\right) \\
&= 100 - \frac{s^6 - 138s^4 + 889s^2 - 900}{s^6 - 14s^4 + 49s^2 - 36} \\
&= \frac{99s^6 - 1262s^4 + 4011s^2 - 2700}{s^6 - 14s^4 + 49s^2 - 36}
\end{aligned}$$

On the other hand,

$$V = \gamma^2 I - D^{\mathrm{T}} D = 100 - 1 = 99$$

$$A + BV^{-1}D^{\mathrm{T}}C = \begin{bmatrix} -1.0101 & -0.0202 & -0.0303 \\ -0.0202 & -2.0404 & -0.0606 \\ -0.0303 & -0.0606 & -3.0909 \end{bmatrix}$$

$$BV^{-1}B^{\mathrm{T}} = \begin{bmatrix} 0.0101 & 0.0202 & 0.0303 \\ 0.0202 & 0.0404 & 0.0606 \\ 0.0303 & 0.0606 & 0.0909 \end{bmatrix}$$

$$-C^{\mathrm{T}}(I + DV^{-1}D^{\mathrm{T}})C = \begin{bmatrix} -1.0101 & -2.0202 & -3.0303 \\ -2.0202 & -4.0404 & -6.0606 \\ -3.0303 & -6.0606 & -9.0909 \end{bmatrix}$$

$$-(A + BV^{-1}D^{\mathrm{T}}C)^{\mathrm{T}} = \begin{bmatrix} 1.0101 & 0.0202 & 0.0303 \\ 0.0202 & 2.0404 & 0.0606 \\ 0.0303 & 0.0606 & 3.0909 \end{bmatrix}$$

$$BV^{-1} = \begin{bmatrix} 0.0101 \\ 0.0202 \\ 0.0303 \end{bmatrix}$$

$$-C^{\mathrm{T}}DV^{-1} = \begin{bmatrix} 0.0101 \\ 0.0202 \\ 0.0303 \end{bmatrix}$$

$$V^{-1}D^{\mathrm{T}}C = \begin{bmatrix} -0.0101 & -0.0202 & -0.0303 \end{bmatrix}$$

$$V^{-1}B^{\mathrm{T}} = \begin{bmatrix} 0.0101\ 0.0202\ 0.0303 \end{bmatrix}$$

Using MATLAB, we can calculate the transfer function of

$$x = \begin{bmatrix} A + BV^{-1}D^{\mathrm{T}}C & BV^{-1}B^{\mathrm{T}} \\ -C^{\mathrm{T}}(I + DV^{-1}D^{\mathrm{T}})C & -(A + BV^{-1}D^{\mathrm{T}}C)^{\mathrm{T}} \end{bmatrix} x + \begin{bmatrix} BV^{-1} \\ -C^{T}DV^{-1} \end{bmatrix} u$$

$$y = \begin{bmatrix} V^{-1}D^{\mathrm{T}}C\ V^{-1}B^{\mathrm{T}} \end{bmatrix} x + V^{-1}u$$

as

$$\Lambda(s) = \frac{0.0101s^6 - 0.1414s^4 + 0.4949s^2 - 0.3636}{s^6 - 12.7475s^4 + 40.5152s^2 - 27.2727}$$

Clearly

$$\Lambda(s) = U(s)^{-1}$$

Since checking $\|G\|_\infty < \gamma$ is much easier than calculating $\|G\|_\infty$ directly, we can use the following algorithm of bisection to 'calculate' the H_∞ norm for a proper real rational transfer function matrix $G(s)$.

Algorithm 8.1

Input: a proper real rational transfer function matrix $G(s)$ and a percentage tolerance δ;
Output: $\|G\|_\infty$ with error less than δ;

Step 1. Find a realization of $G(s)$:

$$G(s) = \begin{pmatrix} A & B \\ C & D \end{pmatrix}$$

Step 2. Pick an upper bound $\overline{\gamma}$ and a lower bound $\underline{\gamma}$ such that $\underline{\gamma} < \|G\|_\infty < \overline{\gamma}$ (for example, we can let $\underline{\gamma}$ be zero and $\overline{\gamma}$ be sufficiently large)

Step 3. If $(\overline{\gamma} - \underline{\gamma})/\overline{\gamma} < 2\delta$,

then let $\|G\|_\infty = (\overline{\gamma} - \underline{\gamma})/2$ and stop

else let $\gamma = (\overline{\gamma} - \underline{\gamma})/2$ and go to Step 4

Step 4. Find all the eigenvalues of

$$\begin{bmatrix} A+BV^{-1}D^TC & BV^{-1}B^T \\ -C^T(I+DV^{-1}D^T)C & -(A+BV^{-1}D^TC)^T \end{bmatrix}$$

Step 5. If there exists eigenvalues on the imaginary axis

$$\text{then let } \underline{\gamma} = \gamma$$

$$\text{else let } \overline{\gamma} = \gamma$$

Step 6. Go to Step 3

Example 8.3

Let us calculate the H_∞ norm of

$$G(s) = \frac{s^3 - 8s^2 - 37s - 30}{s^3 + 6s^2 + 11s + 6}$$

From Example 8.2, we know that $G(s)$ has a realization:

$$A = \begin{bmatrix} -1 & 0 & 0 \\ 0 & -2 & 0 \\ 0 & 0 & -3 \end{bmatrix} \quad B = \begin{bmatrix} 1 \\ 2 \\ 3 \end{bmatrix}$$

$$C = \begin{bmatrix} -1 & -2 & -3 \end{bmatrix} \quad D = 1$$

Let $\overline{\gamma} = 10$ and $\underline{\gamma} = 1$. Then $\gamma = (\overline{\gamma} - \underline{\gamma})/2 = 5.5$, $V = \gamma^2 I - D^T D = 29.25$, and

$$\Lambda = \begin{bmatrix} A+BV^{-1}D^TC & BV^{-1}B^T \\ -C^T(I+DV^{-1}D^T)C & -(A+BV^{-1}D^TC)^T \end{bmatrix}$$

has eigenvalues

$$\lambda(\Lambda) = \{-2.6443, -1.5428, -0.6231, 2.6443, 0.6231, 1.5428\}$$

Since no eigenvalue is on the imaginary axis, the new upper bound is $\overline{\gamma} = \gamma = 5.5$.

Next, let $\gamma = (\overline{\gamma} - \underline{\gamma})/2 = 3.25$. We find the following eigenvalues

$$\lambda(\Lambda) = \{j2.5268, -j2.5268, -2.4491, -1.1914, 2.4491, 1.1914\}$$

Since there are two eigenvalues on the imaginary axis, the new lower bound is $\underline{\gamma} = \gamma = 3.25$. This process can continue until we find $\|G\|_\infty$ with sufficient accuracy.

We can also use MATLAB to find the H_2 norm and the H_∞ norm using the MATLAB commands 'pss2sys', 'h2norm', and 'hinfnorm' as shown in the following example.

Example 8.4

Let $G(s)$ be the transfer function of

$$\dot{x} = \begin{bmatrix} 0 & 1 & 0 \\ 0 & 0 & 1 \\ -2 & -5 & -1 \end{bmatrix} x + \begin{bmatrix} 0 \\ 0 \\ 1 \end{bmatrix} u$$

$$y = x$$

that is

$$G(s) = \begin{bmatrix} \dfrac{1}{s^3+s^2+5s+2} \\ \dfrac{s}{s^3+s^2+5s+2} \\ \dfrac{s^2}{s^3+s^2+5s+2} \end{bmatrix}$$

Then using 'h2norm', we find $\|G\|_2 = 1.041$. Using 'hinfnorm' with tolerance of 1%, we find $\|G\|_\infty$ is between 1.8984 and 1.9174.

8.4 ROBUST CONTROL PROBLEM AS H_2 AND H_∞ CONTROL PROBLEM

To formulate and solve a robust control problem as an H_2 or H_∞ control problem, let us first present a small-gain theorem.

Consider a system with uncertainty. Let us assume that we can separate the uncertainty from the nominal system in a feedback loop, as shown in Figure 8.1.

In Figure 8.1, $G(s)$ is (the transfer function of) the nominal system; and $\Delta(s)$ is the uncertainty. v and z are the input and output of the overall perturbed system. w is the input of the nominal system. The assumption that the uncertainty can be separated from the nominal system as shown in Figure 8.1 is not very restrictive, as illustrated by the following example.

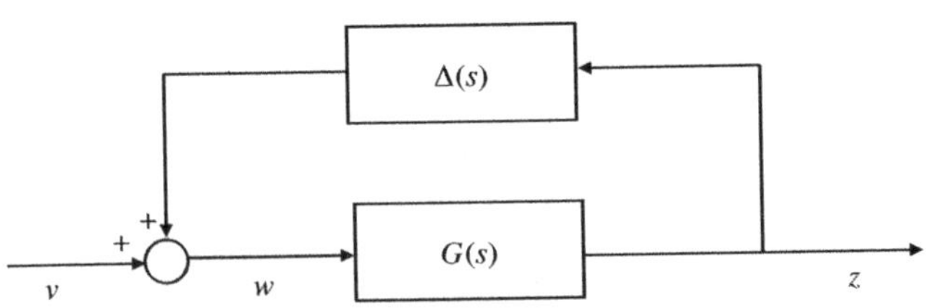

Figure 8.1 Uncertainty and small-gain theorem.

Example 8.5

Consider a system with the following general transfer function:

$$G(s,p) = \frac{b_0 + b_1 s + \ldots + b_{n-1} s^{n-1}}{p_0 + p_1 s + \ldots + p_{n-1} s^{n-1} + s^n}$$

where the uncertainty is described by $p_i \in [p_i^-, p_i^+]$, $i = 0, 1, \ldots, n-1$. We can find its controllable canonical realization as

$$\dot{x} = \begin{bmatrix} 0 & 1 & \ldots & 0 \\ \ldots & & & \\ 0 & 0 & \ldots & 1 \\ -p_0 & -p_1 & \ldots & -p_{n-1} \end{bmatrix} x + \begin{bmatrix} 0 \\ \ldots \\ 0 \\ 1 \end{bmatrix} v$$

$$z = \begin{bmatrix} b_0 & b_1 & \ldots & b_{n-1} \end{bmatrix} x$$

Denote $p_i^o = \frac{p_i^- + p_i^+}{2}$ and $p_i = p_i^o + \Delta p_i$ with $\Delta p_i \in \left[-\frac{p_i^+ - p_i^-}{2}, \frac{p_i^+ - p_i^-}{2} \right]$, $i = 0, 1, \ldots, n-1$. Then we can re-write the state equation as:

$$\dot{x} = \begin{bmatrix} 0 & 1 & \ldots & 0 \\ \ldots & & & \\ 0 & 0 & \ldots & 1 \\ -p_0^o - \Delta p_0 & -p_1^o - \Delta p_1 & \ldots & -p_{n-1}^o - \Delta p_{n-1} \end{bmatrix} x + \begin{bmatrix} 0 \\ \ldots \\ 0 \\ 1 \end{bmatrix} v$$

$$= \begin{bmatrix} 0 & 1 & \ldots & 0 \\ \ldots & & & \\ 0 & 0 & \ldots & 1 \\ -p_0^o & -p_1^o & \ldots & -p_{n-1}^o \end{bmatrix} x + \begin{bmatrix} 0 \\ \ldots \\ 0 \\ 1 \end{bmatrix} v + \begin{bmatrix} 0 & 0 & \ldots & 0 \\ \ldots & & & \\ 0 & 0 & \ldots & 0 \\ -\Delta p_0 & -\Delta p_1 & \ldots & -\Delta p_{n-1} \end{bmatrix} x$$

Let

$$A = \begin{bmatrix} 0 & 1 & \ldots & 0 \\ \ldots & & & \\ 0 & 0 & \ldots & 1 \\ -p_0^o & -p_1^o & \ldots & -p_{n-1}^o \end{bmatrix} \qquad B = \begin{bmatrix} 0 \\ \ldots \\ 0 \\ 1 \end{bmatrix}$$

$$\Delta = \begin{bmatrix} -\Delta p_0 & -\Delta p_1 & \ldots & -\Delta p_{n-1} \end{bmatrix}$$

Define $e = \Delta \cdot x$. Then we have the following equations

$$\dot{x} = Ax + Bu + B\Delta \cdot x = Ax + Bu + Be = Ax + B(u + e)$$

$$z = x$$

In other words, we can translate the system into the one in Figure 8.2.

So, let us now consider the system in Figure 8.1. The problem we want to investigate is as follows. Assume that the nominal system $G(s)$ is stable.

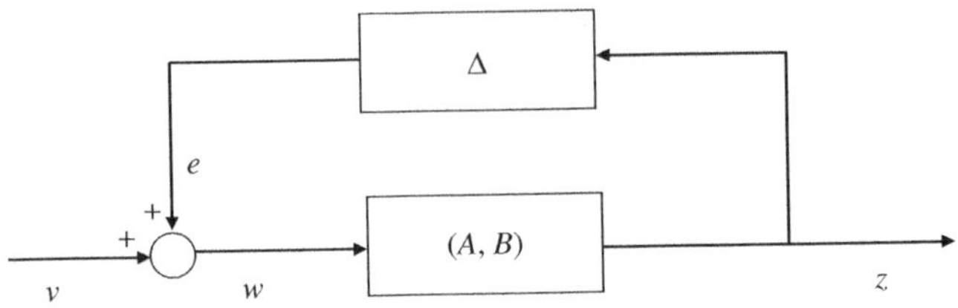

Figure 8.2 Separation of uncertainty from the system.

How big can the uncertainty $\Delta(s)$ be, before the perturbed system becomes unstable? In other words, what is the bound on the uncertainty $\Delta(s)$ that guarantees the stability of the perturbed system? This question is partially answered by the following small-gain theorem.

Theorem 8.2

Consider the system in Figure 8.1. Let $G(s)$ be a proper real rational stable transfer function. Assume that $\|G\|_\infty < \gamma$ for some $\gamma > 0$. Then the perturbed (closed-loop) system is stable for all proper real rational stable transfer functions $\Delta(s)$ such that $\|\Delta\|_\infty \leq 1/\gamma$.

Proof

It is easy to check that in Figure 8.1, the transfer function from v to z is

$$M(s) = (I - G(s)\Delta(s))^{-1}G(s).$$

For the perturbed system to be stable, all the poles of $M(s)$ must be in the open left half plane. Since $G(s)$ has all the poles in the open left half plane, stability requires that $(I - G(s)\Delta(s))^{-1}$ has all the poles in the open left half plane. Equivalently, this means that all the zeros of $(I - G(s)\Delta(s))$ must be in the open left half plane. In other words

$$\inf_{\mathrm{Re}(s)\geq 0} \underline{\sigma}(I - G(s)\Delta(s)) \neq 0$$

Hence, to prove the theorem, we only need to prove the above condition is true for all proper real rational stable transfer functions $\Delta(s)$ such that $\|\Delta\|_\infty \leq 1/\gamma$. The proof is as follows.

$$\begin{aligned}
&\|G\|_\infty < \gamma \wedge \|\Delta\|_\infty < 1/\gamma \\
&\Rightarrow \|G\Delta\|_\infty < 1 \\
&\Rightarrow \sup_{\mathrm{Re}(s)\geq 0} \overline{\sigma}(G(s)\Delta(s)) < 1
\end{aligned}$$

$$\Rightarrow 1 - \sup_{\mathrm{Re}(s)\geq 0} \overline{\sigma}(G(s)\Delta(s)) > 0$$

$$\Rightarrow \inf_{\mathrm{Re}(s)\geq 0} \underline{\sigma}(I - G(s)\Delta(s)) > 0$$

$$\Rightarrow \inf_{\mathrm{Re}(s)\geq 0} \underline{\sigma}(I - G(s)\Delta(s)) \neq 0$$

Q.E.D.

Let us now apply the small-gain theorem to the following example.

Example 8.6

Consider the following system:

$$\dot{x} = \begin{bmatrix} 0 & 1 & 0 \\ 0 & 0 & 1 \\ -3+\Delta_1 & -4+\Delta_2 & -7+\Delta_3 \end{bmatrix} x + \begin{bmatrix} 0 \\ 0 \\ 1 \end{bmatrix} u$$

$$y = x$$

where Δ_1, Δ_2, and Δ_3 are uncertainties. The system can be decomposed as the nominal system (A, B) and the uncertainty Δ as in Figure 8.2. The nominal system is stable. Let $G(s)$ be the transfer function of the nominal system

$$\dot{x} = \begin{bmatrix} 0 & 1 & 0 \\ 0 & 0 & 1 \\ -3 & -4 & -7 \end{bmatrix} x + \begin{bmatrix} 0 \\ 0 \\ 1 \end{bmatrix} u$$

$$y = x$$

Then we can calculate the H_∞ norm using MATLAB and obtain $\|G\|_\infty < 0.5522$. Since the H_∞ norm of $\Delta(s) = [\Delta_1 \ \Delta_2 \ \Delta_3]$ is $\|\Delta\|_\infty = \sqrt{\Delta_1^2 + \Delta_2^2 + \Delta_3^2}$, by Theorem 8.2, the perturbed system is stable for all uncertainties such that

$$\|\Delta\|_\infty = \sqrt{\Delta_1^2 + \Delta_2^2 + \Delta_3^2} \leq 1/0.5522 = 1.811$$

or

$$\Delta_1^2 + \Delta_2^2 + \Delta_3^2 \leq 3.2796 \tag{8.9}$$

Note that this condition is actually very conservative. To see this, let us write the characteristic equation of the perturbed system as

$$s^3 + (7 - \Delta_3)s^2 + (4 - \Delta_2)s + (3 - \Delta_1) = 0$$

By the Routh–Hurwitz criterion, we know that the perturbed system is stable if and only if

$$7-\Delta_3>0\wedge 4-\Delta_2>0\wedge 3-\Delta_1>0\wedge(7-\Delta_3)(4-\Delta_2)>3-\Delta_1 \qquad (8.10)$$

Condition (8.10) is much weaker than Condition (8.9). For example, if $\Delta_1=2.5$, $\Delta_2=3.5$, and $\Delta_3=5$, then Condition (8.10) is satisfied, however

$$\Delta_1^2+\Delta_2^2+\Delta_3^2=43.5000$$

which is much greater than 3.2796.

So far, we have discussed analysis problems; that is, given a perturbed system with bounds on the uncertainty, we can check if the condition in Theorem 8.2 is satisfied. If the condition is satisfied, then the robust stability is guaranteed; otherwise, the system may or may not be robustly stable. In the next section, we will turn to synthesis problem; that is, how to design a controller that will achieve robust stability for the largest bounds on the uncertainty.

8.5 H_2/H_∞ CONTROL SYNTHESIS

Before we discuss H_2/H_∞ control synthesis, let us first mention that the H_2/H_∞ approach is very different from the optimal approach discussed in Chapters 5 and 6. In the optimal control approach, we start with the bounds on uncertainties. We then design a controller based on these bounds. As the result, if the controller exists, then it is guaranteed to robustly stabilize the perturbed system. On the other hand, in the H_2/H_∞ approach, the bounds on uncertainties are not given in advance. The synthesis will try to achieve the largest tolerance range on uncertainty. However, there is no guarantee that the range is large enough to cover all possible uncertainties. In other words, the H_2/H_∞ approach cannot guarantee the robustness of the resulting controller. The approach will do its best to make the resulting controller robust. Whether the best is good enough depends on the nature of the uncertainties.

To formulate the H_2/H_∞ approach, let us consider the setting in Figure 8.1, but assume the $G(s)$ can now be modified by introducing a controller as shown in Figure 8.3. In Figure 8.3, $F(s)$ is (the transfer function of) the plant; $K(s)$ is (the transfer function of) the controller to be designed; u is the input for control; and y is the output (measurement) for control. Comparing Figure 8.3 with Figure 8.1, we see that nominal system $G(s)$ is now equivalent to the controlled system consisting of the plant $F(s)$ and the controller $K(s)$. The plant $F(s)$ is given while the controller $K(s)$ is to be designed. Form the previous discussions, we know that in order to maximize the tolerance range on uncertainty, we need to design

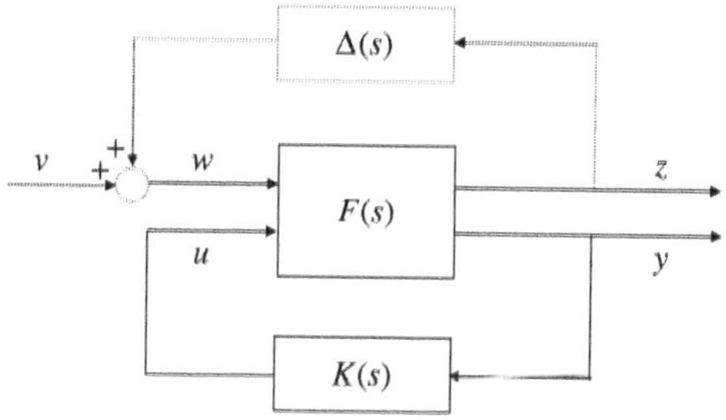

Figure 8.3 H_2/H_∞ approach: introduction of a controller to minimize the H_2/H_∞ norm.

a feasible controller that minimizes the norm of the transfer function from w to z.

Formally, let us assume that the plant $F(s)$ has the following realization

$$\dot{x} = Ax + B_1 w + B_2 u$$
$$z = C_1 x + D_{11} w + D_{12} u$$
$$y = C_2 x + D_{21} w + D_{22} u$$

Sometimes, we denote the above realization of $F(s)$ as

$$F(s) = \begin{pmatrix} A & B_1 & B_2 \\ C_1 & D_{11} & D_{12} \\ C_2 & D_{21} & D_{22} \end{pmatrix} = \begin{bmatrix} F_{11}(s) & F_{12}(s) \\ F_{21}(s) & F_{22}(s) \end{bmatrix}$$

where $F_{11}(s) = C_1(sI - A)^{-1}B_1 + D_{11}$ is the transfer function from w to z; $F_{12}(s) = C_1(sI - A)^{-1}B_2 + D_{12}$ is the transfer function from u to z; $F_{21}(s) = C_2(sI - A)^{-1}B_1 + D_{21}$ is the transfer function from w to y; and $F_{22}(s) = C_2(sI - A)^{-1}B_2 + D_{22}$ is the transfer function from u to y. Clearly, $F(s)$ is a proper real rational transfer function. The transfer function $G(s)$ of the controlled system with controller $K(s)$ can be derived as follows.

$$\begin{bmatrix} Z(s) \\ Y(s) \end{bmatrix} = \begin{bmatrix} F_{11}(s) & F_{12}(s) \\ F_{21}(s) & F_{22}(s) \end{bmatrix} \begin{bmatrix} W(s) \\ U(s) \end{bmatrix}$$

implies

$$Z(s) = F_{11}(s)W(s) + F_{12}(s)U(s)$$
$$Y(s) = F_{21}(s)W(s) + F_{22}(s)U(s).$$

Since $U(s) = K(s)Y(s)$, we have

$$\begin{aligned} Y(s) &= F_{21}(s)W(s) + F_{22}(s)K(s)Y(s) \\ \Rightarrow (I - F_{22}(s)K(s))Y(s) &= F_{21}(s)W(s) \\ \Rightarrow Y(s) &= (I - F_{22}(s)K(s))^{-1}F_{21}(s)W(s) \\ \Rightarrow U(s) &= K(s)(I - F_{22}(s)K(s))^{-1}F_{21}(s)W(s) \end{aligned}$$

Hence

$$\begin{aligned} Z(s) &= F_{11}(s)W(s) + F_{12}(s)U(s) \\ &= F_{11}(s)W(s) + F_{12}(s)K(s)(I - F_{22}(s)K(s))^{-1}F_{21}(s)W(s) \\ &= (F_{11}(s) + F_{12}(s)K(s)(I - F_{22}(s)K(s))^{-1}F_{21}(s))W(s) \end{aligned}$$

That is, the transfer function from w to z is given by

$$G(s) = F_{11}(s) + F_{12}(s)K(s)(I - F_{22}(s)K(s))^{-1}F_{21}(s) \tag{8.11}$$

One basic requirement on $K(s)$ is that it must internally stabilize the controlled system. This in turn requires that (A, B_2) is stabilizable and (A, C_2) is detectable; that is, there exist L_1, L_2 such that $A + B_2L_1$ and $A + L_2C_2$ are stable. Assuming this is true, we can characterize all stabilizing controllers $K(s)$ as shown in the following theorem.

Theorem 8.3

Let the plant

$$F(s) = \begin{pmatrix} A & B_1 & B_2 \\ C_1 & D_{11} & D_{12} \\ C_2 & D_{21} & D_{22} \end{pmatrix} = \begin{bmatrix} F_{11}(s) & F_{12}(s) \\ F_{21}(s) & F_{22}(s) \end{bmatrix}$$

and L_1, L_2 be such that $A + B_2L_1$ and $A + L_2C_2$ are stable. Then all controllers that internally stabilize the controlled system can be parameterized as

$$K(s) = M_{11}(s) + M_{12}(s)Q(s)(I - M_{22}(s)Q(s))^{-1}M_{21}(s)$$

where $Q(s)$ is any proper real rational transfer function such that $Q(s) \in H_\infty$ and $M_{ij}(s)$ is given by

$$\begin{bmatrix} M_{11}(s) & M_{12}(s) \\ M_{21}(s) & M_{22}(s) \end{bmatrix} = \begin{pmatrix} A + B_2L_1 + L_2C_2 + L_2D_{22}L_1 & -L_2 & B_2 + L_2D_{22} \\ L_1 & 0 & I \\ -(C_2 + D_{22}L_1) & I & -D_{22} \end{pmatrix}$$

Proof

We will only prove the special case when all A, B_1, B_2, C_1, C_2, D_{11}, D_{12}, D_{21}, D_{22} are scalars because it illustrates the idea and avoids tedious matrix manipulations. In the scalar case

$$\begin{bmatrix} M_{11}(s) & M_{12}(s) \\ M_{21}(s) & M_{22}(s) \end{bmatrix}$$

$$= \begin{bmatrix} L_1 \\ -(C_2 + D_{22}L_1) \end{bmatrix} \frac{1}{s - (A + B_2L_1 + L_2C_2 + L_2D_{22}L_1)} [-L_2 \;\; B_2 + L_2D_{22}]$$

$$+ \begin{bmatrix} 0 & 1 \\ 1 & -D_{22} \end{bmatrix}$$

$$= \frac{1}{s - (A + B_2L_1 + L_2C_2 + L_2D_{22}L_1)} \begin{bmatrix} -L_1L_2 & L_1(B_2 + L_2D_{22}) \\ (C_2 + D_{22}L_1)L_2 & -(C_2 + D_{22}L_1)(B_2 + L_2D_{22}) \end{bmatrix} + \begin{bmatrix} 0 & 1 \\ 1 & -D_{22} \end{bmatrix}$$

$$= \frac{1}{s - A - B_2L_1 - L_2C_2 - L_2D_{22}L_1} \begin{bmatrix} -L_1L_2 & s - A - L_2C_2 \\ s - A - B_2L_1 & -C_2B_2 - D_{22}(s - A) \end{bmatrix}$$

Hence

$$I - M_{22}(s)Q(s) = 1 - \frac{-C_2B_2 - D_{22}(s - A)}{s - A - B_2L_1 - L_2C_2 - L_2D_{22}L_1} Q(s)$$

$$= \frac{s - A - B_2L_1 - L_2C_2 - L_2D_{22}L_1 + (C_2B_2 + D_{22}(s - A))Q(s)}{s - A - B_2L_1 - L_2C_2 - L_2D_{22}L_1}$$

and

$$K(s) = M_{11}(s) + M_{12}(s)Q(s)(I - M_{22}(s)Q(s))^{-1}M_{21}(s)$$

$$= \frac{-L_1L_2}{s - A - B_2L_1 - L_2C_2 - L_2D_{22}L_1} + \frac{s - A - L_2C_2}{s - A - B_2L_1 - L_2C_2 - L_2D_{22}L_1}$$
$$\times Q(s)(I - M_{22}(s)Q(s))^{-1} \frac{s - A - B_2L_1}{s - A - B_2L_1 - L_2C_2 - L_2D_{22}L_1}$$

$$= \frac{-L_1L_2}{s - A - B_2L_1 - L_2C_2 - L_2D_{22}L_1} + \frac{s - A - L_2C_2}{s - A - B_2L_1 - L_2C_2 - L_2D_{22}L_1} Q(s)$$
$$\times \frac{s - A - B_2L_1 - L_2C_2 - L_2D_{22}L_1}{s - A - B_2L_1 - L_2C_2 - L_2D_{22}L_1 + (C_2B_2 + D_{22}(s - A))Q(s)}$$
$$\times \frac{s - A - B_2L_1}{s - A - B_2L_1 - L_2C_2 - L_2D_{22}L_1}$$

$$
\begin{aligned}
&= \frac{-L_1L_2}{s-A-B_2L_1-L_2C_2-L_2D_{22}L_1} + \frac{(s-A-L_2C_2)Q(s)}{s-A-B_2L_1-L_2C_2-L_2D_{22}L_1} \\
&\quad \times \frac{s-A-B_2L_1}{s-A-B_2L_1-L_2C_2-L_2D_{22}L_1+(C_2B_2+D_{22}(s-A))Q(s)} \\
&= \frac{-L_1L_2+(s-A)Q(s)}{s-A-B_2L_1-L_2C_2-L_2D_{22}L_1+(C_2B_2+D_{22}(s-A))Q(s)}
\end{aligned}
$$

Let

$$
\begin{aligned}
K_1(s) &= -L_1L_2+(s-A)Q(s) \\
K_2(s) &= s-A-B_2L_1-L_2C_2-L_2D_{22}L_1+(C_2B_2+D_{22}(s-A))Q(s)
\end{aligned}
$$

then we can write

$$
K(s) = \frac{K_1(s)}{K_2(s)}
$$

By Equation (8.11), the controlled system is given by

$$
G(s) = F_{11}(s)+F_{12}(s)K(s)(I-F_{22}(s)K(s))^{-1}F_{21}(s)
$$

We can calculate $G(s)$ as follows.

$$
\begin{aligned}
\begin{bmatrix} F_{11}(s) & F_{12}(s) \\ F_{21}(s) & F_{22}(s) \end{bmatrix} &= \begin{bmatrix} C_1 \\ C_2 \end{bmatrix} \frac{1}{s-A} \begin{bmatrix} B_1 & B_2 \end{bmatrix} + \begin{bmatrix} D_{11} & D_{12} \\ D_{21} & D_{22} \end{bmatrix} \\
&= \frac{1}{s-A} \begin{bmatrix} C_1B_1 & C_1B_2 \\ C_2B_1 & C_2B_2 \end{bmatrix} + \begin{bmatrix} D_{11} & D_{12} \\ D_{21} & D_{22} \end{bmatrix} \\
&= \frac{1}{s-A} \begin{bmatrix} C_1B_1+D_{11}(s-A) & C_1B_2+D_{12}(s-A) \\ C_2B_1+D_{21}(s-A) & C_2B_2+D_{22}(s-A) \end{bmatrix}
\end{aligned}
$$

Therefore

$$
\begin{aligned}
I-F_{22}(s)K(s) &= 1 - \frac{C_2B_2+D_{22}(s-A)}{s-A}K(s) \\
&= \frac{s-A-(C_2B_2+D_{22}(s-A))K(s)}{s-A} \\
&= \frac{(s-A)K_2(s)-(C_2B_2+D_{22}(s-A))K_1(s)}{(s-A)K_2(s)}
\end{aligned}
$$

Since

$$
\begin{aligned}
&(s-A)K_2(s)-(C_2B_2+D_{22}(s-A))K_1(s) \\
&= (s-A)(s-A-B_2L_1-L_2C_2-L_2D_{22}L_1+(C_2B_2+D_{22}(s-A))Q(s))
\end{aligned}
$$

$$
\begin{aligned}
&- (C_2B_2 + D_{22}(s - A))(-L_1L_2 + (s - A)Q(s)) \\
&= (s - A)(s - A - B_2L_1 - L_2C_2) + C_2B_2L_1L_2 \\
&= (s - A - L_2C_2)(s - A - B_2L_1)
\end{aligned}
$$

we have

$$
\begin{aligned}
K(s)(I - F_{22}(s)K(s))^{-1} &= \frac{K_1(s)}{K_2(s)} \times \frac{(s - A)K_2(s)}{(s - A - L_2C_2)(s - A - B_2L_1)} \\
&= \frac{(s - A)K_1(s)}{(s - A - L_2C_2)(s - A - B_2L_1)}
\end{aligned}
$$

and

$$
\begin{aligned}
G(s) &= F_{11}(s) + F_{12}(s)K(s)(I - F_{22}(s)K(s))^{-1}F_{21}(s) \\
&= \frac{C_1B_1 + D_{11}(s - A)}{s - A} + \frac{C_1B_2 + D_{12}(s - A)}{s - A} \\
&\quad \times \frac{(s - A)K_1(s)}{(s - A - L_2C_2)(s - A - B_2L_1)} \times \frac{C_2B_1 + D_{21}(s - A)}{s - A} \\
&= \frac{C_1B_1 + D_{11}(s - A)}{s - A} + \frac{(C_1B_2 + D_{12}(s - A))(C_2B_1 + D_{21}(s - A))K_1(s)}{(s - A)(s - A - L_2C_2)(s - A - B_2L_1)} \\
&= \frac{C_1B_1 + D_{11}(s - A)}{s - A} \\
&\quad + \frac{(C_1B_2 + D_{12}(s - A))(C_2B_1 + D_{21}(s - A))(-L_1L_2 + (s - A)Q(s))}{(s - A)(s - A - L_2C_2)(s - A - B_2L_1)} \\
&= \frac{C_1B_1 + D_{11}(s - A)}{s - A} - \frac{(C_1B_2 + D_{12}(s - A))(C_2B_1 + D_{21}(s - A))L_1L_2}{(s - A)(s - A - L_2C_2)(s - A - B_2L_1)} \\
&\quad + \frac{(C_1B_2 + D_{12}(s - A))(C_2B_1 + D_{21}(s - A))}{(s - A - L_2C_2)(s - A - B_2L_1)}Q(s) \\
&= \frac{C_1B_1(s - A - L_2C_2)(s - A - B_2L_1) - (C_1B_2 + D_{12}(s - A))(C_2B_1 + D_{21}(s - A))L_1L_2}{(s - A)(s - A - L_2C_2)(s - A - B_2L_1)} \\
&\quad + D_{11} + \frac{(C_1B_2 + D_{12}(s - A))(C_2B_1 + D_{21}(s - A))}{(s - A - L_2C_2)(s - A - B_2L_1)}Q(s) \\
&= \frac{(C_1B_1 - D_{12}D_{21}L_1L_2)(s - A) - C_1B_1(L_2C_2 + B_2L_1) - (D_{12}C_2B_1 + D_{21}C_1B_2)L_1L_2}{(s - A - L_2C_2)(s - A - B_2L_1)} \\
&\quad + D_{11} + \frac{(C_1B_2 + D_{12}(s - A))(C_2B_1 + D_{21}(s - A))}{(s - A - L_2C_2)(s - A - B_2L_1)}Q(s)
\end{aligned}
$$

Since $A + B_2L_1$ and $A + L_2C_2$ are stable, by observing the eigenvalues of $G(s)$, it is clear that $G(s)$ is stable if and only if $Q(s)$ is a proper real rational transfer function such that $Q(s) \in H_\infty$.

Q.E.D.

Let us consider the following example.

Example 8.7

Consider the system

$$F(s) = \begin{pmatrix} A & B_1 & B_2 \\ C_1 & D_{11} & D_{12} \\ C_2 & D_{21} & D_{22} \end{pmatrix}$$

with

$$A = \begin{bmatrix} -5 & 2 & -4 \\ 0 & -3 & 0 \\ 0 & 7 & -1 \end{bmatrix} \quad B_1 = \begin{bmatrix} 7 \\ -3 \\ 1 \end{bmatrix} \quad B_2 = \begin{bmatrix} 6 \\ 8 \\ -5 \end{bmatrix} \quad C_1 = \begin{bmatrix} -2 & 9 & 4 \end{bmatrix}$$

$$C_2 = \begin{bmatrix} 6 & 3 & -1 \end{bmatrix}$$

$D_{11} = 0$, $D_{12} = 1$, $D_{21} = 2$, and $D_{22} = 0$. The elements of the transfer function $F(s)$ can be found as follows.

$$F_{11}(s) = \frac{-37s^2 - 2509s - 669}{s^3 + 9s^2 + 23s + 15}$$

$$F_{12}(s) = \frac{s^3 + 49s^2 + 339s + 1455}{s^3 + 9s^2 + 23s + 15}$$

$$F_{21}(s) = \frac{2s^3 + 50s^2 + 113s + 597}{s^3 + 9s^2 + 23s + 15}$$

$$F_{22}(s) = \frac{65s^2 + 488s - 865}{s^3 + 9s^2 + 23s + 15}$$

Since A is stable, we can take $L_1 = 0$ and $L_2 = 0$. Therefore, the matrix

$$\begin{bmatrix} M_{11}(s) & M_{12}(s) \\ M_{21}(s) & M_{22}(s) \end{bmatrix} = \begin{pmatrix} A & 0 & B_2 \\ 0 & 0 & 1 \\ -C_2 & 1 & 0 \end{pmatrix}$$

can be computed as follows.

$$M_{11}(s) = 0$$

$$M_{12}(s) = 1$$

$$M_{21}(s) = 1$$

$$M_{22}(s) = \frac{-65s^2 - 488s + 865}{s^3 + 9s^2 + 23s + 15}$$

Hence, the controller can be written as

$$\begin{aligned}
K(s) &= M_{11}(s) + M_{12}(s)Q(s)(I - M_{22}(s)Q(s))^{-1}M_{21}(s)\\
&= 0 + Q(s)(I - M_{22}(s)Q(s))^{-1}\\
&= Q(s)\left(1 - \frac{-65s^2 - 488s + 865}{s^3 + 9s^2 + 23s + 15}Q(s)\right)^{-1}\\
&= Q(s)\left(\frac{s^3 + 9s^2 + 23s + 15 + (65s^2 + 488s - 865)Q(s)}{s^3 + 9s^2 + 23s + 15}\right)^{-1}\\
&= \frac{(s^3 + 9s^2 + 23s + 15)Q(s)}{s^3 + 9s^2 + 23s + 15 + (65s^2 + 488s - 865)Q(s)}
\end{aligned}$$

To find the controlled system, let us first calculate

$$\begin{aligned}
&I - F_{22}(s)K(s)\\
&=1 - \frac{65s^2 + 488s - 865}{s^3 + 9s^2 + 23s + 15} \times \frac{(s^3 + 9s^2 + 23s + 15)Q(s)}{s^3 + 9s^2 + 23s + 15 + (65s^2 + 488s - 865)Q(s)}\\
&=1 - \frac{(65s^2 + 488s - 865)Q(s)}{s^3 + 9s^2 + 23s + 15 + (65s^2 + 488s - 865)Q(s)}\\
&=\frac{s^3 + 9s^2 + 23s + 15}{s^3 + 9s^2 + 23s + 15 + (65s^2 + 488s - 865)Q(s)}
\end{aligned}$$

and

$$\begin{aligned}
&K(s)(I - F_{22}(s)K(s))^{-1}\\
&=\frac{(s^3 + 9s^2 + 23s + 15)Q(s)}{s^3 + 9s^2 + 23s + 15 + (65s^2 + 488s - 865)Q(s)}\\
&\quad\times\frac{s^3 + 9s^2 + 23s + 15 + (65s^2 + 488s - 865)Q(s)}{s^3 + 9s^2 + 23s + 15}\\
&=Q(s)
\end{aligned}$$

Therefore, the controlled system is given by

$$\begin{aligned}
G(s) &= F_{11}(s) + F_{12}(s)K(s)(I - F_{22}(s)K(s))^{-1}F_{21}(s)\\
&= \frac{-37s^2 - 2509s - 669}{s^3 + 9s^2 + 23s + 15} + \frac{s^3 + 49s^2 + 339s + 1455}{s^3 + 9s^2 + 23s + 15}\\
&\quad\times Q(s)\frac{2s^3 + 50s^2 + 113s + 597}{s^3 + 9s^2 + 23s + 15}
\end{aligned}$$

The above example shows that the transfer function of the controlled systems can be written in a specific form. In general, using the controller

$$K(s) = M_{11}(s) + M_{12}(s)Q(s)(I - M_{22}(s)Q(s))^{-1}M_{21}(s)$$

The transfer function of the controlled system can be derived as follows.

Theorem 8.4

Let the controller be

$$K(s) = M_{11}(s) + M_{12}(s)Q(s)(I - M_{22}(s)Q(s))^{-1}M_{21}(s)$$

The controlled system has the following transfer function

$$G(s) = \Psi_{11}(s) + \Psi_{12}(s)Q(s)\Psi_{21}(s),$$

where $\Psi_{ij}(s)$ is given by

$$\begin{bmatrix} \Psi_{11}(s) & \Psi_{12}(s) \\ \Psi_{21}(s) & \Psi_{22}(s) \end{bmatrix} = \left(\begin{array}{cccc} A+B_2L_1 & -B_2L_1 & B_1 & B_2 \\ 0 & A+L_2C_2 & B_1+L_2D_{21} & 0 \\ C_1+D_{12}L_1 & -D_{12}L_1 & D_{11} & D_{12} \\ 0 & C_2 & D_{21} & 0 \end{array}\right)$$

Proof

We will only prove the special case when all A, B_1, B_2, C_1, C_2, D_{11}, D_{12}, D_{21}, D_{22} are scalars. From the proof of Theorem 8.3, we have

$$\begin{aligned} G(s) = & \frac{(C_1B_1 - D_{12}D_{21}L_1L_2)(s-A) - C_1B_1(L_2C_2 + B_2L_1) - (D_{12}C_2B_1 + D_{21}C_1B_2)L_1L_2}{(s-A-L_2C_2)(s-A-B_2L_1)} \\ & + D_{11} + \frac{(C_1B_2 + D_{12}(s-A))(C_2B_1 + D_{21}(s-A))}{(s-A-L_2C_2)(s-A-B_2L_1)}Q(s) \end{aligned} \tag{8.12}$$

On the other hand

$$\begin{aligned} & \begin{bmatrix} \Psi_{11}(s) & \Psi_{12}(s) \\ \Psi_{21}(s) & \Psi_{22}(s) \end{bmatrix} \\ & = \begin{bmatrix} C_1+D_{12}L_1 & -D_{12}L_1 \\ 0 & C_2 \end{bmatrix} \begin{bmatrix} s-A-B_2L_1 & B_2L_1 \\ 0 & s-A-L_2C_2 \end{bmatrix}^{-1} \\ & \times \begin{bmatrix} B_1 & B_2 \\ B_1+L_2D_{21} & 0 \end{bmatrix} + \begin{bmatrix} D_{11} & D_{12} \\ D_{21} & 0 \end{bmatrix} \end{aligned}$$

$$
\begin{aligned}
&= \begin{bmatrix} C_1 + D_{12}L_1 & -D_{12}L_1 \\ 0 & C_2 \end{bmatrix} \begin{bmatrix} s - A - L_2C_2 & -B_2L_1 \\ 0 & s - A - B_2L_1 \end{bmatrix} \\
&\quad \times \frac{1}{(s - A - L_2C_2)(s - A - B_2L_1)} \begin{bmatrix} B_1 & B_2 \\ B_1 + L_2D_{21} & 0 \end{bmatrix} + \begin{bmatrix} D_{11} & D_{12} \\ D_{21} & 0 \end{bmatrix} \\
&= \begin{bmatrix} (C_1 + D_{12}L_1)(s - A - L_2C_2) & -(C_1 + D_{12}L_1)B_2L_1 - D_{12}L_1(s - A - B_2L_1) \\ 0 & C_2(s - A - B_2L_1) \end{bmatrix} \\
&\quad \times \frac{1}{(s - A - L_2C_2)(s - A - B_2L_1)} \begin{bmatrix} B_1 & B_2 \\ B_1 + L_2D_{21} & 0 \end{bmatrix} + \begin{bmatrix} D_{11} & D_{12} \\ D_{21} & 0 \end{bmatrix} \\
&= \begin{bmatrix} (C_1 + D_{12}L_1)(s - A - L_2C_2) & -C_1B_2L_1 - D_{12}L_1(s - A) \\ 0 & C_2(s - A - B_2L_1) \end{bmatrix} \begin{bmatrix} B_1 & B_2 \\ B_1 + L_2D_{21} & 0 \end{bmatrix} \\
&\quad \times \frac{1}{(s - A - L_2C_2)(s - A - B_2L_1)} + \begin{bmatrix} D_{11} & D_{12} \\ D_{21} & 0 \end{bmatrix} \\
&= \begin{bmatrix} (C_1 + D_{12}L_1)(s - A - L_2C_2)B_1 - (C_1B_2L_1 + D_{12}L_1(s - A))(B_1 + L_2D_{21}) & (C_1 + D_{12}L_1)(s - A - L_2C_2)B_2 \\ C_2(s - A - B_2L_1)(B_1 + L_2D_{21}) & 0 \end{bmatrix} \\
&\quad \times \frac{1}{(s - A - L_2C_2)(s - A - B_2L_1)} + \begin{bmatrix} D_{11} & D_{12} \\ D_{21} & 0 \end{bmatrix}
\end{aligned}
$$

Therefore

$$
\begin{aligned}
&\Psi_{11}(s) \\
&= \frac{(C_1 + D_{12}L_1)(s - A - L_2C_2)B_1 - (C_1B_2L_1 + D_{12}L_1(s - A))(B_1 + L_2D_{21})}{(s - A - L_2C_2)(s - A - B_2L_1)} + D_{11} \\
&= \frac{(s - A)(C_1B_1 - D_{12}L_1L_2D_{21}) - (C_1 + D_{12}L_1)L_2C_2B_1 - C_1B_2L_1(B_1 + L_2D_{21})}{(s - A - L_2C_2)(s - A - B_2L_1)} + D_{11} \\
&= \frac{(C_1B_1 - D_{12}D_{21}L_1L_2)(s - A) - C_1B_1(L_2C_2 + B_2L_1) - (D_{12}C_2B_1 + D_{21}C_1B_2)L_1L_2}{(s - A - L_2C_2)(s - A - B_2L_1)} + D_{11}
\end{aligned}
$$

Also

$$
\begin{aligned}
&\Psi_{12}(s)\Psi_{21}(s) \\
&= \left(\frac{(C_1 + D_{12}L_1)(s - A - L_2C_2)B_2}{(s - A - L_2C_2)(s - A - B_2L_1)} + D_{12} \right) \left(\frac{C_2(s - A - B_2L_1)(B_1 + L_2D_{21})}{(s - A - L_2C_2)(s - A - B_2L_1)} + D_{21} \right) \\
&= \left(\frac{(C_1 + D_{12}L_1)B_2}{s - A - B_2L_1} + D_{12} \right) \left(\frac{C_2(B_1 + L_2D_{21})}{s - A - L_2C_2} + D_{21} \right) \\
&= \frac{(C_1 + D_{12}L_1)B_2 + D_{12}(s - A - B_2L_1)}{s - A - B_2L_1} \times \frac{C_2(B_1 + L_2D_{21}) + D_{21}(s - A - L_2C_2)}{s - A - L_2C_2} \\
&= \frac{(C_1B_2 + D_{12}(s - A))(C_2B_1 + D_{21}(s - A))}{(s - A - L_2C_2)(s - A - B_2L_1)}
\end{aligned}
$$

Compare these expressions with Equation (8.12), clearly

$$G(s) = \Psi_{11}(s) + \Psi_{12}(s)Q(s)\Psi_{21}(s)$$

Q.E.D.

With Theorems 8.3 and 8.4, we can now present the H_2/H_∞ approach. We first consider H_2 control. We make the following assumption on the plant

$$F(s) = \begin{pmatrix} A & B_1 & B_2 \\ C_1 & D_{11} & D_{12} \\ C_2 & D_{21} & D_{22} \end{pmatrix} = \begin{bmatrix} F_{11}(s) & F_{12}(s) \\ F_{21}(s) & F_{22}(s) \end{bmatrix} \tag{8.13}$$

Assumption 8.1

The plant has the following properties.

1. (A, B_2) is stabilizable and (A, C_2) is detectable.
2. $D_{11} = 0, \quad D_{22} = 0, \quad D_{12}^T D_{12} > 0$, and $D_{21} D_{21}^T > 0$.
3. For all ω, $rank\left(\begin{bmatrix} A - j\omega I & B_2 \\ C_1 & D_{12} \end{bmatrix}\right) = \text{number of columns}$

$$rank\left(\begin{bmatrix} A - j\omega I & B_1 \\ C_2 & D_{21} \end{bmatrix}\right) = \text{number of rows}$$

Under these assumptions, our goal is to solve the following H_2 control problem.

H_2 Control Problem 8.1

For the plant given in Equation (8.13) satisfying Assumption 8.1, find a controller with a proper real rational transfer function $K(s)$ that internally stabilizes the controlled system and minimizes the H_2 norm $\|G\|_2$ of the transfer function $G(s)$ from w to z.

We will call such a controller optimal H_2 controller. To find the optimal H_2 controller, let us first solve the following two algebraic Riccati equations.

$$\begin{aligned} &(A - B_2(D_{12}^T D_{12})^{-1} D_{12}^T C_1)^T S_1 + S_1(A - B_2(D_{12}^T D_{12})^{-1} D_{12}^T C_1) \\ &- S_1 B_2 (D_{12}^T D_{12})^{-1} B_2^T S_1 + C_1^T (I - D_{12}(D_{12}^T D_{12})^{-1} D_{12}^T) C_1 = 0 \end{aligned} \tag{8.14}$$

$$\begin{aligned}&(A - B_1 D_{21}^{\mathrm{T}}(D_{21}^{\mathrm{T}} D_{21})^{-1} C_2) S_2 + S_2 (A - B_1 D_{21}^{\mathrm{T}}(D_{21}^{\mathrm{T}} D_{21})^{-1} C_2)^{\mathrm{T}} \\ &- S_2 C_2^{\mathrm{T}} (D_{21}^{\mathrm{T}} D_{21})^{-1} C_2 S_2 + B_1 (I - D_{21}^{\mathrm{T}} (D_{21}^{\mathrm{T}} D_{21})^{-1} D_{21}) B_1^{\mathrm{T}} = 0\end{aligned} \tag{8.15}$$

The optimal H_2 controller is given in the following theorem.

Theorem 8.5

Let the plant be given in Equation (8.13) which satisfies Assumption 8.1. Let S_1, S_2 be the solutions to the two algebraic Riccati Equations (8.14) and (8.15) respectively. The optimal H_2 controller for H_2 Control Problem 8.1 is given by

$$K(s) = -L_1 (sI - A - B_2 L_1 - L_2 C_2)^{-1} L_2$$

where

$$\begin{aligned}L_1 &= -(D_{12}^{\mathrm{T}} D_{12})^{-1} (B_2^{\mathrm{T}} S_1 + D_{12}^{\mathrm{T}} C_1) \\ L_2 &= -(S_2 C_2^{\mathrm{T}} + B_1 D_{21}^{\mathrm{T}})(D_{21}^{\mathrm{T}} D_{21})^{-1}\end{aligned}$$

Proof

Since (A, B_2) is stabilizable and (A, C_2) is detectable, the solutions to the two algebraic Riccati Equations (8.14) and (8.15) exist and the corresponding $A + B_2 L_1$ and $A + L_2 C_2$ are stable. By Theorem 8.3, all controllers that internally stabilize the controlled system can be parameterized as

$$K(s) = M_{11}(s) + M_{12}(s) Q(s) (I - M_{22}(s) Q(s))^{-1} M_{21}(s)$$

where $Q(s)$ is any proper real rational transfer function such that $Q(s) \in H_\infty$ and $M_{ij}(s)$ is given by

$$\begin{bmatrix} M_{11}(s) & M_{12}(s) \\ M_{21}(s) & M_{22}(s) \end{bmatrix} = \begin{pmatrix} A + B_2 L_1 + L_2 C_2 & -L_2 & B_2 \\ L_1 & 0 & I \\ -C_2 & I & 0 \end{pmatrix}$$

Furthermore, the transfer function of the controlled system is

$$G(s) = \Psi_{11}(s) + \Psi_{12}(s) Q(s) \Psi_{21}(s)$$

where $\Psi_{ij}(s)$ is given by

$$\begin{bmatrix} \Psi_{11}(s) & \Psi_{12}(s) \\ \Psi_{21}(s) & \Psi_{22}(s) \end{bmatrix} = \begin{pmatrix} A + B_2 L_1 & -B_2 L_1 & B_1 & B_2 \\ 0 & A + L_2 C_2 & B_1 + L_2 D_{21} & 0 \\ C_1 + D_{12} L_1 & -D_{12} L_1 & D_{11} & D_{12} \\ 0 & C_2 & D_{21} & 0 \end{pmatrix}$$

Since the controller is obtained by solving algebraic Riccati equations (8.14) and (8.15), it can be shown that Ψ_{11} and $\Psi_{12}Q\Psi_{21}$ are orthogonal. Hence

$$\|G\|_2 = \|\Psi_{11}\|_2 + \|\Psi_{12}Q\Psi_{21}\|_2$$

Clearly, to minimize $\|G\|_2$, we must select $Q(s) = 0$. Therefore

$$K(s) = M_{11}(s) = -L_1(sI - A - B_2L_1 - L_2C_2)^{-1}L_2$$

Q.E.D.

The following example illustrates the design of the optimal H_2 controller.

Example 8.8

Consider the system in Example 8.7. Let us first check if Assumption 8.1 is satisfied. Clearly

1. Since A is stable, (A, B_2) is stabilizable and (A, C_2) is detectable
2. $D_{11} = 0, \quad D_{22} = 0, \quad D_{12}^TD_{12} = 1 > 0$, and $D_{21}D_{21}^T = 4 > 0$
3. For all ω

$$rank\left(\begin{bmatrix} A - j\omega I & B_2 \\ C_1 & D_{12} \end{bmatrix}\right) = rank\left(\begin{bmatrix} -5 - j\omega & 2 & -4 & 6 \\ 0 & -3 - j\omega & 0 & 8 \\ 0 & 7 & -1 - j\omega & -5 \\ -2 & 9 & 4 & 1 \end{bmatrix}\right) = 4$$

$$rank\left(\begin{bmatrix} A - j\omega I & B_1 \\ C_2 & D_{21} \end{bmatrix}\right) = rank\left(\begin{bmatrix} -5 - j\omega & 2 & -4 & 7 \\ 0 & -3 - j\omega & 0 & -3 \\ 0 & 7 & -1 - j\omega & 1 \\ 6 & 3 & -1 & 1 \end{bmatrix}\right) = 4$$

So Assumption 8.1 is satisfied. It is not difficult to see that since

$$C_1^T(I - D_{12}(D_{12}^TD_{12})^{-1}D_{12}^T)C_1 = C_1^T(1 - 1)C_1 = 0$$

and

$$B_1(I - D_{21}^T(D_{21}^TD_{21})D_{21})B_1^T = B_1(1 - 1)B_1^T = 0$$

the solutions to the algebraic Riccati equations

$$\begin{aligned}(A - B_2(D_{12}^TD_{12})^{-1}D_{12}^TC_1)^TS_1 + S_1(A - B_2(D_{12}^TD_{12})^{-1}D_{12}^TC_1) \\ - S_1B_2(D_{12}^TD_{12})^{-1}B_2^TS_1 + C_1^T(I - D_{12}(D_{12}^TD_{12})^{-1}D_{12}^T)C_1 = 0\end{aligned}$$

and

$$\begin{aligned}&(A - B_1 D_{21}^{\mathrm{T}}(D_{21}^{\mathrm{T}}D_{21})^{-1}C_2)S_2 + S_2(A - B_1 D_{21}^{\mathrm{T}}(D_{21}^{\mathrm{T}}D_{21})^{-1}C_2)^{\mathrm{T}}\\ &- S_2 C_2^{\mathrm{T}}(D_{21}^{\mathrm{T}}D_{21})^{-1}C_2 S_2 + B_1(I - D_{21}^{\mathrm{T}}(D_{21}^{\mathrm{T}}D_{21})D_{21})B_1^{\mathrm{T}} = 0\end{aligned}$$

are

$$S_1 = 0$$
$$S_2 = 0$$

Hence

$$\begin{aligned}L_1 &= -(D_{12}^{\mathrm{T}}D_{12})^{-1}(B_2^{\mathrm{T}}S_1 + D_{12}^{\mathrm{T}}C_1) = -(D_{12}^{\mathrm{T}}D_{12})^{-1}D_{12}^{\mathrm{T}}C_1\\ &= -C_1 = \begin{bmatrix}-2 & -9 & 4\end{bmatrix}\\ L_2 &= -(S_2 C_2^{\mathrm{T}} + B_1 D_{21}^{\mathrm{T}})(D_{21}^{\mathrm{T}}D_{21})^{-1} = -B_1 D_{21}^{\mathrm{T}}(D_{21}^{\mathrm{T}}D_{21})^{-1}\\ &= -B_1/2 = \begin{bmatrix}-3.5\\ 1.5\\ -0.5\end{bmatrix}\end{aligned}$$

The optimal H_2 controller is then obtained as follows.

$$\begin{aligned}A + B_2L_1 + L_2C_2 &= \begin{bmatrix}-5 & 2 & -4\\ 0 & -3 & 0\\ 0 & 7 & -1\end{bmatrix} + \begin{bmatrix}6\\ 8\\ -5\end{bmatrix}\begin{bmatrix}-2 & -9 & 4\end{bmatrix} + \begin{bmatrix}-3.5\\ 1.5\\ -0.5\end{bmatrix}\begin{bmatrix}6 & 3 & -1\end{bmatrix}\\ &= \begin{bmatrix}-14 & -62 & -24.5\\ 25 & -70.5 & -33.5\\ -13 & 50.5 & 19.5\end{bmatrix}\end{aligned}$$

$$\begin{aligned}K(s) &= -L_1(sI - A - B_2L_1 - L_2C_2)^{-1}L_2\\ &= \begin{bmatrix}-2 & -9 & 4\end{bmatrix}\begin{bmatrix}s+14 & 62 & 24.5\\ -25 & s+70.5 & 33.5\\ 13 & -50.5 & s-19.5\end{bmatrix}^{-1}\begin{bmatrix}-3.5\\ 1.5\\ -0.5\end{bmatrix}\\ &= \frac{18.5s^2 + 125s + 334.5}{s^3 + 65s^2 + 2275s + 9665}\end{aligned}$$

$K(s)$ can be realized as

$$\begin{aligned}\dot{\eta} &= \begin{bmatrix}-14 & -62 & -24.5\\ 25 & -70.5 & -33.5\\ -13 & 50.5 & 19.5\end{bmatrix}\eta + \begin{bmatrix}-3.5\\ 1.5\\ -0.5\end{bmatrix}y\\ u &= \begin{bmatrix}-2 & -9 & 4\end{bmatrix}\eta\end{aligned}$$

MATLAB command 'h2syn' can also be used to synthesize an optimal H_2 controller as illustrated in the following example.

Example 8.9

Consider the system

$$F(s) = \begin{pmatrix} A & B_1 & B_2 \\ C_1 & D_{11} & D_{12} \\ C_2 & D_{21} & D_{22} \end{pmatrix}$$

where

$$A = \begin{bmatrix} -5 & 2 & -4 & -7 \\ 0 & -3 & 2 & -6 \\ 0 & 7 & -1 & 4 \\ -2 & 3 & 5 & 8 \end{bmatrix} \quad B_1 = \begin{bmatrix} 0 & 7 \\ 5 & -3 \\ 0 & 1 \\ -2 & 6 \end{bmatrix} \quad B_2 = \begin{bmatrix} 6 \\ 8 \\ -5 \\ 2 \end{bmatrix}$$

$$C_1 = \begin{bmatrix} -2 & 9 & 4 & -1 \\ 3 & 0 & -7 & 0 \end{bmatrix} \quad C_2 = \begin{bmatrix} 6 & 3 & -1 & 7 \end{bmatrix}$$

$$D_{11} = \begin{bmatrix} 0 & 0 \\ 0 & 0 \end{bmatrix} \quad D_{12} = \begin{bmatrix} 0 \\ 1 \end{bmatrix} \quad D_{21} = \begin{bmatrix} 1 & 0 \end{bmatrix} \quad \text{and} \quad D_{22} = 0$$

We can check that Assumption 8.1 is satisfied. Using MATLAB command 'h2syn', we obtain the following optimal H_2 controller

$$\dot{\eta} = \begin{bmatrix} -34387 & -1864 & 385 & -4576 \\ -4550 & -2478 & 516 & -6067 \\ -2661 & -1199 & 5295 & -2630 \\ 4556 & 2232 & -816 & 5137 \end{bmatrix} \eta + \begin{bmatrix} 583 \\ 773 \\ 434 \\ -756 \end{bmatrix} y$$
$$u = \begin{bmatrix} 108 & -195 & -324 & -815 \end{bmatrix} \eta$$

The above examples show how to solve the H_2 control problem. Next, we discuss an H_∞ control problem. The H_∞ control problem is similar, but more complicated. We first make the following assumption on the plant given in Equation (8.13).

Assumption 8.2

The plant given in Equation (8.13) has the following properties.

1. (A, B_1) is controllable and (A, C_1) is observable
2. (A, B_2) is stabilizable and (A, C_2) is detectable
3. $D_{11} = 0$, and $D_{22} = 0$
4. $D_{12}^T \begin{bmatrix} C_1 & D_{12} \end{bmatrix} = \begin{bmatrix} 0 & I \end{bmatrix}$
5. $\begin{bmatrix} B_1 \\ D_{21} \end{bmatrix} D_{21}^T = \begin{bmatrix} 0 \\ I \end{bmatrix}$

Actually, it is possible to relax the above assumption and still solve the H_∞ control problem. However, the solution will be more complex. Therefore, in this book, we will present the solution to the H_∞ control problem when Assumption 8.2 is satisfied. Formally, the H_∞ control problem is as follows.

H_∞ Control Problem 8.2

For the plant given in Equation (8.13) satisfying Assumption 8.2 and for a given γ, find all controllers $K(s)$, if there are any, such that: (1) $K(s)$ internally stabilizes the controlled system; and (2) the transfer function $G(s)$ from w to z has the H_∞ norm $\|G\|_\infty < \gamma$.

Note the difference between H_2 Control Problem 8.1 and H_∞ Control Problem 8.2. The H_2 control problem is to find a controller that minimizes the H_2 norm of $G(s)$. The H_∞ control problem is to find controllers that ensure that the H_∞ norm of $G(s)$ is less than a particular constant. The reason for this difference is that, as shown in Section 8.3, while it is easy to compute the H_2 norm, it is much more difficult to compute the H_∞ norm.

We will call a controller solving H_∞ Control Problem 8.2 a H_∞ controller. To find H_∞ controllers, let us first solve the following two algebraic Riccati equations.

$$A^{\mathrm{T}}S_1 + S_1A + S_1(\gamma^{-2}B_1B_1^{\mathrm{T}} - B_2B_2^{\mathrm{T}})S_1 + C_1^TC_1 = 0 \tag{8.16}$$

$$AS_1 + S_1A^{\mathrm{T}} + S_1(\gamma^{-2}C_1^{\mathrm{T}}C_1 - C_2^TC_2)S_1 + B_1B_1^{\mathrm{T}} = 0 \tag{8.17}$$

Theorem 8.6

Let the plant be given in Equation (8.13) which satisfies Assumption 8.2. Let S_1, S_2 be the solutions to the two algebraic Riccati Equations (8.16) and (8.17) respectively. There exists an H_∞ controller such that $\|G\|_\infty < \gamma$ if and only if: (1) $S_1 > 0$, $S_2 > 0$; and (2) the spectrum radius of S_1S_2, $\rho(S_1S_2) = \max|\lambda(S_1S_2)| < \gamma^2$. If these two conditions are satisfied, then the set of all H_∞ controllers is given by

$$K(s) = M_{11}(s) + M_{12}(s)Q(s)(I - M_{22}(s)Q(s))^{-1}M_{21}(s)$$

where $Q(s)$ is any proper real rational transfer function such that $Q(s) \in H_\infty$, $\|Q\|_\infty < \gamma$ and $M_{ij}(s)$ is

$$\begin{bmatrix} M_{11}(s) & M_{12}(s) \\ M_{21}(s) & M_{22}(s) \end{bmatrix} = \begin{pmatrix} A + \gamma^{-2}B_1B_1^{\mathrm{T}}S_1 + B_2L_1 + L_2C_2 & -L_2 & (I - \gamma^{-2}S_2S_1)^{-1}B_2 \\ L_1 & 0 & I \\ -C_2 & I & 0 \end{pmatrix}$$

with $L_1 = -B_2^TS_1$ and $L_2 = -(I - \gamma^{-2}S_2S_1)^{-1}S_2C_2^{\mathrm{T}}$.

The proof of Theorem 8.6 is rather tedious and hence is omitted.

Using MATLAB command 'hinfsyn', we can synthesize an 'optimal' H_∞ controller that internally stabilizes the controlled system and 'minimizes' the H_∞ norm $\|G\|_\infty$ of the transfer function $G(s)$ from w to z. This is done by using an algorithm of bisection similar to the one used in calculating the H_∞ norm. It can be used for systems not satisfying Assumption 8.2 as the assumption is only sufficient, but not necessary.

Example 8.10

Consider the system in Example 8.7. Using MATLAB command 'hinfsyn' with $\overline{\gamma} = 10000$, $\underline{\gamma} = 0$, and tolerance of 1%, we obtain the following 'optimal' H_∞ controller

$$\dot{\eta} = \begin{bmatrix} -14 & -62 & -24.5 \\ 25 & -70.5 & -33.5 \\ -13 & 50.5 & 19.5 \end{bmatrix} \eta + \begin{bmatrix} 3.7596 \\ -1.6113 \\ 0.5371 \end{bmatrix} y$$

$$u = \begin{bmatrix} 1.8619 & -8.3785 & -3.7238 \end{bmatrix} \eta$$

Let us conclude this chapter by considering a more complex example of a system with multiple inputs and multiple outputs.

Example 8.11

The system is given by

$$A = \begin{bmatrix} -2.9252 & -1.5104 & 0.4013 & -4.8025 & 1.7004 \\ 3.5064 & -3.3745 & 1.9835 & -4.0158 & -0.4383 \\ 2.1311 & 0.3975 & -4.9795 & -1.2584 & 3.3804 \\ -4.1958 & 4.2333 & 3.3660 & 4.6748 & 3.2310 \\ 3.1870 & -0.6288 & 3.8942 & 1.1959 & 0.4211 \end{bmatrix}$$

$$B_1 = \begin{bmatrix} -4.0761 & 0.4889 \\ -2.0472 & -3.4384 \\ 1.0224 & 4.2507 \\ -3.1716 & -2.9784 \\ 2.3317 & -4.9155 \end{bmatrix}, \quad B_2 = \begin{bmatrix} 1.8911 & -0.2602 \\ -0.9903 & -4.8224 \\ -1.2042 & 3.1995 \\ 4.0319 & -4.4261 \\ 4.4335 & -4.0987 \end{bmatrix}$$

$$C_1 = \begin{bmatrix} 1.9665 & -4.4242 & 2.2283 & 1.9572 & 1.2539 \\ -3.5285 & -4.3664 & -4.6629 & -2.7882 & 0.1329 \end{bmatrix}$$

$$C_1 = \begin{bmatrix} 0.7571 & 3.9268 & -2.0752 & 4.9402 & 4.7309 \\ -2.1276 & -4.0741 & 3.2190 & -4.3218 & 4.2279 \end{bmatrix}$$

$$D_{11} = \begin{bmatrix} 0 & 0 \\ 0 & 0 \end{bmatrix} \quad D_{12} = \begin{bmatrix} 1 & 0 \\ 0 & 2 \end{bmatrix} \quad D_{21} = \begin{bmatrix} 1 & 0 \\ 0 & 3 \end{bmatrix} \quad \text{and} \quad D_{22} = \begin{bmatrix} 0 & 0 \\ 0 & 0 \end{bmatrix}$$

Using MATLAB, we obtain the 'optimal' H_∞ controller as follows.

$$\dot{\eta} = \begin{bmatrix} 13192 & 1162 & 6859 & -8014 & 3137 \\ 15069 & 1283 & 7845 & -9176 & 3577 \\ -15053 & -1342 & -7808 & 9106 & -3587 \\ 35705 & 3179 & 18529 & -21621 & 8503 \\ 37162 & 3273 & 19304 & -22548 & 8812 \end{bmatrix} \eta + \begin{bmatrix} 1 & -2 \\ 4 & -4 \\ 4 & 0 \\ -5 & -1 \\ 3 & -1 \end{bmatrix} y$$

$$u = \begin{bmatrix} 4782 & 426 & 2483 & -2899 & 1134 \\ -3229 & -283 & -1676 & 1958 & -767 \end{bmatrix} \eta.$$

8.6 NOTES AND REFERENCES

The H_∞ control was first introduced by Zames [196]. Since then, many papers and book have been published, including references [15, 36, 50, 51, 54, 58–61, 67, 76, 87, 94, 153, 158, 200, 201]. Initially, the H_∞/H_2 approach is based in transfer function model. Results are obtained using transfer functions in the frequency domain. Late, it was found the H_∞/H_2 approach can be effectively presented using the state space model of systems. The state space model is what we use in this book because it is simpler to use the state space model to handle with multivariable systems with multiple inputs and multiple outputs. It also makes the presentation uniform since all other chapters use the state space model.

8.7 PROBLEMS

8.1 For the following systems, calculate their H_2 norms.

(a)

$$\dot{x} = \begin{bmatrix} 0 & 1 & 0 \\ 0 & 0 & 1 \\ -3 & -7 & -1 \end{bmatrix} x + \begin{bmatrix} 7 & 0 \\ 0 & -4 \\ 1 & 0 \end{bmatrix} u$$

$$y = \begin{bmatrix} 0 & -3 & 6 \\ 7 & -1 & 0 \end{bmatrix} x$$

(b)

$$\dot{x} = \begin{bmatrix} -2 & 0 & 2 & -3 \\ 5 & -1 & 5 & -6 \\ 0 & 0 & -7 & 3 \\ 0 & 0 & 0 & -4 \end{bmatrix} x + \begin{bmatrix} 0 \\ 4 \\ -4 \\ 2 \end{bmatrix} u$$
$$y = \begin{bmatrix} -9 & 0 & 4 & -2 \end{bmatrix} x$$

(c)

$$\dot{x} = \begin{bmatrix} 0 & 1 & 0 & 0 & 0 \\ 0 & 0 & 1 & 0 & 0 \\ -4 & -7 & -1 & 0 & 0 \\ 3 & -6 & 9 & -5 & 0 \\ 7 & 2 & -5 & 1 & -6 \end{bmatrix} x + \begin{bmatrix} 0 \\ 0 \\ 6 \\ 2 \\ -4 \end{bmatrix} u$$
$$y = \begin{bmatrix} 2 & -8 & -6 & 1 & 3 \end{bmatrix} x$$

8.2 Given a function $f(t) \in L_2[0, \infty)$, its Laplace transform is denoted by $F(s) = L[f(t)]$. Prove

$$\frac{1}{2\pi}\int_{-\infty}^{\infty} |F(j\omega)|^2 d\omega = \int_{-\infty}^{\infty} |f(t)|^2 dt$$

8.3 For systems with the following transfer functions, calculate the H_2 norms.

(a) $G(s) = \dfrac{-3s^2 + 2s - 6}{s^3 + 7s^2 + 4s + 5}$

(b) $G(s) = \dfrac{3s^3 + 3s^2 - 5s - 9}{s^4 + 3s^3 + 2s^2 + 4s + 7}$

(c) $G(s) = \dfrac{-6s^5 + 3s^4 - 4s^3 + 3s^2 - 5s - 9}{s^6 + 3s^5 + 5s^4 + 9s^3 + 2s^2 + 4s + 7}$

8.4 For the following systems, calculate their H_∞ norms.

(a)

$$\dot{x} = \begin{bmatrix} 0 & 1 & 0 \\ 0 & 0 & 1 \\ -3 & -7 & -1 \end{bmatrix} x + \begin{bmatrix} 7 & 0 \\ 0 & -4 \\ 1 & 0 \end{bmatrix} u$$
$$y = \begin{bmatrix} 0 & -3 & 6 \\ 7 & -1 & 0 \end{bmatrix} x + \begin{bmatrix} 0 & 1 \\ 2 & 0 \end{bmatrix}$$

(b)

$$\dot{x} = \begin{bmatrix} -2 & 0 & 2 & -3 \\ 5 & -1 & 5 & -6 \\ 0 & 0 & -7 & 3 \\ 0 & 0 & 0 & -4 \end{bmatrix} x + \begin{bmatrix} 0 \\ 4 \\ -4 \\ 2 \end{bmatrix} u$$

$$y = \begin{bmatrix} -9 & 0 & 4 & -2 \end{bmatrix} x$$

(c)

$$\dot{x} = \begin{bmatrix} 0 & 1 & 0 & 0 & 0 \\ 0 & 0 & 1 & 0 & 0 \\ -4 & -7 & -1 & 0 & 0 \\ 3 & -6 & 9 & -5 & 0 \\ 7 & 2 & -5 & 1 & -6 \end{bmatrix} x + \begin{bmatrix} 0 \\ 0 \\ 6 \\ 2 \\ -4 \end{bmatrix} u$$

$$y = \begin{bmatrix} 2 & -8 & -6 & 1 & 3 \end{bmatrix} x - 4$$

8.5 For the following general second-order system, calculate its H_∞ and H_2 norms.

$$G(s) = \frac{\omega_n^2}{s^2 + 2\xi\omega_n s + \omega_n^2}$$

8.6 For systems with the following transfer functions, calculate their H_2 norms.

(a) $G(s) = \dfrac{2s^3 - 3s^2 + 2s - 6}{s^3 + 7s^2 + 4s + 5}$

(b) $G(s) = \dfrac{3s^3 + 3s^2 - 5s - 9}{s^4 + 3s^3 + 2s^2 + 4s + 7}$

(c) $G(s) = \dfrac{3s^6 - 6s^5 + 3s^4 - 4s^3 + 3s^2 - 5s - 9}{s^6 + 3s^5 + 5s^4 + 9s^3 + 2s^2 + 4s + 7}$

8.7 Consider the following stable system:

$$\dot{x} = \begin{bmatrix} 0 & 1 & 0 \\ 0 & 0 & 1 \\ -6 + \Delta_1 & -9 + \Delta_2 & -3 + \Delta_3 \end{bmatrix} x + \begin{bmatrix} 0 \\ 0 \\ 1 \end{bmatrix} u$$

$$y = x$$

where Δ_1, Δ_2, and Δ_3 are uncertainties. Decompose the system as the nominal system (A, B) and the uncertainty Δ as in Figure 8.2. Use the small-gain theorem to find the bound on the uncertainty Δ that guarantees the stability of the system.

8.8 For the system

$$F(s) = \begin{pmatrix} A & B_1 & B_2 \\ C_1 & D_{11} & D_{12} \\ C_2 & D_{21} & D_{22} \end{pmatrix}$$

with

$$A = \begin{bmatrix} -5 & -8 & 9 \\ 0 & -2 & 0 \\ 0 & 5 & -1 \end{bmatrix} \qquad B_1 = \begin{bmatrix} 6 \\ -3 \\ 2 \end{bmatrix} \qquad B_2 = \begin{bmatrix} -3 \\ 8 \\ 4 \end{bmatrix}$$

$C_1 = [-2 \quad 3 \quad 7]$, $C_2 = [4 \quad -3 \quad -1]$, $D_{11} = 0$, $D_{12} = 5$, $D_{21} = 2$, and $D_{22} = 0$, find the elements of the transfer function $F(s)$. Also find all controllers that internally stabilize the controlled system.

8.9 For the system in Problem 8.8, find the optimal H_2 controller.

8.10 Simulate the controlled system obtained in Problem 8.9.

8.11 Consider the system

$$F(s) = \begin{pmatrix} A & B_1 & B_2 \\ C_1 & D_{11} & D_{12} \\ C_2 & D_{21} & D_{22} \end{pmatrix}$$

where

$$A = \begin{bmatrix} -3 & 0 & -4 & 0 \\ 4 & -5 & 2 & 6 \\ -3 & 7 & -2 & 4 \\ 0 & 3 & 5 & 7 \end{bmatrix} \qquad B_1 = \begin{bmatrix} 0 & -1 \\ -2 & 5 \\ 0 & 4 \\ 7 & -3 \end{bmatrix} \qquad B_2 = \begin{bmatrix} 6 \\ -3 \\ 7 \\ 2 \end{bmatrix}$$

$$C_1 = \begin{bmatrix} 7 & -6 & 4 & -1 \\ -3 & 0 & 5 & 0 \end{bmatrix} \qquad C_2 = [-1 \quad 3 \quad 8 \quad 7]$$

$$D_{11} = \begin{bmatrix} 0 & 0 \\ 0 & 0 \end{bmatrix} \qquad D_{12} = \begin{bmatrix} 0 \\ 1 \end{bmatrix} \qquad D_{21} = [1 \quad 0] \qquad \text{and} \qquad D_{22} = 0$$

Find the H_∞ controller.

8.12 Simulate the controlled system obtained in Problem 8.11.

9

Robust Active Damping

In this chapter, we present the first of three applications of the optimal control approach developed in Chapters 5 and 6. We will design a robust active damping control law for stability enhancement of vibration systems. Many practical systems exhibit vibration: buildings, flexible structures, vehicles, etc. How to control and reduce (damp) vibration is an important problem in terms of safety, comfort, and durability. Vibration damping can be classified into two types: passive and active. Passive damping tries to add some dampers to the system, while active damping will use external force to actively control the system to reduce the vibration. We will consider active damping.

To facilitate the discussion, we will introduce a special inner product and the associated energy norm. The control law can then be obtained by solving an LQR problem. Interestingly, the resulting control system is no longer a 'classical system' in the sense that the stiffness and damping matrices are no longer symmetric. We apply the results to active vehicle suspension systems.

9.1 INTRODUCTION

Let us first consider the following example.

Example 9.1

Consider an active vehicle suspension system shown in Figure 9.1. In the system, the mass M represents the body of the vehicle and m the unsprung part of the vehicle including the tyres and axles. The connection between the body and the unsprung part, or the suspension system, is modelled by a spring K_1 and a dashpot D. The spring K_2 acts between the axle and the road and models the stiffness of the tires. The active suspension is achieved by applying a force u between M and m.

Assuming that gravity is balanced by the resting forces of the springs, we can model the active suspension system as follows. Let x_1, x_2 be the resting positions of masses M, m. The input to the system is the force u. The forces due to the springs are linear with respect to the corresponding displacement. The force of the dashpot is linear with respect to the corresponding velocity.

The free body diagrams of two masses are shown in Figure 9.2. The minus signs before some forces reflect the fact that the actual direction of a force is opposite to the reference direction specified by the arrow. Applying Newton's second law, we obtain the following dynamic equations.

$$M\ddot{x}_1 = u - K_1(x_1 - x_2) - D(\dot{x}_1 - \dot{x}_2)$$

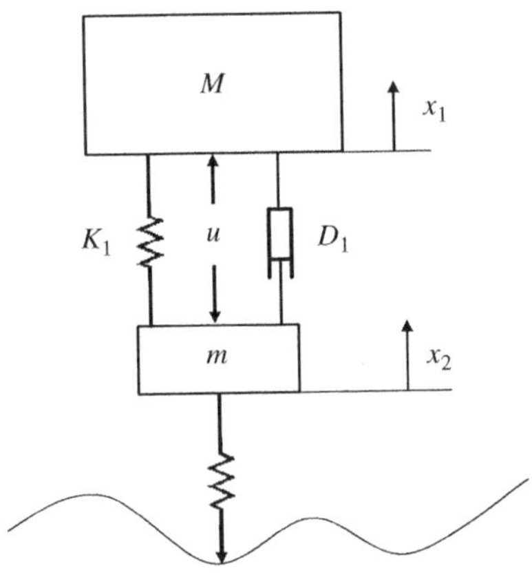

Figure 9.1 An active vehicle suspension system.

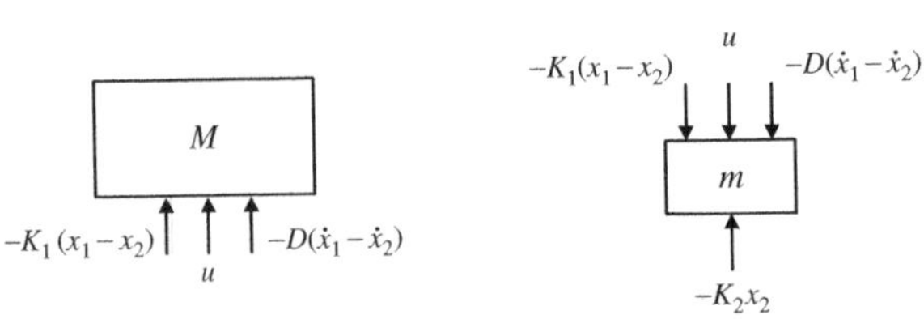

Figure 9.2 Free body diagrams of the suspension system in Figure 9.1.

$$m\ddot{x}_2 = -u + K_1(x_1 - x_2) + D(\dot{x}_1 - \dot{x}_2) - K_2 x_2$$

The uncertainty of the system appears in the dashpot coefficient D as we do not know its exact value, we assume that we know its bounds, that is, $D \in [D_{\min}, D_{\max}]$.

In summary, we can write the dynamic equation of the system as follows.

$$\begin{bmatrix} M & 0 \\ 0 & m \end{bmatrix} \begin{bmatrix} \ddot{x}_1 \\ \ddot{x}_1 \end{bmatrix} + \begin{bmatrix} K_1 & -K_1 \\ -K_1 & K_1 + K_2 \end{bmatrix} \begin{bmatrix} x_1 \\ x_1 \end{bmatrix} = \begin{bmatrix} 1 \\ -1 \end{bmatrix} u + \begin{bmatrix} 1 \\ -1 \end{bmatrix} D(\dot{x}_1 - \dot{x}_2)$$

9.2 PROBLEM FORMULATION

To formulate the robust damping problem, let us consider the following n-DOF (degree of freedom) linear vibration system with uncertainties.

$$M_o\ddot{x} + A_o x = B_o u + C_o f_o(x, \dot{x}) \tag{9.1}$$

where

x is the n-dimensional displacement vector

u is the m-dimensional control vector

M_o is the $n \times n$ mass matrix (symmetric and positive definite)

A_o is the $n \times n$ stiffness matrix (symmetric and positive definite)

B_o is a $n \times m$ matrix

C_o is a $n \times p$ matrix

$f_o(x, \dot{x})$ is the uncertainty

In the model, we have neglected internal damping for the sake of simplicity, as it can be added without changing the problem significantly.

Our goal is to enhance the stability of the system under uncertainty. In other words, we would like to design a feedback control to stabilize the system under uncertainty. This is achieved by adding more damping to the system.

In order to simplify the notation used in this chapter, we introduce the following variables and matrices.

$$y = M_o^{1/2} x$$

$$A = M_o^{-1/2} A_o M_o^{-1/2}$$

$$B = M_o^{-1/2} B_o$$
$$C = M_o^{-1/2} C_o$$
$$f(y, \dot{y}) = f_o(M_o^{-1/2} y, M_o^{-1/2} \dot{y})$$

Then equation (9.1) can be rewritten as

$$M_o^{1/2}\ddot{x} + M_o^{-1/2} A_o M_o^{-1/2} M_o^{1/2} x = M_o^{-1/2} B_o u + M_o^{-1/2} C_o f_o (M_o^{-1/2} M_o^{1/2} x, M_o^{-1/2} M_o^{1/2} \dot{x})$$

that is,

$$\ddot{y} + Ay = Bu + Cf(y, \dot{y})$$

Note that the uncertainty $f(y, \dot{y})$ depends on y and $\dot{y}$. If the uncertainty satisfies the matching condition, then $B = C$. Otherwise, we have $B \neq C$. As in Chapters 5 and 6, if the matching condition is not satisfied, then the problem becomes more complex. We will consider both cases.

We make the following assumptions.

Assumption 9.1

The uncertainty $f(y, \dot{y})$ is bounded, that is, there exists a non-negative function $g_{\max}(y, \dot{y})$ such that

$$\|f(y, \dot{y})\| \leq g_{\max}(y, \dot{y})$$

Let us now formulate our robust active damping problem. We stack the displacement and its velocity to obtain the following first-order model.

$$\frac{d}{dt}\begin{bmatrix} y \\ \dot{y} \end{bmatrix} = \begin{bmatrix} 0 & I \\ -A & o \end{bmatrix}\begin{bmatrix} y \\ \dot{y} \end{bmatrix} + \begin{bmatrix} 0 \\ B \end{bmatrix} u + \begin{bmatrix} 0 \\ C \end{bmatrix} f(y, \dot{y}) = \widetilde{A}\begin{bmatrix} y \\ \dot{y} \end{bmatrix} + \begin{bmatrix} 0 \\ B \end{bmatrix} u + \begin{bmatrix} 0 \\ C \end{bmatrix} f(y, \dot{y})$$

where

$$\widetilde{A} = \begin{bmatrix} 0 & I \\ -A & 0 \end{bmatrix}$$

In order to formulate robust active damping problem in a rigorous fashion, we introduce the following inner product $\langle .,. \rangle_E$ and the associated energy norm for matrix $A = M_o^{-1/2} A_o M_o^{-1/2}$, which is symmetric and describes the mass and stiffness in the system. Given two vectors

$$v = \begin{bmatrix} v_1 \\ v_2 \end{bmatrix} \in R^{2n}, \quad w = \begin{bmatrix} w_1 \\ w_2 \end{bmatrix} \in R^{2n}$$

we define the inner product on R^{2n} by

$$\langle v, w\rangle_E = \langle Av_1, w_1\rangle + \langle v_2, w_2\rangle$$

Hence, the corresponding energy norm is given by

$$\|v\|_E^2 = \|\sqrt{A}v_1\|^2 + \|v_2\|^2.$$

Note that the kinetic energy of a system is given by

$$\frac{1}{2}\langle \dot{y}, \dot{y}\rangle = \frac{1}{2}(M_o^{1/2}x)^T M_o^{1/2}\dot{x} = \frac{1}{2}x^T M_o \dot{x}$$

Similarly, the potential energy is given by

$$\frac{1}{2}\langle Ay, y\rangle = \frac{1}{2}(M_o^{-1/2}A_o M_o^{-1/2}M_o^{1/2}x)^T M_o^{1/2}x = \frac{1}{2}(M_o^{-1/2}A_o x)^T M_o^{1/2}x = \frac{1}{2}x^T A_o x$$

Therefore, the total energy at time t is given by

$$E(t) = \frac{1}{2}\left\| \begin{bmatrix} y(t) \\ \dot{y}(t) \end{bmatrix} \right\|_E^2$$

Under the inner product defined above, the adjoint operator $\tilde{A}^*$ of A^* is given by

$$\tilde{A}^* = \begin{bmatrix} 0 & -I \\ A & 0 \end{bmatrix}$$

This is because

$$\begin{aligned}
\left\langle \tilde{A}^* v, w \right\rangle_E &= \left\langle \begin{bmatrix} -v_2 \\ Av_1 \end{bmatrix}, \begin{bmatrix} w_1 \\ w_2 \end{bmatrix} \right\rangle_E \\
&= \langle -Av_2, w_1\rangle + \langle Av_1, w_2\rangle \\
&= \langle Av_1, w_2\rangle + \langle v_2, -Aw_1\rangle \\
&= \left\langle \begin{bmatrix} v_1 \\ v_2 \end{bmatrix}, \begin{bmatrix} w_2 \\ -Aw_1 \end{bmatrix} \right\rangle_E \\
&= \left\langle v, \tilde{A}w \right\rangle_E
\end{aligned}$$

Note that $\tilde{A}^* + \tilde{A} = 0$.

We can now state the robust active damping problem as follows.

Robust Active Damping Problem 9.1

For the following system

$$\frac{d}{dt}\begin{bmatrix} y \\ \dot{y} \end{bmatrix} = \begin{bmatrix} 0 & I \\ -A & 0 \end{bmatrix}\begin{bmatrix} y \\ \dot{y} \end{bmatrix} + \begin{bmatrix} 0 \\ B \end{bmatrix} u + \begin{bmatrix} 0 \\ C \end{bmatrix} f(y, \dot{y})$$

find a feedback control $u = u_o(y, \dot{y})$ such that the energy of the closed-loop system decays to zero, that is

$$E(t) = \frac{1}{2}\left\| \begin{bmatrix} y(t) \\ \dot{y}(t) \end{bmatrix} \right\|_E^2 \to 0 \qquad \text{as} \qquad t \to \infty$$

for all uncertainties $f(y, \dot{y})$ satisfying $\|f(y, \dot{y})\| \leq g_{\max}(y, \dot{y})$.

9.3 ROBUST ACTIVE DAMPING DESIGN

We will design the robust active damping by translating the robust control problem into an optimal control problem. In order to construct the corresponding optimal control problem, we first perform the orthogonal decomposition of the uncertainties as outlined in Chapter 5.

$$\begin{bmatrix} 0 \\ C \end{bmatrix} f(y, \dot{y}) = \begin{bmatrix} 0 \\ B \end{bmatrix}\begin{bmatrix} 0 \\ B \end{bmatrix}^+ \begin{bmatrix} 0 \\ C \end{bmatrix} f(y, \dot{y}) + \left(\begin{bmatrix} I & 0 \\ 0 & I \end{bmatrix} - \begin{bmatrix} 0 \\ B \end{bmatrix}\begin{bmatrix} 0 \\ B \end{bmatrix}^+\right)\begin{bmatrix} 0 \\ C \end{bmatrix} f(y, \dot{y})$$

It can be shown that pseudo-inverse

$$\begin{bmatrix} 0 \\ B \end{bmatrix}^+$$

satisfies the following property

$$\begin{bmatrix} 0 \\ B \end{bmatrix}^+ = [0\; B^+]$$

Therefore, the orthogonal decomposition becomes

$$\begin{bmatrix} 0 \\ C \end{bmatrix} f(y, \dot{y}) = \begin{bmatrix} 0 \\ BB^+C \end{bmatrix} f(y, \dot{y}) + \begin{bmatrix} 0 \\ (I - BB^+)C \end{bmatrix} f(y, \dot{y})$$

With the above decomposition of the uncertainties, the dynamic equation of the optimal control problem is given by

$$\frac{d}{dt}\begin{bmatrix} y \\ \dot{y} \end{bmatrix} = \begin{bmatrix} 0 & I \\ -A & o \end{bmatrix}\begin{bmatrix} y \\ \dot{y} \end{bmatrix} + \begin{bmatrix} 0 \\ B \end{bmatrix} u + \begin{bmatrix} 0 \\ (I - BB^+)C \end{bmatrix} v$$

Define

$$\widetilde{B} = \begin{bmatrix} 0 & 0 \\ B & (I - BB^{+})C \end{bmatrix},$$

then the dynamic equation becomes

$$\frac{\mathrm{d}}{\mathrm{d}t}\begin{bmatrix} y \\ \dot{y} \end{bmatrix} = \widetilde{A}\begin{bmatrix} y \\ \dot{y} \end{bmatrix} + \widetilde{B}\begin{bmatrix} u \\ v \end{bmatrix}$$

In order to reduce the optimal control problem to a LQR problem, we assume that the uncertainties are bounded linear in the energy norm.

$$\|f(y, \dot{y})\|^2 \le g_{\max}(y, \dot{y})^2 \le \left\langle H\begin{bmatrix} y \\ \dot{y} \end{bmatrix}, \begin{bmatrix} y \\ \dot{y} \end{bmatrix}\right\rangle_E$$

$$\left\| \begin{bmatrix} 0 \\ B \end{bmatrix}^{+} \begin{bmatrix} 0 \\ C \end{bmatrix} f(y, \dot{y}) \right\|^2 = \|B^{+}C\, f(y, \dot{y})\|^2 \le f_{\max}(y, \dot{y})^2 \le \left\langle G\begin{bmatrix} y \\ \dot{y} \end{bmatrix}, \begin{bmatrix} y \\ \dot{y} \end{bmatrix}\right\rangle_E$$

where G and H are some positive semi-definite matrices. The corresponding LQR problem is as follows.

LQR Problem 9.2

For the following system

$$\frac{\mathrm{d}}{\mathrm{d}t}\begin{bmatrix} y \\ \dot{y} \end{bmatrix} = \widetilde{A}\begin{bmatrix} y \\ \dot{y} \end{bmatrix} + \widetilde{B}\begin{bmatrix} u \\ v \end{bmatrix}$$

find a feedback control (u_o, v_o) that minimizes the cost functional

$$\int_0^{\infty} \left(\left\langle P\begin{bmatrix} y \\ \dot{y} \end{bmatrix}, \begin{bmatrix} y \\ \dot{y} \end{bmatrix}\right\rangle_E + \|u\|^2 + \rho^2 \|v\|^2 \right) \mathrm{d}t$$

where $P = \beta^2 I + G + \rho^2 H$ with design parameters β and ρ (in this chapter, we assume that $\alpha = 1$).

This LQR problem can be solved by solving the following Riccati type equation.

$$S\widetilde{A}+\widetilde{A}^{*}S+P-S\widetilde{B}R^{-1}\widetilde{B}^{\mathrm{T}}S=0 \tag{9.2}$$

where

$$R=\begin{bmatrix} I & 0 \\ 0 & \rho^2 I \end{bmatrix}$$

The control is then given by

$$\begin{bmatrix} u_o \\ v_o \end{bmatrix}=-R^{-1}\widetilde{B}^{\mathrm{T}}S\begin{bmatrix} y \\ \dot{y} \end{bmatrix} \tag{9.3}$$

The difference between Equation (9.2) and a standard algebraic Riccati equation is that we have A^* here instead of A^{T}. Note that in most computer software, the routine for solving an algebraic Riccati equation is written for standard finite dimensional inner product space in which $\langle v, w\rangle = v^{\mathrm{T}}w$, instead of $\langle v, w\rangle_{\mathrm{E}}$. In standard finite dimensional inner product space, the adjoint of a real matrix is the same as its transpose. However, with inner product $\langle v, w\rangle_{\mathrm{E}}$, the adjoint and transpose of a real matrix are different in general, that is, $\widetilde{A}^* = -\widetilde{A} \neq \widetilde{A}^{\mathrm{T}}$. Therefore, Equation (9.2) for our problem must be transformed into an equivalent 'standard version', in order to obtain a numerical solution of the LQR problem.

To this end, we introduce the following two matrices:

$$\widetilde{S}=\begin{bmatrix} A & 0 \\ 0 & I \end{bmatrix}S \qquad \widetilde{P}=\begin{bmatrix} A & 0 \\ 0 & I \end{bmatrix}P$$

We can show that $\widetilde{S}$ and $\widetilde{P}$ have the following property.

Proposition 9.1

Matrices $\widetilde{S}$ and $\widetilde{P}$ are symmetric, that is

$$\widetilde{S}^{\mathrm{T}}=\widetilde{S} \qquad \widetilde{P}^{\mathrm{T}}=\widetilde{P}$$

Proof

From the definition, P is self-adjoint with respect to inner product $\langle .,. \rangle_{\mathrm{E}}$, that is, $\langle Pv, w\rangle_{\mathrm{E}} = \langle v, Pw\rangle_{\mathrm{E}}$. Let

$$P=\begin{bmatrix} P_{11} & P_{12} \\ P_{21} & P_{22} \end{bmatrix}$$

then we have

$$\begin{aligned}\langle Pv, w\rangle_E &= \left\langle \begin{bmatrix} P_{11} & P_{12} \\ P_{21} & P_{22}\end{bmatrix}\begin{bmatrix} v_1 \\ v_2\end{bmatrix}, \begin{bmatrix} w_1 \\ w_2\end{bmatrix}\right\rangle_E \\ &= \left\langle \begin{bmatrix} P_{11}v_1 + P_{12}v_2 \\ P_{21}v_1 + P_{22}v_2\end{bmatrix}, \begin{bmatrix} w_1 \\ w_2\end{bmatrix}\right\rangle_E \\ &= \langle A(P_{11}v_1 + P_{12}v_2), w_1\rangle + \langle P_{21}v_1 + P_{22}v_2, w_2\rangle \\ &= v_1^{\mathrm{T}}P_{11}^{\mathrm{T}}Aw_1 + v_2^{\mathrm{T}}P_{12}^{\mathrm{T}}Aw_1 + v_1^{\mathrm{T}}P_{21}^{\mathrm{T}}w_2 + v_2^{\mathrm{T}}P_{22}^{\mathrm{T}}w_2\end{aligned}$$

and

$$\begin{aligned}\langle v, Pw\rangle_E &= \left\langle \begin{bmatrix} v_1 \\ v_2\end{bmatrix}, \begin{bmatrix} P_{11} & P_{12} \\ P_{21} & P_{22}\end{bmatrix}\begin{bmatrix} w_1 \\ w_2\end{bmatrix}\right\rangle_E \\ &= \left\langle \begin{bmatrix} v_1 \\ v_2\end{bmatrix}, \begin{bmatrix} P_{11}w_1 + P_{12}w_2 \\ P_{21}w_1 + P_{22}w_2\end{bmatrix}\right\rangle_E \\ &= \langle Av_1, P_{11}w_1 + P_{12}w_2\rangle + \langle v_2, P_{21}w_1 + P_{22}w_2\rangle \\ &= v_1^{\mathrm{T}}AP_{11}w_1 + v_2^{\mathrm{T}}P_{21}w_1 + v_1^{\mathrm{T}}AP_{12}w_2 + v_2^{\mathrm{T}}P_{22}w_2\end{aligned}$$

Therefore

$$P_{11}^{\mathrm{T}}A = AP_{11} \qquad P_{12}^{\mathrm{T}}A = P_{21} \qquad P_{21}^{\mathrm{T}} = AP_{12} \qquad P_{22}^{\mathrm{T}} = P_{22}$$

This implies

$$\begin{aligned}\widetilde{P}^{\mathrm{T}} &= \begin{bmatrix} P_{11} & P_{12} \\ P_{21} & P_{22}\end{bmatrix}^{\mathrm{T}}\begin{bmatrix} A & 0 \\ 0 & I\end{bmatrix} = \begin{bmatrix} P_{11}^{\mathrm{T}}A & P_{21}^{\mathrm{T}} \\ P_{12}^{\mathrm{T}}A & P_{22}^{\mathrm{T}}\end{bmatrix} = \begin{bmatrix} AP_{11} & AP_{12} \\ P_{21} & P_{22}\end{bmatrix} \\ &= \begin{bmatrix} A & 0 \\ 0 & I\end{bmatrix}\begin{bmatrix} P_{11} & P_{12} \\ P_{21} & P_{22}\end{bmatrix} = \widetilde{P}\end{aligned}$$

Similarly, we can prove that

$$\widetilde{S}^{\mathrm{T}} = \widetilde{S}$$

Q.E.D.

Using the result and left-multiplying

$$\begin{bmatrix} A & 0 \\ 0 & I\end{bmatrix}$$

on both sides of Equation (9.2), we obtain

$$\begin{bmatrix} A & 0 \\ 0 & I\end{bmatrix}S\widetilde{A} + \begin{bmatrix} A & 0 \\ 0 & I\end{bmatrix}\widetilde{A}^{*}S + \begin{bmatrix} A & 0 \\ 0 & I\end{bmatrix}P - \begin{bmatrix} A & 0 \\ 0 & I\end{bmatrix}S\widetilde{B}R^{-1}\widetilde{B}^{\mathrm{T}}S = 0$$

Since

$$\begin{bmatrix} A & 0 \\ 0 & I \end{bmatrix} \widetilde{A}^* = \begin{bmatrix} A & 0 \\ 0 & I \end{bmatrix} \begin{bmatrix} 0 & -I \\ A & 0 \end{bmatrix} = \begin{bmatrix} 0 & -A \\ A & 0 \end{bmatrix} = \begin{bmatrix} 0 & -A \\ I & 0 \end{bmatrix} \begin{bmatrix} A & 0 \\ 0 & I \end{bmatrix}$$

$$= \widetilde{A}^{\mathrm{T}} \begin{bmatrix} A & 0 \\ 0 & I \end{bmatrix}$$

$$\widetilde{B}^{\mathrm{T}} = \left(\begin{bmatrix} A & 0 \\ 0 & I \end{bmatrix} \widetilde{B} \right)^{\mathrm{T}} = \widetilde{B}^{\mathrm{T}} \begin{bmatrix} A & 0 \\ 0 & I \end{bmatrix}$$

we have

$$\begin{bmatrix} A & 0 \\ 0 & I \end{bmatrix} S\widetilde{A} + \widetilde{A}^{\mathrm{T}} \begin{bmatrix} A & 0 \\ 0 & I \end{bmatrix} S + \begin{bmatrix} A & 0 \\ 0 & I \end{bmatrix} P - \begin{bmatrix} A & 0 \\ 0 & I \end{bmatrix} S\widetilde{B}R^{-1}\widetilde{B}^{\mathrm{T}} \begin{bmatrix} A & 0 \\ 0 & I \end{bmatrix} S = 0$$

or

$$\widetilde{S}\widetilde{A} + \widetilde{A}^{\mathrm{T}}\widetilde{S} + \widetilde{P} - \widetilde{S}\widetilde{B}R^{-1}\widetilde{B}^{\mathrm{T}}\widetilde{S} = 0 \tag{9.4}$$

This is a standard algebraic Riccati equation that can be solved using MATLAB, for example. Let the solution be

$$\widetilde{S} = \begin{bmatrix} \widetilde{S}_{11} & \widetilde{S}_{12} \\ \widetilde{S}_{21} & \widetilde{S}_{22} \end{bmatrix}$$

Then the corresponding optimal control (9.3) can be written as

$$\begin{bmatrix} u_o \\ v_o \end{bmatrix} = -R^{-1}\widetilde{B}^{\mathrm{T}} S \begin{bmatrix} y \\ \dot{y} \end{bmatrix} = -R^{-1}\widetilde{B}^{\mathrm{T}} \begin{bmatrix} A & 0 \\ 0 & I \end{bmatrix} S \begin{bmatrix} y \\ \dot{y} \end{bmatrix} = -R^{-1}\widetilde{B}^{\mathrm{T}}\widetilde{S} \begin{bmatrix} y \\ \dot{y} \end{bmatrix}$$

or

$$\begin{bmatrix} u_o \\ v_o \end{bmatrix} = -\begin{bmatrix} I & 0 \\ 0 & \rho^2 I \end{bmatrix}^{-1} \begin{bmatrix} 0 & 0 \\ B & (I - BB^{+})C \end{bmatrix}^{\mathrm{T}} \begin{bmatrix} \widetilde{S}_{11} & \widetilde{S}_{12} \\ \widetilde{S}_{21} & \widetilde{S}_{22} \end{bmatrix} \begin{bmatrix} y \\ \dot{y} \end{bmatrix}$$

$$= -\begin{bmatrix} B^{\mathrm{T}} \\ \rho^{-2} C^{\mathrm{T}}(I - BB^{+}) \end{bmatrix} [\widetilde{S}_{21} \quad \widetilde{S}_{22}] \begin{bmatrix} y \\ \dot{y} \end{bmatrix}$$

The sufficient condition $\beta^2 I - 2\rho^2 L^{\mathrm{T}} L > 0$ of Theorem 5.3 then becomes

$$\beta^2 I - 2\rho^{-2} \begin{bmatrix} \widetilde{S}_{21} & \widetilde{S}_{22} \end{bmatrix}^{\mathrm{T}} (I - BB^{+})CC^{\mathrm{T}}(I - BB^{+}) \begin{bmatrix} \widetilde{S}_{21} & \widetilde{S}_{22} \end{bmatrix} > 0$$

If this condition is satisfied, then the robust control is given by

$$\begin{aligned} u_o &= -B_o^{\mathrm{T}} M_o^{-1/2} (\widetilde{S}_{21} y + \widetilde{S}_{22} \dot{y}) \\ &= -B_o^{\mathrm{T}} M_o^{-1/2} (\widetilde{S}_{21} M_o^{1/2} x + \widetilde{S}_{22} M_o^{1/2} \dot{x}) \end{aligned} \tag{9.5}$$

If the matching condition is satisfied, that is, $B_o = C_o$, then

$$\tilde{B} = \begin{bmatrix} 0 \\ B \end{bmatrix}$$

and we will take $\beta = 1$ and $\rho = 0$. In this case, the LQR problem is as follows. For the following system

$$\frac{d}{dt}\begin{bmatrix} y \\ \dot{y} \end{bmatrix} = \tilde{A}\begin{bmatrix} y \\ \dot{y} \end{bmatrix} + \tilde{B}u$$

find a feedback control $u = u_o(y, \dot{y})$ that minimizes the cost functional

$$\int_0^\infty \left(\left\langle (I+G)\begin{bmatrix} y \\ \dot{y} \end{bmatrix}, \begin{bmatrix} y \\ \dot{y} \end{bmatrix} \right\rangle_E + \|u\|^2 \right) dt$$

The solution to the above LQR problem always exists, and so is Robust Active Damping Problem 9.1

9.4 ACTIVE VEHICLE SUSPENSION SYSTEM

Let us now apply the results to the active vehicle suspension system discussed in Section 9.1. Its dynamics is given by

$$M\ddot{x}_1 = u - K_1(x_1 - x_2) - D(\dot{x}_1 - \dot{x}_2)$$
$$m\ddot{x}_2 = -u + K_1(x_1 - x_2) + D(\dot{x}_1 - \dot{x}_2) - K_2x_2$$

where

$M = 3000\,\text{kg}$

$m = 500\,\text{kg}$

$K_1 = 3000\,\text{N/m}$

$K_2 = 30000\,\text{N/m}$

$D \in [500, 1000]\ \text{N/ms}^{-1}$ is the uncertainty

In other words

$$M_o = \begin{bmatrix} 3000 & 0 \\ 0 & 500 \end{bmatrix}$$

$$A_o = \begin{bmatrix} 3000 & -3000 \\ -3000 & 33000 \end{bmatrix}$$

$$B_o = \begin{bmatrix} 1 \\ -1 \end{bmatrix}$$

$$C_o = B_o$$

$$f_o(x, \dot{x}) = D(\dot{x}_2 - \dot{x}_1)$$

Therefore

$$y = M_o^{1/2} x = \begin{bmatrix} 54.77 & 0 \\ 0 & 22.36 \end{bmatrix} x$$

$$A = M_o^{-1/2} A_o M_o^{-1/2} = \begin{bmatrix} 1 & -2.45 \\ -2.45 & 66 \end{bmatrix}$$

$$B = M_o^{-1/2} B_o = \begin{bmatrix} 0.0183 \\ -0.0447 \end{bmatrix}$$

$$C = B$$

$$f(y, \dot{y}) = f_o(M_o^{-1/2} y, M_o^{-1/2} \dot{y}) = D(0.0447\dot{y}_2 - 0.0183\dot{y}_1)$$

Since the matching condition is satisfied, we solve the LQR problem for

$$\frac{d}{dt}\begin{bmatrix} y \\ \dot{y} \end{bmatrix} = \widetilde{A}\begin{bmatrix} y \\ \dot{y} \end{bmatrix} + \widetilde{B}u$$

with

$$\widetilde{A} = \begin{bmatrix} 0 & I \\ -A & 0 \end{bmatrix} = \begin{bmatrix} 0 & 0 & 1 & 0 \\ 0 & 0 & 0 & 1 \\ -1 & 2.45 & 0 & 0 \\ 2.45 & -66 & 0 & 0 \end{bmatrix}$$

$$\widetilde{B} = \begin{bmatrix} 0 \\ B \end{bmatrix} = \begin{bmatrix} 0 \\ 0 \\ 0.0183 \\ -0.0447 \end{bmatrix}$$

The cost functional is

$$\int_0^\infty \left(\left\langle (I + G)\begin{bmatrix} y \\ \dot{y} \end{bmatrix}, \begin{bmatrix} y \\ \dot{y} \end{bmatrix} \right\rangle_E + \|u\|^2 \right) dt$$

To find $G = \begin{bmatrix} G_1 & G_3 \\ G_2 & G_4 \end{bmatrix}$, we note

$$\|B^+ C f(y, \dot{y})\|^2 = \dot{y}^T \begin{bmatrix} -0.0183 \\ 0.0447 \end{bmatrix} D^2 \begin{bmatrix} -0.0183 & 0.0447 \end{bmatrix} \dot{y}$$

$$\leq \dot{y}^T \begin{bmatrix} 333 & -816 \\ -816 & 2000 \end{bmatrix} \dot{y}$$

and

$$\left\langle G\begin{bmatrix} y \\ \dot{y} \end{bmatrix}, \begin{bmatrix} y \\ \dot{y} \end{bmatrix}\right\rangle_E = \begin{bmatrix} y^T & \dot{y}^T \end{bmatrix}\begin{bmatrix} G_1^T A & G_3^T \\ G_2^T A & G_4^T \end{bmatrix}\begin{bmatrix} y \\ \dot{y} \end{bmatrix}$$

Hence, $G_1 = G_2 = G_3 = 0$ and

$$G_4 = \begin{bmatrix} 333 & -816 \\ -816 & 2000 \end{bmatrix}$$

Consequently

$$P = I + G = \begin{bmatrix} 1 & 0 & 0 & 0 \\ 0 & 1 & 0 & 0 \\ 0 & 0 & 334 & -816 \\ 0 & 0 & -816 & 2001 \end{bmatrix}$$

$$\widetilde{P} = \begin{bmatrix} A & 0 \\ 0 & I \end{bmatrix} P = \begin{bmatrix} 1 & -2.45 & 0 & 0 \\ -2.45 & 66 & 0 & 0 \\ 0 & 0 & 334 & -817 \\ 0 & 0 & -817 & 2001 \end{bmatrix}$$

Solving the algebraic Riccati Equation (9.4)

$$\widetilde{S}\widetilde{A} + \widetilde{A}^T\widetilde{S} + \widetilde{P} - \widetilde{S}\widetilde{B}R^{-1}\widetilde{B}^T\widetilde{S} = 0$$

with $R = 1$ using MATLAB, we obtain

$$\widetilde{S} = \begin{bmatrix} 1003.51 & -2450.98 & 0.62 & 0.05 \\ -2450.98 & 66035.77 & -3.14 & 0.38 \\ 0.62 & -3.14 & 1003.67 & 0.13 \\ 0.05 & 0.38 & 0.13 & 1000.49 \end{bmatrix}$$

The optimal control is given by Equation (9.5)

$$\begin{aligned} u_o &= -B_o^T M_o^{-1/2}(\widetilde{S}_{21} M_o^{1/2} x + \widetilde{S}_{22} M_o^{1/2} \dot{x}) \\ &= -0.500x_1 + 1.665x_2 - 1003.344\dot{x}_1 + 1000.442\dot{x}_2 \end{aligned}$$

To test the robustness of the control thus obtained, we simulate the actual responses of the system for different D. We use the following initial conditions.

$$x_1(0) = 0.1$$

$$x_2(0) = -0.1$$

$$\dot{x}_1(0) = 0$$

$$\dot{x}_2(0) = 0$$

For $D = 500$, x_1, x_2 are given in Figure 9.3 and $\dot{x}_1$, $\dot{x}_2$ are given in Figure 9.4. (The slow dynamics occurs for x_1.)

For $D = 750$, x_1, x_2 are given in Figure 9.5 and $\dot{x}_1$, $\dot{x}_2$ are given in Figure 9.6.

For $D = 1000$, x_1, x_2 are given in Figure 9.7 and $\dot{x}_1$, $\dot{x}_2$ are given in Figure 9.8.

From the figures, we can see that for the same control, the system responses to different D are very similar.

In comparison, we also simulated the system without control (that is, without active damping). The resulting x_1, x_2 are shown in Figure 9.9 and $\dot{x}_1$, $\dot{x}_2$ are shown in Figure 9.10. Obviously, without active damping, the car vibrates much more.

9.5 DISCUSSION

Consider the vibration system

$$\ddot{y} + Ay = Bu + Cf(y, \dot{y})$$

If the uncertainty is from actuators only, then the matching condition will be satisfied. That is, $C = BC'$ for some C'. Therefore, the sufficient

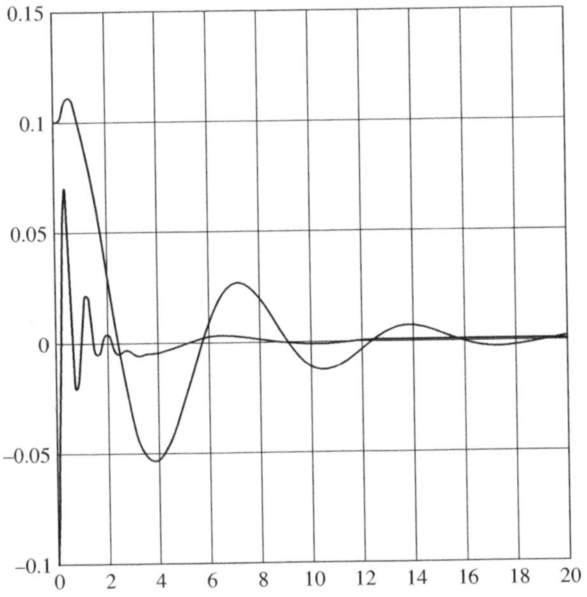

Figure 9.3 Simulation of displacements for $D = 500$.

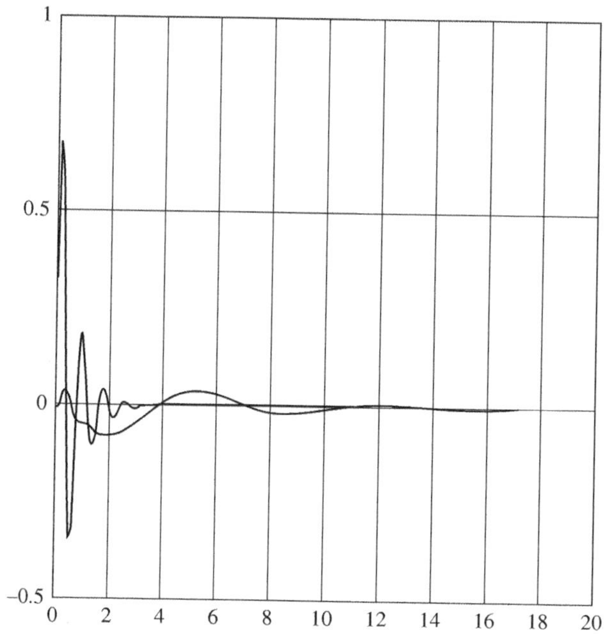

Figure 9.4 Simulation of velocities for $D = 500$.

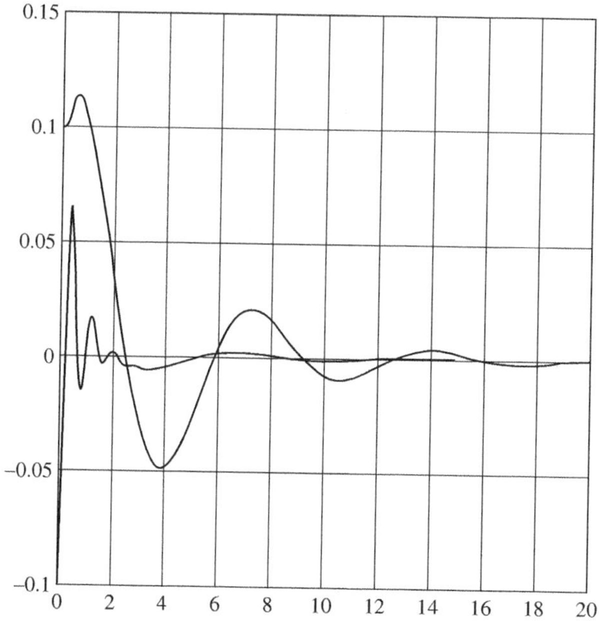

Figure 9.5 Simulation of displacements for $D = 750$.

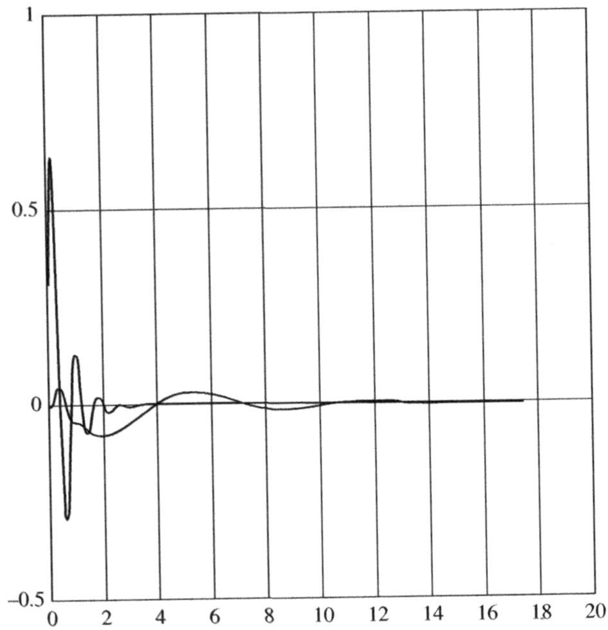

Figure 9.6 Simulation of velocities for $D = 750$.

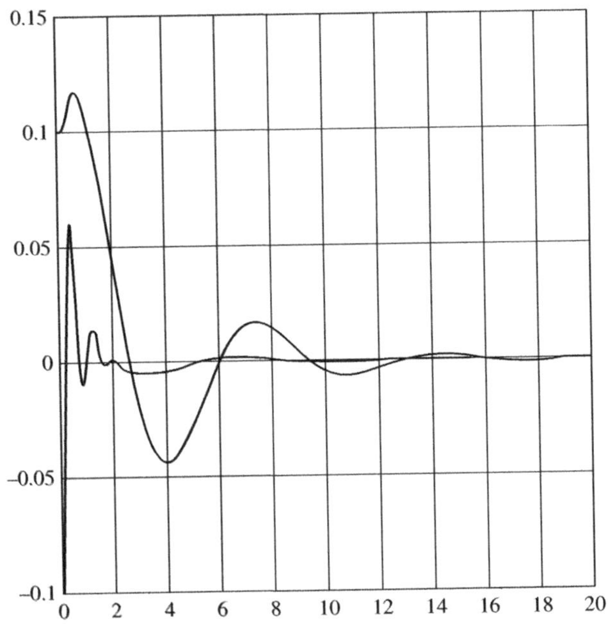

Figure 9.7 Simulation of displacements for $D = 1000$.

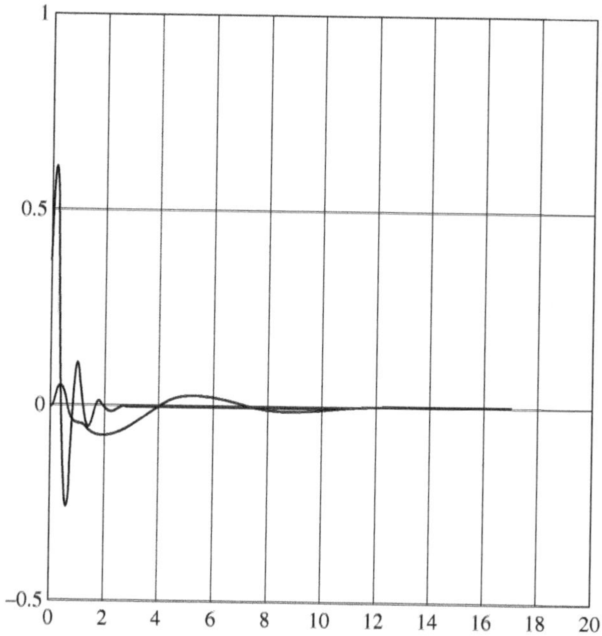

Figure 9.8 Simulation of velocities for $D = 1000$.

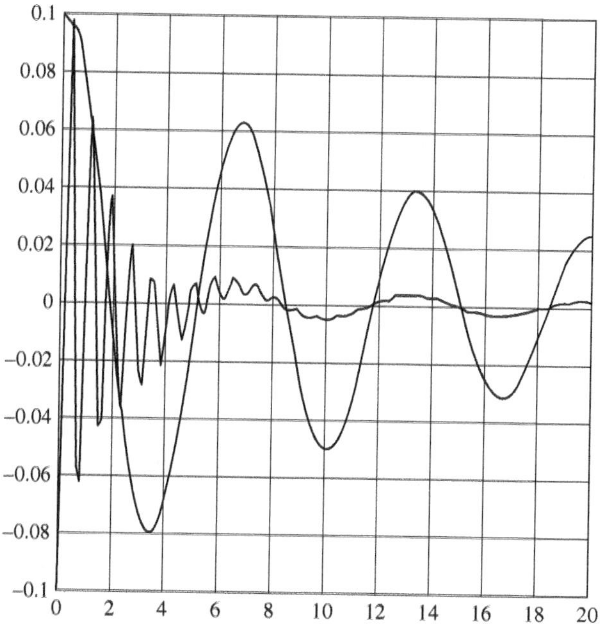

Figure 9.9 Simulation of displacements without active damping.

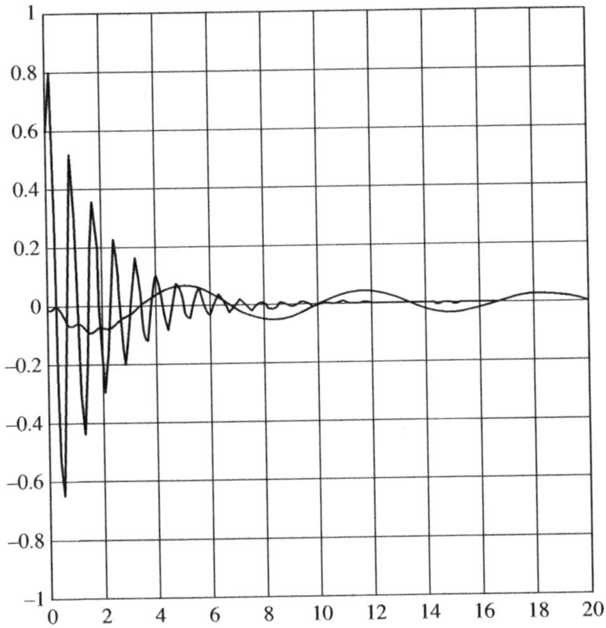

Figure 9.10 Simulation of velocities without active damping.

condition for robust control is automatically satisfied. In other words, the robust active damping control always guarantees robust stability.

In the absence of uncertainty, that is $f = 0$, it is well known that a natural choice is to use collocated actuators/sensors to construct the following rate feedback through the so-called direct connection

$$u = -kB^{\mathrm{T}}\dot{y} \tag{9.6}$$

where k stands for the gain and the measurements are $z = B^{\mathrm{T}}\dot{y}$.The resulting closed-loop system is

$$\ddot{y} + kBB^{\mathrm{T}}\dot{y} + Ay = 0 \tag{9.7}$$

Here, we point out that, the direct rate feedback (9.6) can also be derived through our approach. In fact, by letting $P = k^2\widetilde{B}\widetilde{B}^{\mathrm{T}}, k > 0$ Equation (9.2) becomes

$$S\widetilde{A} + \widetilde{A}^{*}S + k^2\widetilde{B}\widetilde{B}^{\mathrm{T}} - S\widetilde{B}R^{-1}\widetilde{B}^{\mathrm{T}}S = 0$$

It can be shown that $S = kI$ is the solution to the above problem

$$S\widetilde{A} + \widetilde{A}^{*}S + k^2\widetilde{B}\widetilde{B}^{\mathrm{T}} - S\widetilde{B}R^{-1}\widetilde{B}^{\mathrm{T}}S = k\widetilde{A} + k\widetilde{A}^{*} + k^2\widetilde{B}\widetilde{B}^{\mathrm{T}} - k^2\widetilde{B}R^{-1}\widetilde{B}^{\mathrm{T}}$$

Since $\widetilde{A}+\widetilde{A}^*=0$ and $R=I$, we have

$$S\widetilde{A}+\widetilde{A}^*S+k^2\widetilde{B}\widetilde{B}^{\mathrm{T}}-S\widetilde{B}R^{-1}\widetilde{B}^{\mathrm{T}}S=0$$

By Equation (9.3), the corresponding control is

$$\begin{bmatrix} u_o \\ v_o \end{bmatrix} = -R^{-1}\widetilde{B}^{\mathrm{T}}S\begin{bmatrix} y \\ \dot{y} \end{bmatrix} = -\begin{bmatrix} 0 & 0 \\ B & (I-BB^+)C \end{bmatrix}^T \begin{bmatrix} y \\ \dot{y} \end{bmatrix} = -\begin{bmatrix} B^{\mathrm{T}}\dot{y} \\ C^T(I-BB^+)\dot{y} \end{bmatrix}$$

which is the direct rate feedback.

In general, with the robust active damping control law given by Equation (9.5)

$$u_o=-B_o^{\mathrm{T}}M_o^{-1/2}(\widetilde{S}_{21}M_o^{1/2}x+\widetilde{S}_{22}M_o^{1/2}\dot{x})$$

the closed-loop system becomes

$$M_o\ddot{x}+A_ox+B_oB_o^{\mathrm{T}}M_o^{-1/2}(\widetilde{S}_{21}M_o^{1/2}x+\widetilde{S}_{22}M_o^{1/2}\dot{x})=C_of_o(x,\dot{x})$$

or

$$M_o\ddot{x}+\Omega_1\dot{x}+\Omega_2x=C_of_o(x,\dot{x})$$

where $\Omega_1 = B_oB_o^{\mathrm{T}}M_o^{-1/2}\widetilde{S}_{22}M_o^{1/2}$ and $\Omega_2 = A_o + B_oB_o^{\mathrm{T}}M_o^{-1/2}\widetilde{S}_{21}M_o^{1/2}$. In general, neither Ω_1 nor Ω_2 is symmetric. Thus, the closed-loop system is no longer a 'classical' vibration system. As for a classical vibration system, it is usually assumed that M_o is symmetric and positive-definite, and that Ω_1 and Ω_2 are symmetric and positive semi-definite. These assumptions have solid footing in the theory of Lagrangian dynamics. In the last several decades, classical systems have constituted a subject of intense investigation in vibration theory.

9.6 NOTES AND REFERENCES

We have discussed robust active damping of vibration systems. We first formulated the active damping problem as a robust control problem. We then introduced an inner product that corresponds to the energy stored in the system, including both the kinetic energy and potential energy. This norm allows us to solve the robust damping problem efficiently. The solution to the robust damping problem is obtained by translating it into an optimal control problem. Since the inner product used here is different from that used in Chapter 5, the method used here to solve the robust damping problem is also different. We also applied the method to an active vehicle suspension system. The simulation results show some very nice performance. Our initial work on robust active damping is published in reference [109].

There are other approaches to the stability enhancement of vibration systems. For example, in reference [11] a stochastic linear quadratic Guassian (LQG)-based approach to the design of compensators for stability enhancement applicable to flexible multibody systems with collocated rate sensors/actuators was presented. Frequency-domain approaches to compensator design for stability enhancement were also presented in references [16, 47]. Our optimal control approach provides another avenue to stability enhancement of vibration systems.

The use of modern control devices such as microprocessors can easily lead to equations for which matrices no longer have any symmetry or definiteness property. The asymmetry is sometimes addressed in the context of gyroscopic and follower forces. Linear systems governed by equations for which matrices lack any specific symmetry or definiteness will be termed nonclassical systems. These systems arise especially frequently in the emerging area of microdynamics. Nonclassical systems will be encountered more frequently in the future as microdevices need to be designed with higher precision. In reference [31] a necessary and sufficient condition under which nonclassical linear systems can be decoupled or become solvable was given. It is also the first attempt at an organized investigation of nonclassical systems. More details can be found in [31] and reference cited therein.

10

Robust Control of Manipulators

In this chapter, we apply the optimal control approach to robust control of robot manipulators. The manipulator control problem to be solved can be described as follows. Suppose that a robot manipulator is been used to move an unknown object. To control the manipulator, the following uncertainties must be dealt with: (1) the weight of the object is unknown because the object itself is not known beforehand; and (2) the friction and other parameters in the manipulator dynamics may be uncertain because it is difficult to model and measure them precisely. Our goal is to design a robust control that can handle these uncertainties.

We can formulate this robust control problem in our framework. It turns out that, for robot manipulators, the matching condition is satisfied, but there is uncertainty in the input matrix. We will derive a general robust control law using the optimal control approach and apply it to a two-joint SCARA-type robot.

10.1 ROBOT DYNAMICS

A common way to derive robot dynamics is to use Lagrange's equation of motion, which relates generalized coordinates with generalized forces via the kinetic and potential energies of a conservative system.

Robust Control Design: An Optimal Control Approach F. Lin

To present Lagrange's equation, let us first recall that the kinetic energy of a mass m moving with a linear velocity of v is given by

$$K = \frac{1}{2} m v^2$$

Similarly, an object with moment of inertia J rotating at angular velocity ω is given by

$$K = \frac{1}{2} J \omega^2$$

The potential energy of a mass m at a height h in a gravitational field is given by

$$P = mgh$$

Lagrange's equation can then be written as

$$\frac{\mathrm{d}}{\mathrm{d}t}\frac{\partial L}{\partial \dot{q}} - \frac{\partial L}{\partial q} = \tau \tag{10.1}$$

where

q is an n-dimensional vector of generalized coordinates q_i
τ is an n-dimensional vector of generalized forces τ_i
$L = K - P$

the difference between the kinetic and potential energies, is the Lagrangian.

Example 10.1

Let us consider a two-link planar revolution/prismatic (RP) robot arm shown in Figure 10.1. For simplicity, we assume that the link masses are concentrated at the centers of masses. The parameters associated with the first link are

l_1 the length of link 1
r_1 the distance from joint 1 to the centre of mass
m_1 the mass of link 1

The parameters associated with the second link are

l_2 the length of link 2
r_2 the distance from joint 2 to the centre of mass
m_2 the mass of link 2

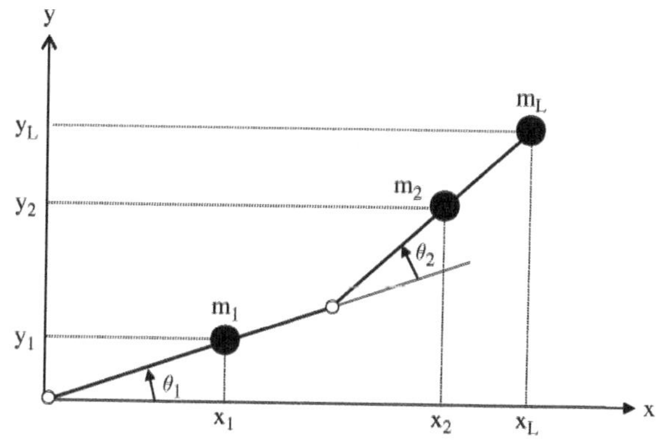

Figure 10.1 Two-link planar RP robot arm.

The robot will pick a load described by

m_L the mass of the load

To describe the dynamics of the robot, we define the following generalized coordinates and generalized forces.

θ_1 the angle of link 1 relative to the horizontal line
θ_2 the angle of link 2 relative to link 1
τ_1 the torque applied at joint 1
τ_2 the torque applied at joint 2

From Figure 10.1, it is easy to see that the x–y coordinates of masses m_1, m_2, m_L are given as follows.

$$x_1 = r_1 \cos\theta_1$$
$$x_2 = l_1 \cos\theta_1 + r_2 \cos(\theta_1 + \theta_2)$$
$$x_L = l_1 \cos\theta_1 + l_2 \cos(\theta_1 + \theta_2)$$

$$y_1 = r_1 \sin\theta_1$$
$$y_2 = l_1 \sin\theta_1 + r_2 \sin(\theta_1 + \theta_2)$$
$$y_L = l_1 \sin\theta_1 + l_2 \sin(\theta_1 + \theta_2)$$

Their derivatives are

$$\dot{x}_1 = -r_1\dot{\theta}_1 \sin\theta_1$$
$$\dot{x}_2 = -l_1\dot{\theta}_1 \sin\theta_1 - r_2(\dot{\theta}_1 + \dot{\theta}_2)\sin(\theta_1 + \theta_2)$$
$$\dot{x}_L = -l_1\dot{\theta}_1 \sin\theta_1 - l_2(\dot{\theta}_1 + \dot{\theta}_2)\sin(\theta_1 + \theta_2)$$

$$\dot{y}_1 = r_1\dot{\theta}_1 \cos\theta_1$$
$$\dot{y}_2 = l_1\dot{\theta}_1 \cos\theta_1 + r_2(\dot{\theta}_1 + \dot{\theta}_2)\cos(\theta_1 + \theta_2)$$
$$\dot{y}_L = l_1\dot{\theta}_1 \cos\theta_1 + l_2(\dot{\theta}_1 + \dot{\theta}_2)\cos(\theta_1 + \theta_2)$$

Hence, the kinetic and potential energies of link 1 are

$$K_1 = \frac{1}{2}m_1 v_1^2 = \frac{1}{2}m_1(\dot{x}_1^2 + \dot{y}_1^2) = \frac{1}{2}m_1 r_1^2 \dot{\theta}_1^2$$
$$P_1 = m_1 g h_1 = m_1 g y_1 = m_1 g r_1 \sin\theta_1$$

The kinetic and potential energies of link 2 are

$$\begin{aligned}
K_2 &= \frac{1}{2}m_2 v_2^2 = \frac{1}{2}m_2(\dot{x}_2^2 + \dot{y}_2^2) \\
&= \frac{1}{2}m_2((l_1\dot{\theta}_1 \sin\theta_1 + r_2(\dot{\theta}_1 + \dot{\theta}_2)\sin(\theta_1 + \theta_2))^2 \\
&\quad + (l_1\dot{\theta}_1 \cos\theta_1 + r_2(\dot{\theta}_1 + \dot{\theta}_2)\cos(\theta_1 + \theta_2))^2) \\
&= \frac{1}{2}m_2(l_1^2\dot{\theta}_1^2 + r_2^2(\dot{\theta}_1 + \dot{\theta}_2)^2 + 2l_1\dot{\theta}_1 r_2(\dot{\theta}_1 + \dot{\theta}_2)\cos\theta_2) \\
P_2 &= m_2 g h_2 = m_2 g y_2 = m_2 g(l_1 \sin\theta_1 + r_2 \sin(\theta_1 + \theta_2))
\end{aligned}$$

The kinetic and potential energies of load are

$$\begin{aligned}
K_L &= \frac{1}{2}m_L v_L^2 = \frac{1}{2}m_2(\dot{x}_L^2 + \dot{y}_L^2) \\
&= \frac{1}{2}m_L((l_1\dot{\theta}_1 \sin\theta_1 + l_2(\dot{\theta}_1 + \dot{\theta}_2)\sin(\theta_1 + \theta_2))^2 \\
&\quad + (l_1\dot{\theta}_1 \cos\theta_1 + l_2(\dot{\theta}_1 + \dot{\theta}_2)\cos(\theta_1 + \theta_2))^2) \\
&= \frac{1}{2}m_L(l_1^2\dot{\theta}_1^2 + l_2^2(\dot{\theta}_1 + \dot{\theta}_2)^2 + 2l_1\dot{\theta}_1 l_2(\dot{\theta}_1 + \dot{\theta}_2)\cos\theta_2) \\
P_L &= m_L g h_L = m_L g y_L = m_L g(l_1 \sin\theta_1 + l_2 \sin(\theta_1 + \theta_2))
\end{aligned}$$

Therefore, the Lagrangian for the entire arm is

$$\begin{aligned}
L &= K_1 + K_2 + K_L - P_1 - P_2 - P_L \\
&= \frac{1}{2}m_1 r_1^2\dot{\theta}_1^2 + \frac{1}{2}m_2(l_1^2\dot{\theta}_1^2 + r_2^2(\dot{\theta}_1 + \dot{\theta}_2)^2 + 2l_1\dot{\theta}_1 r_2(\dot{\theta}_1 + \dot{\theta}_2)\cos\theta_2) \\
&\quad + \frac{1}{2}m_L(l_1^2\dot{\theta}_1^2 + l_2^2(\dot{\theta}_1 + \dot{\theta}_2)^2 + 2l_1\dot{\theta}_1 l_2(\dot{\theta}_1 + \dot{\theta}_2)\cos\theta_2) \\
&\quad - m_1 g r_1 \sin\theta_1 - m_2 g(l_1 \sin\theta_1 + r_2 \sin(\theta_1 + \theta_2)) \\
&\quad - m_L g(l_1 \sin\theta_1 + l_2 \sin(\theta_1 + \theta_2))
\end{aligned}$$

To derive Lagrange's equation, let us first calculate

$$\frac{\partial L}{\partial \dot{\theta}_1} = m_1 r_1^2 \dot{\theta}_1 + m_2(l_1^2 \dot{\theta}_1 + r_2^2(\dot{\theta}_1 + \dot{\theta}_2) + l_1 r_2(2\dot{\theta}_1 + \dot{\theta}_2)\cos\theta_2)$$

$$+ m_L(l_1^2 \dot{\theta}_1 + l_2^2(\dot{\theta}_1 + \dot{\theta}_2) + l_1 l_2(2\dot{\theta}_1 + \dot{\theta}_2)\cos\theta_2)$$

$$\frac{\mathrm{d}}{\mathrm{d}t}\frac{\partial L}{\partial \dot{\theta}_1} = m_1 r_1^2 \ddot{\theta}_1$$

$$+ m_2(l_1^2 \ddot{\theta}_1 + r_2^2(\ddot{\theta}_1 + \ddot{\theta}_2) + l_1 r_2(2\ddot{\theta}_1 + \ddot{\theta}_2)\cos\theta_2 - l_1 r_2(2\dot{\theta}_1 + \dot{\theta}_2)\dot{\theta}_2 \sin\theta_2)$$

$$+ m_L(l_1^2 \ddot{\theta}_1 + l_2^2(\ddot{\theta}_1 + \ddot{\theta}_2) + l_1 l_2(2\ddot{\theta}_1 + \ddot{\theta}_2)\cos\theta_2 - l_1 l_2(2\dot{\theta}_1 + \dot{\theta}_2)\dot{\theta}_2 \sin\theta_2)$$

$$\frac{\partial L}{\partial \theta_1} = -m_1 g r_1 \cos\theta_1 - m_2 g(l_1 \cos\theta_1 + r_2\cos(\theta_1 + \theta_2))$$

$$- m_L g(l_1 \cos\theta_1 + l_2 \cos(\theta_1 + \theta_2))$$

$$\frac{\partial L}{\partial \dot{\theta}_2} = m_2(r_2^2(\dot{\theta}_1 + \dot{\theta}_2) + l_1 r_2 \dot{\theta}_1 \cos\theta_2) + m_L(l_2^2(\dot{\theta}_1 + \dot{\theta}_2) + l_1 l_2 \dot{\theta}_1 \cos\theta_2)$$

$$\frac{\mathrm{d}}{\mathrm{d}t}\frac{\partial L}{\partial \dot{\theta}_2} = m_2(r_2^2(\ddot{\theta}_1 + \ddot{\theta}_2) + l_1 r_2 \ddot{\theta}_1 \cos\theta_2 - l_1 r_2 \dot{\theta}_1 \dot{\theta}_2 \sin\theta_2)$$

$$+ m_L(l_2^2(\ddot{\theta}_1 + \ddot{\theta}_2) + l_1 l_2 \ddot{\theta}_1 \cos\theta_2 - l_1 l_2 \dot{\theta}_1 \dot{\theta}_2 \sin\theta_2)$$

$$\frac{\partial L}{\partial \theta_2} = -m_2 l_1 \dot{\theta}_1 r_2(\dot{\theta}_1 + \dot{\theta}_2)\sin\theta_2 - m_L l_1 \dot{\theta}_1 l_2(\dot{\theta}_1 + \dot{\theta}_2)\sin\theta_2$$

$$- m_2 g r_2 \cos(\theta_1 + \theta_2) - m_L g l_2 \cos(\theta_1 + \theta_2)$$

Lagrange's Equation (10.1) now becomes

$$\frac{\mathrm{d}}{\mathrm{d}t}\frac{\partial L}{\partial \dot{\theta}_1} - \frac{\partial L}{\partial \theta_1} = \tau_1 \Rightarrow$$

$$m_1 r_1^2 \ddot{\theta}_1 + m_2(l_1^2 \ddot{\theta}_1 + r_2^2(\ddot{\theta}_1 + \ddot{\theta}_2) + l_1 r_2(2\ddot{\theta}_1 + \ddot{\theta}_2)\cos\theta_2 - l_1 r_2(2\dot{\theta}_1 + \dot{\theta}_2)\dot{\theta}_2 \sin\theta_2)$$

$$+ m_L(l_1^2 \ddot{\theta}_1 + l_2^2(\ddot{\theta}_1 + \ddot{\theta}_2) + l_1 l_2(2\ddot{\theta}_1 + \ddot{\theta}_2)\cos\theta_2 - l_1 l_2(2\dot{\theta}_1 + \dot{\theta}_2)\dot{\theta}_2 \sin\theta_2)$$

$$+ m_1 g r_1 \cos\theta_1 + m_2 g(l_1\cos\theta_1 + r_2\cos(\theta_1 + \theta_2)) + m_L g(l_1 \cos\theta_1$$

$$+ l_2\cos(\theta_1 + \theta_2)) = \tau_1$$

$$\frac{\mathrm{d}}{\mathrm{d}t}\frac{\partial L}{\partial \dot{\theta}_2} - \frac{\partial L}{\partial \theta_2} = \tau_2 \Rightarrow$$

$$m_2(r_2^2(\ddot{\theta}_1 + \ddot{\theta}_2) + l_1 r_2 \ddot{\theta}_1 \cos\theta_2 - l_1 r_2 \dot{\theta}_1 \dot{\theta}_2 \sin\theta_2) + m_L(l_2^2(\ddot{\theta}_1 + \ddot{\theta}_2) + l_1 l_2 \ddot{\theta}_1 \cos\theta_2$$

$$- l_1 l_2 \dot{\theta}_1 \dot{\theta}_2 \sin\theta_2) + m_2 l_1 \dot{\theta}_1 r_2(\dot{\theta}_1 + \dot{\theta}_2)\sin\theta_2 + m_L l_1 \dot{\theta}_1 l_2(\dot{\theta}_1 + \dot{\theta}_2)\sin\theta_2$$

$$+ m_2 g r_2 \cos(\theta_1 + \theta_2) + m_L g r_2 \cos(\theta_1 + \theta_2) = \tau_2$$

Or, in matrix form

$$\begin{bmatrix} m_1r_1^2+m_2l_1^2+m_2r_2^2+2m_2l_1r_2\cos\theta_2+m_Ll_1^2+m_Ll_2^2+2m_Ll_1l_2\cos\theta_2 & m_2r_2^2+m_2l_1r_2\cos\theta_2+m_Lr_2^2+m_Ll_1l_2\cos\theta_2 \\ m_2r_2^2+m_2l_1r_2\cos\theta_2+m_Ll_2^2+m_Ll_1l_2\cos\theta_2 & m_2r_2^2+m_Ll_2^2 \end{bmatrix}\begin{bmatrix}\ddot\theta_1\\ \ddot\theta_2\end{bmatrix}$$

$$+\begin{bmatrix} -m_2l_1r_2(2\dot\theta_1+\dot\theta_2)\dot\theta_2\sin\theta_2-m_Ll_1l_2(2\dot\theta_1+\dot\theta_2)\dot\theta_2\sin\theta_2 \\ m_2l_1\dot\theta_1^2r_2\sin\theta_2+m_Ll_1\dot\theta_1^2l_2\sin\theta_2 \end{bmatrix}$$

$$+\begin{bmatrix} m_1gr_1\cos\theta_1+m_2g(l_1\cos\theta_1+r_2\cos(\theta_1+\theta_2))+m_Lg(l_1\cos\theta_1 \\ +l_2\cos(\theta_1+\theta_2))m_2gr_2\cos(\theta_1+\theta_2)+m_Lgr_2\cos(\theta_1+\theta_2) \end{bmatrix}=\begin{bmatrix}\tau_1\\ \tau_2\end{bmatrix}$$

10.2 PROBLEM FORMULATION

As illustrated in Example 10.1, using Lagrange's equation, we can derive the dynamics of robot manipulator as

$$M(q)\ddot q+V(q,\dot q)+U(\dot q)+W(q)=\tau$$

where

q is the generalized coordinate vector
τ is the generalized force vector
$M(q)$ is the inertia matrix
$V(q,\dot q)$ is the Coriolis/centripetal vector
$W(q)$ is the gravity vector
$U(\dot q)$ is the friction vector

For simplicity, we denote

$$N(q,\dot q)=V(q,\dot q)+U(\dot q)+W(q)$$

There are uncertainties in $M(q)$ and $V(q,\dot q)$ due to issues such as the unknown load to be picked and unmodelled frictions. We assume the following bounds on the uncertainties.

Assumption 10.1

There exist positive definite matrices $M_o(q)$ and $M_{\min}(q)$ such that

$$M_o(q)\ge M(q)\ge M_{\min}(q)$$

Assumption 10.2

There exist a vector $N_o(q, \dot{q})$ and a nonnegative function $n_{\max}(q, \dot{q})$ such that

$$\|N(q, \dot{q}) - N_o(q, \dot{q})\| \leq n_{\max}(q, \dot{q})$$

Our robust control problem is to design a control law to control the robot manipulator from some initial position to $(q, \dot{q}) = (0, 0)$. A more general robust tracking problem can be studied if we introduce a desired trajectory.

To derive the state equation for the robust control problem, we define the state variables in the usual manner

$$x_1 = q$$
$$x_2 = \dot{q}$$

Define the control variable as

$$u = M_o(q)^{-1}(\tau - N_o(q, \dot{q}))$$

Then the dynamics of the robot manipulator become

$$\begin{aligned}
\dot{x}_1 &= \dot{q} = x_2 \\
\dot{x}_2 &= \ddot{q} = M(q)^{-1}(\tau - N(q, \dot{q})) \\
&= M(q)^{-1}(\tau - N_o(q, \dot{q})) + M(q)^{-1}(N_o(q, \dot{q}) - N(q, \dot{q})) \\
&= M(q)^{-1}M_o(q)M_o^{-1}(q)(\tau - N_o(q, \dot{q})) + M(q)^{-1}(N_o(q, \dot{q}) - N(q, \dot{q})) \\
&= M(x_1)^{-1}M_o(x_1)u + M(x_1)^{-1}(N_o(x_1, x_2) - N(x_1, x_2))
\end{aligned}$$

Let us define

$$\begin{aligned}
h(x) &= M(x_1)^{-1}M_o(x_1) - I \\
f(x) &= M(x_1)^{-1}(N_o(x_1, x_2) - N(x_1, x_2))
\end{aligned} \tag{10.2}$$

The state equation becomes

$$\dot{x}_1 = x_2$$
$$\dot{x}_2 = (u - h(x)u) + f(x)$$

which, in matrix form, reads as follows

$$\dot{x} = Ax + B(u + h(x)u) + Bf(x)$$

where

$$x = \begin{bmatrix} x_1 \\ x_2 \end{bmatrix}$$

$$A = \begin{bmatrix} 0 & I \\ 0 & 0 \end{bmatrix}$$

$$B = \begin{bmatrix} 0 \\ I \end{bmatrix}$$

Therefore, our robust control problem can be stated as follows.

Robust Manipulator Control Problem 10.1

Find a feedback control law $u = u_o(x)$ such that the closed-loop system

$$\dot{x} = Ax + B(u_o(x) + h(x)u_o(x)) + Bf(x)$$

is globally asymptotically stable for all uncertainties $f(x)$, $h(x)$ satisfying Assumptions 10.1 and 10.2.

10.3 ROBUST CONTROL DESIGN

Since the uncertainties $h(x)$, $f(x)$ satisfy Assumptions 10.1 and 10.2, they are bounded as follows.

$$\begin{aligned} h(x) &= M(x_1)^{-1}M_o(x_1) - I \geq 0 \\ \|f(x)\| &= \|M(x_1)^{-1}(N_o(x_1, x_2) - N(x_1, x_2))\| \\ &\leq \|M(x_1)\|^{-1}\|(N_o(x_1, x_2) - N(x_1, x_2))\| \\ &\leq \|M_{min}(x_1)\|^{-1} n_{max}(x_1, x_2) \\ &= f_{max}(x) \end{aligned}$$

Using the results of Chapter 6, we can translate Robust Control Problem 10.1 into the following problem.

Optimal Control Problem 10.2

For the system

$$\dot{x} = Ax + Bu$$

find a feedback control law $u = u_o(x)$ that minimizes the following cost functional

$$\int_0^\infty (f_{max}(x)^2 + x^T x + u^T u)dt$$

Note that the matching condition holds for robot manipulators. By Theorem 4.3, if the solution to the Optimal Control Problem 10.2 exists, then it is a solution to the Robust Control Problem 10.1.

Although for the $f(x)$ given in Equation (10.2), $\|f(x)\|^2$ may not be quadratically bounded, in many cases, we can find the largest physically feasible region of x and determine a quadratic bound for $\|f(x)^2\|$. Assume such a bound is given by

$$f(x)^T f(x) \leq x^T Fx$$

for some positive definite matrix F. Then Optimal Control Problem 10.2 becomes the following LQR problem.

LQR Problem 10.3

For the system

$$\dot{x} = Ax + Bu$$

find a feedback control law $u = Kx$ that minimizes the cost functional

$$\int_0^\infty (x^T Fx + x^T x + u^T u)dt$$

The solution can be obtained by first solving the algebraic Riccati equation (note that $R = R^{-1} = I$)

$$A^T S + SA + F + I - SBB^T S = 0$$

The optimal control is then given by

$$u = -B^T Sx$$

Because of the special structure of A and B, the solution to the algebraic Riccati equation exhibits a simple form. To see this, let

$$F = \begin{bmatrix} F_1 & F_2 \\ F_2^T & F_3 \end{bmatrix}$$

$$S = \begin{bmatrix} S_1 & S_2 \\ S_2^T & S_3 \end{bmatrix}$$

Substitute A, B, F and S into the algebraic Riccati equation, we get

$$\begin{bmatrix} 0 & I \\ 0 & 0 \end{bmatrix}^{\mathrm{T}} \begin{bmatrix} S_1 & S_2 \\ S_2^T & S_3 \end{bmatrix} + \begin{bmatrix} S_1 & S_2 \\ S_2^T & S_3 \end{bmatrix} \begin{bmatrix} 0 & I \\ 0 & 0 \end{bmatrix} + \begin{bmatrix} F_1 & F_2 \\ F_2^T & F_3 \end{bmatrix} + \begin{bmatrix} I & 0 \\ 0 & I \end{bmatrix}$$
$$- \begin{bmatrix} S_1 & S_2 \\ S_2^T & S_3 \end{bmatrix} \begin{bmatrix} 0 \\ I \end{bmatrix} \begin{bmatrix} 0 \\ I \end{bmatrix}^{\mathrm{T}} \begin{bmatrix} S_1 & S_2 \\ S_2^T & S_3 \end{bmatrix} = \begin{bmatrix} 0 & 0 \\ 0 & 0 \end{bmatrix}$$

This implies

$$F_1 + I - S_2 S_2^{\mathrm{T}} = 0$$
$$S_1 + F_2 - S_2 S_3 = 0$$
$$S_2 + S_2^{\mathrm{T}} + F_3 + I - S_3^2 = 0$$

Therefore

$$S_1 = (F_1 + I)^{1/2}(2(F_1 + I)^{1/2} + F_3 + I)^{1/2} - F_2$$
$$S_2 = (F_1 + I)^{1/2}$$
$$S_3 = (2(F_1 + I)^{1/2} + F_3 + I)^{1/2}$$

The optimal control is

$$u = -B^{\mathrm{T}} S x = -\begin{bmatrix} 0 \\ I \end{bmatrix}^{\mathrm{T}} \begin{bmatrix} S_1 & S_2 \\ S_2^T & S_3 \end{bmatrix} \begin{bmatrix} x_1 \\ x_2 \end{bmatrix}$$
$$= -(F_1 + I)^{1/2} x_1 - (2(F_1 + I)^{1/2} + F_3 + I)^{1/2} x_2$$

10.4 SIMULATIONS

We now illustrate the performance of the control law by simulation of a two-joint SCARA-type robot. The dynamics of the robot is similar to the one we discussed in Section 10.2, except for the following. We no longer assume that the link masses are concentrated at the centres of masses. Hence, each link has a moment of inertia. We also assume that there are frictions in the joints. The configuration of the robot manipulator is shown in Figure 10.2.

We use the same notation as in Section 10.2 for the parameters and variables of the manipulator. However, we need to add the following notation.

J_1	the moment of inertia of link 1 with respect to its center of mass
J_2	the moment of inertia of link 2 with respect to its center of mass
$b_1\dot{\theta}_1$	the friction at joint 1
$b_2\dot{\theta}_2$	the friction at joint 2

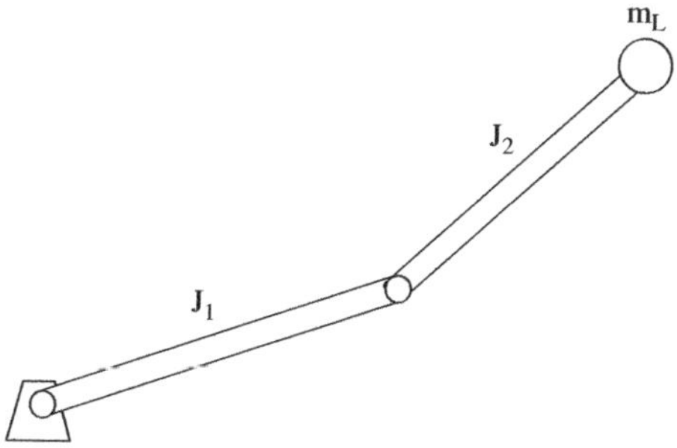

Figure 10.2 Two-joint SCARA-type robot manipulator.

The following values are used in the simulation.

$$m_1 = 13.86\,\text{oz}$$
$$m_2 = 3.33\,\text{oz}$$
$$J_1 = 62.39\,\text{oz in}^2$$
$$J_2 = 110.70\,\text{oz in}^2$$
$$l_1 = 8\,\text{in}$$
$$l_2 = 6\,\text{in}$$
$$r_1 = 4.12\,\text{in}$$
$$r_2 = 3.22\,\text{in}$$
$$b_1 = 20\,\text{oz in s}$$
$$b_2 = 50\,\text{oz in s}$$
$$m_L \in [5, 20]\,\text{oz}$$

We simulate the system with the goal of moving the manipulator from any initial position to the upward position, that is, $\theta_1 = 90°$, $\theta_2 = 0°$. For convenience, let us define

$$q = \begin{bmatrix} q_1 \\ q_2 \end{bmatrix} = \begin{bmatrix} 90° - \theta_1 \\ \theta_2 \end{bmatrix}, \tau = \begin{bmatrix} \tau_1 \\ \tau_2 \end{bmatrix}$$

then the dynamic equation of the manipulator is given by

$$\begin{bmatrix} M_{11} & M_{12} \\ M_{21} & M_{22} \end{bmatrix} \begin{bmatrix} \ddot{q}_1 \\ \ddot{q}_2 \end{bmatrix} + \begin{bmatrix} V_1 \\ V_2 \end{bmatrix} + \begin{bmatrix} U_1 \\ U_2 \end{bmatrix} + \begin{bmatrix} W_1 \\ W_2 \end{bmatrix} = \begin{bmatrix} \tau_1 \\ \tau_2 \end{bmatrix}$$

The elements in the above equation can be calculated as follows.

$$M_{11} = J_1 + J_2 + m_1 r_1^2 + m_2 l_1^2 + m_2 r_2^2 + 2m_2 l_1 r_2 \cos\theta_2 + m_L l_1^2 + m_L l_2^2 \\ + 2m_L l_1 l_2 \cos q_2$$

$$= J_1 + J_2 + m_1 r_1^2 + m_2 l_1^2 + m_2 r_2^2 + 2m_2 l_1 r_2 \cos q_2 + m_L l_1^2 + m_L l_2^2$$
$$+ 2m_L l_1 l_2 \cos q_2$$
$$M_{12} = J_2 + m_2 r_2^2 + m_2 l_1 r_2 \cos\theta_2 + m_L l_2^2 + m_L l_1 l_2 \cos\theta_2$$
$$= J_2 + m_2 r_2^2 + m_2 l_1 r_2 \cos q_2 + m_L l_2^2 + m_L l_1 l_2 \cos q_2$$
$$M_{21} = J_2 + m_2 r_2^2 + m_2 l_1 r_2 \cos\theta_2 + m_L l_2^2 + m_L l_1 l_2 \cos\theta_2$$
$$= J_2 + m_2 r_2^2 + m_2 l_1 r_2 \cos q_2 + m_L l_2^2 + m_L l_1 l_2 \cos q_2$$
$$M_{22} = J_2 + m_2 r_2^2 + m_L l_2^2$$

$$V_1 = - m_2 l_1 r_2 (2\dot\theta_1 + \dot\theta_2)\dot\theta_2 \sin\theta_2 - m_L l_1 l_2 (2\dot\theta_1 + \dot\theta_2)\dot\theta_2 \sin\theta_2$$
$$= - m_2 l_1 r_2 (-2\dot q_1 + \dot q_2)\dot q_2 \sin q_2 - m_L l_1 l_2 (-2\dot q + \dot q_2)\dot q_2 \sin q_2$$
$$= (m_2 l_1 r_2 + m_L l_1 l_2)(2\dot q_1 - \dot q_2)\dot q_2 \sin q_2$$
$$V_2 = m_2 l_1 \dot\theta_1^2 r_2 \sin\theta_2 + m_L l_1 \dot\theta_1^2 l_2 \sin\theta_2$$
$$= m_2 l_1 \dot q_1^2 r_2 \sin q_2 + m_L l_1 \dot q_1^2 l_2 \sin q_2$$
$$= (m_2 l_1 r_2 + m_L l_1 l_2)\dot q_1^2 \sin q_2$$
$$U_1 = b_1 \dot\theta_1 = -b_1 \dot q_1$$
$$U_2 = b_2 \dot\theta_2 = b_2 \dot q_2$$

$$W_1 = m_1 g r_1 \cos\theta_1 + m_2 g(l_1 \cos\theta_1 + r_2 \cos(\theta_1 + \theta_2)) + m_L g(l_1 \cos\theta_1$$
$$+ l_2 \cos(\theta_1 + \theta_2))$$
$$= m_1 g r_1 \sin q_1 + m_2 g(l_1 \sin q_1 + r_2 \sin(q_1 + q_2)) + m_L g(l_1 \sin q_1$$
$$+ l_2 \sin(q_1 + q_2))$$
$$= (m_1 g r_1 + m_2 g l_1 + m_L g l_1)\sin q_1 + (m_2 g r_2 + m_L g l_2)\sin(q_1 + q_2)$$
$$W_2 = m_2 g r_2 \cos(\theta_1 + \theta_2) + m_L g r_2 \cos(\theta_1 + \theta_2)$$
$$= m_2 g r_2 \sin(q_1 + q_2) + m_L g l_2 \sin(q_1 + q_2)$$
$$= (m_2 g r_2 + m_L g l_2)\sin(q_1 + q_2)$$

Inserting the values of the parameters, we obtain

$$M_{11} = 562.0 + 171.6 \cos q_2 + 100 m_L + 96 m_L \cos q_2$$
$$M_{12} = 51.2 + 85.8 \cos q_2 + 36 m_L + 48 m_L \cos q_2$$

$$M_{21} = 51.2 + 85.8\cos q_2 + 36m_L + 48m_L\cos q_2$$
$$M_{22} = 51.2 + 36m_L$$

$$V_1 = (85.8 + 48m_L)(2\dot{q}_1 - \dot{q}_2)\dot{q}_2 \sin q_2$$
$$V_2 = (85.8 + 48m_L)\dot{q}_1^2 \sin q_2$$
$$U_1 = -20\dot{q}_1$$
$$U_2 = 50\dot{q}_2$$
$$W_1 = (820.68 + 78.4m_L)\sin q_1 + (105.1 + 58.8m_L)\sin(q_1 + q_2)$$
$$W_2 = (105.1 + 58.8m_L)\sin(q_1 + q_2)$$

In terms of the above expressions, $M(q)$ and $N(q, \dot{q})$ can be expressed as

$$M(q) = \begin{bmatrix} M_{11} & M_{12} \\ M_{21} & M_{22} \end{bmatrix}$$
$$N(q, \dot{q}) = \begin{bmatrix} V_1 + U_1 + W_1 \\ V_2 + U_2 + W_2 \end{bmatrix}$$

We can find the bounds on the matrices as follows.

$$M_o(q) = \begin{bmatrix} 2562.0 + 2091.6\cos q_2 & 771.2 + 1045.8\cos q_2 \\ 771.2 + 1045.8\cos q_2 & 771.2 \end{bmatrix}$$
$$\geq M(q) = \begin{bmatrix} M_{11} & M_{12} \\ M_{21} & M_{22} \end{bmatrix}$$
$$\geq M_{\min}(q) = \begin{bmatrix} 562.0 + 171.6\cos q_2 & 51.2 + 85.8\cos q_2 \\ 51.2 + 85.8\cos q_2 & 51.2 \end{bmatrix}$$

Hence

$$\|M_{\min}(q)\| = \left\| \begin{bmatrix} 562.0 + 171.6\cos q_2 & 51.2 + 85.8\cos q_2 \\ 51.2 + 85.8\cos q_2 & 51.2 \end{bmatrix} \right\|$$
$$\geq \left\| \begin{bmatrix} 562.0 & 51.2 \\ 51.2 & 51.2 \end{bmatrix} \right\| = 567.1$$

Let $N_o(q, \dot{q})$ be the value of $N(q, \dot{q})$ at $m_L = 0$

$$N_o(q, \dot{q}) = \begin{bmatrix} 85.8(2\dot{q}_1 - \dot{q}_2)\dot{q}_2 \sin q_2 - 20\dot{q}_1 + 820.68\sin q_1 \\ +105.1\sin(q_1 + q_2) \\ 85.8\dot{q}_1^2 \sin q_2 + 50\dot{q}_2 + 105.1\sin(q_1 + q_2) \end{bmatrix}$$

Then

$$
\begin{aligned}
&N(q,\dot q)-N_o(q,\dot q)\\
&=\begin{bmatrix}48m_L(2\dot q_1-\dot q_2)\dot q_2\sin q_2+78.4m_L\sin q_1+58.8m_L\sin(q_1+q_2)\\ 48m_L\dot q_1^2\sin q_2+58.8m_L\sin(q_1+q_2)\end{bmatrix}
\end{aligned}
$$

We assume the speed of the rotation is limited by $\|\dot q_1\|\le 10$ and $\|\dot q_2\|\le 10$, then

$$
\begin{aligned}
&\|N(q,\dot q)-N_o(q,\dot q)\|\\
&=\left\|\begin{bmatrix}48m_L(2\dot q_1-\dot q_2)\dot q_2\sin q_2+78.4m_L\sin q_1+58.8m_L\sin(q_1+q_2)\\ 48m_L\dot q_1^2\sin q_2+58.8m_L\sin(q_1+q_2)\end{bmatrix}\right\|\\
&\le\left\|m_L\begin{bmatrix}480(2\dot q_1-\dot q_2)+78.4q_1+58.8(q_1+q_2)\\ 480\dot q_1+58.8(q_1+q_2)\end{bmatrix}\right\|\\
&\le\left\|20\begin{bmatrix}136.9 & 58.8 & 960 & -480\\ 58.8 & 58.8 & 480 & 0\end{bmatrix}\begin{bmatrix}q_1\\ q_2\\ \dot q_1\\ \dot q_2\end{bmatrix}\right\|\\
&=n_{\max}(q,\dot q)
\end{aligned}
$$

Since $\|M_{\min}(q)\|\ge 567.1$ and $\|f(q,\dot q)\|^2\le\|M_{\min}(q)\|^{-2}n_{\max}(q,\dot q)^2\le x^{\mathrm T}Fx$, we can obtain F as follows.

$$
\begin{aligned}
F&=\left(\frac{20}{567.1}\right)^2\begin{bmatrix}136.9 & 58.8 & 960 & -480\\ 58.8 & 58.8 & 480 & 0\end{bmatrix}^T\begin{bmatrix}136.9 & 58.8 & 960 & -480\\ 58.8 & 58.8 & 480 & 0\end{bmatrix}\\
&=\begin{bmatrix}27.6 & 14.3 & 198.6 & -81.7\\ 14.3 & 8.6 & 105.3 & -35.1\\ 198.6 & 105.3 & 1432.9 & -573.2\\ -81.7 & -35.1 & -573.2 & 286.6\end{bmatrix}
\end{aligned}
$$

The control is given by

$$
\begin{aligned}
u&=-(F_1+I)^{1/2}q-(2(F_1+I)^{1/2}+F_3+I)^{1/2}\dot q\\
&=-\begin{bmatrix}4.9863 & 1.9331\\ 1.9331 & 2.4214\end{bmatrix}q-\begin{bmatrix}36.1281 & -11.7310\\ -11.7310 & 12.4011\end{bmatrix}\dot q
\end{aligned}
$$

Hence

$$
\begin{aligned}
\tau&=M_o(q)u+N_o(q,\dot q)\\
&=\begin{bmatrix}2562.0+2091.6\cos q_2 & 771.2+1045.8\cos q_2\\ 771.2+1045.8\cos q_2 & 771.2\end{bmatrix}u\\
&\quad+\begin{bmatrix}85.8(2\dot q_1-\dot q_2)\dot q_2\sin q_2-20\dot q_1+820.68\sin q_1+105.1\sin(q_1+q_2)\\ 85.8\dot q_1^2\sin q_2+50\dot q_2+105.1\sin(q_1+q_2)\end{bmatrix}
\end{aligned}
$$

We will simulate the system

$$M(q)\ddot{q} + N(q, \dot{q}) = \tau$$

under the above control; that is

$$\begin{aligned}\ddot{q} &= M(q)^{-1}(\tau - N(q, \dot{q})) \\ &= \begin{bmatrix} M_{11} & M_{12} \\ M_{21} & M_{22} \end{bmatrix}^{-1} (\tau - N(q, \dot{q}))\end{aligned}$$

To test the robustness of the control thus obtained, we simulate the actual responses of the system for different m_L. In all simulations, we use the following initial conditions.

$$\begin{aligned} q_1 &= \frac{\pi}{2}(= 90^\circ) \\ q_2 &= \frac{\pi}{2}(= 90^\circ) \\ \dot{q}_1 &= 0 \\ \dot{q}_2 &= 0 \end{aligned}$$

For $m_L = 5$ oz, the angle positions and angle velocities are shown in Figures 10.3 and 10.4 respectively. For convenience, the angle positions are plotted in degrees.

For $m_L = 10$ oz, the angle positions and angle velocities are shown in Figures 10.5 and 10.6, respectively.

For $m_L = 15$ oz, the angle positions and angle velocities are shown in Figures 10.7 and 10.8, respectively.

For $m_L = 20$ oz, the angle positions and angle velocities are shown in Figures 10.9 and 10.10, respectively.

From the figures, we can see that our control is very robust for different values of the load m_L. Also note that the assumption of $\|\dot{q}_1\| \leq 10$ and $\|\dot{q}_2\| \leq 10$ is indeed satisfied.

The response (settling) times and the magnitudes of control inputs depend on the relative weights of states and control inputs in the cost function. Note that we can introduce a relative weight γ in the cost functional

$$\int_0^\infty (x^{\mathrm{T}}Fx + x^{\mathrm{T}}x + \gamma u^{\mathrm{T}}u)dt$$

without changing the robustness of the resulting control. By using a small values of γ, we will achieve fast response times (at the expense of large control inputs and large overshoots). Such performance considerations are subjects of future research.

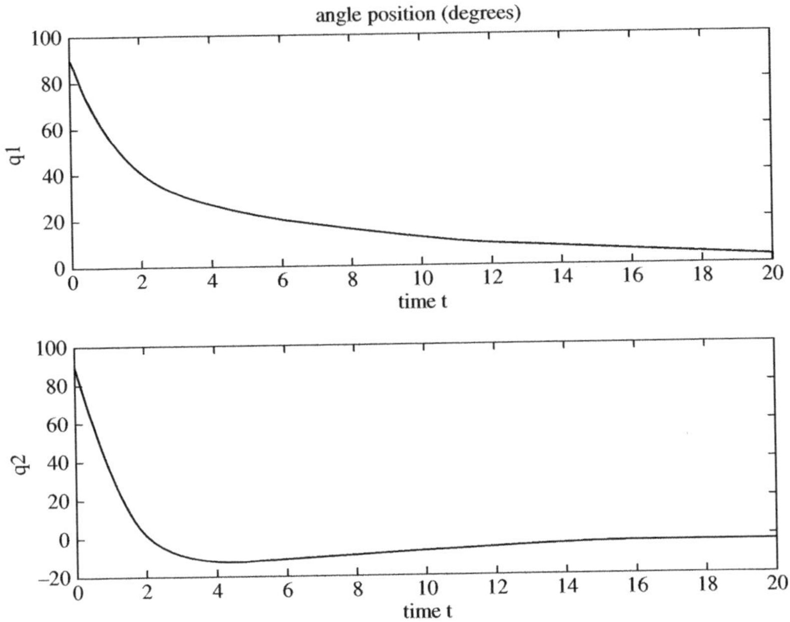

Figure 10.3 Simulation of angle positions for $m_L = 5$.

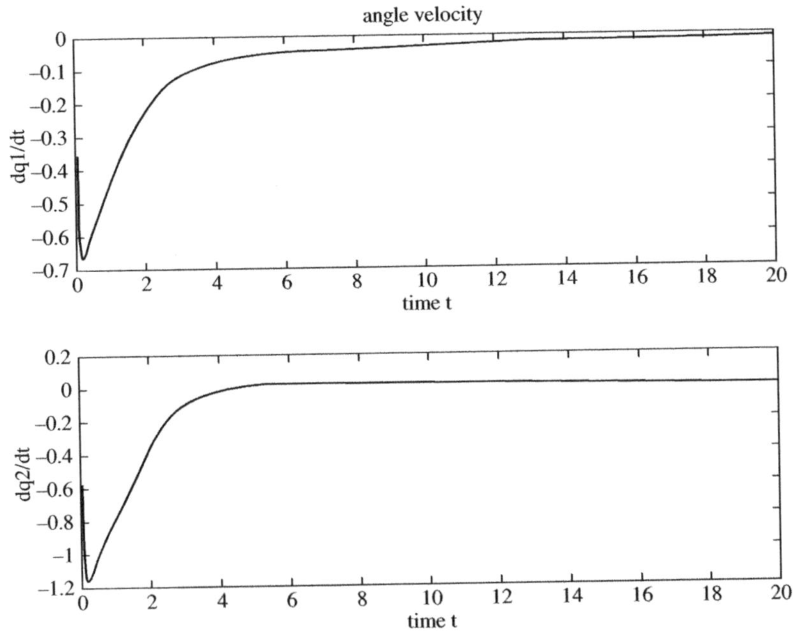

Figure 10.4 Simulation of angle velocities for $m_L = 5$.

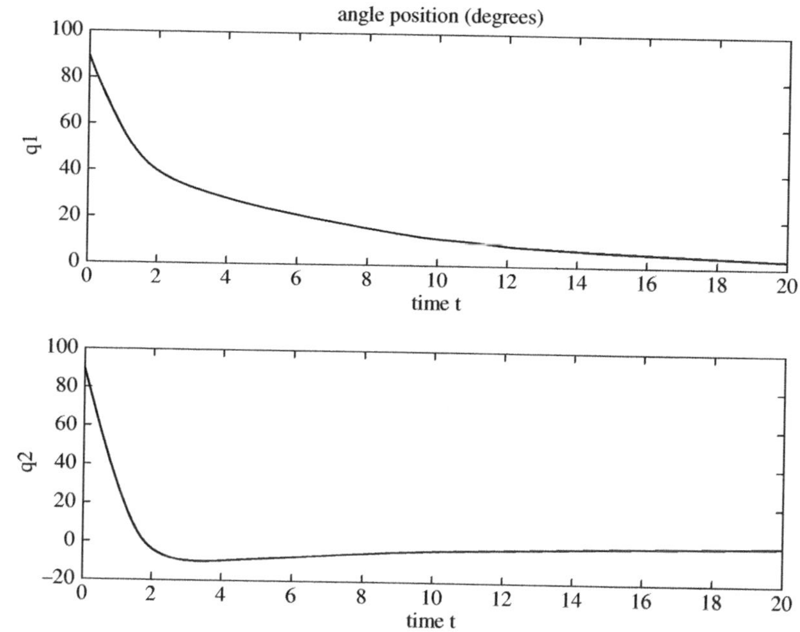

Figure 10.5 Simulation of angle positions for $m_L = 10$.

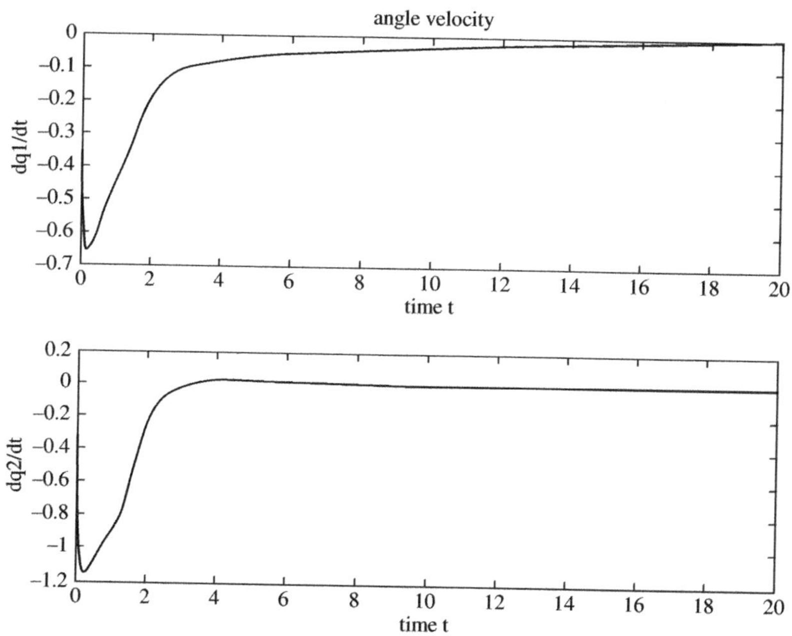

Figure 10.6 Simulation of angle velocities for $m_L = 10$.

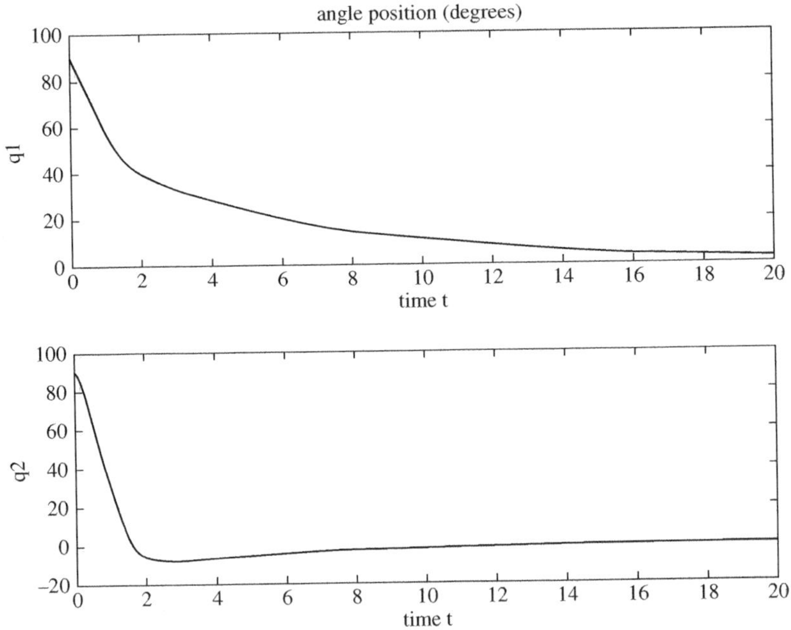

Figure 10.7 Simulation of angle positions for $m_L = 15$.

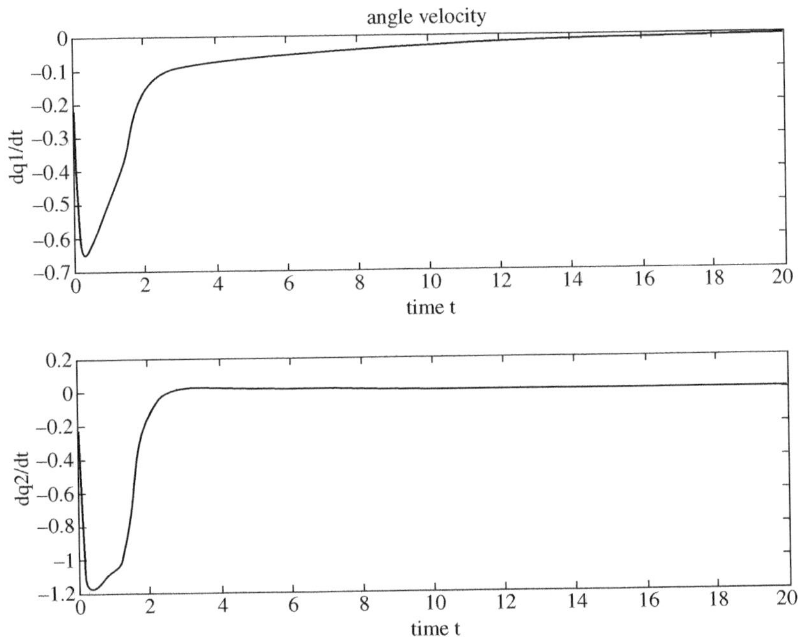

Figure 10.8 Simulation of angle velocities for $m_L = 15$.

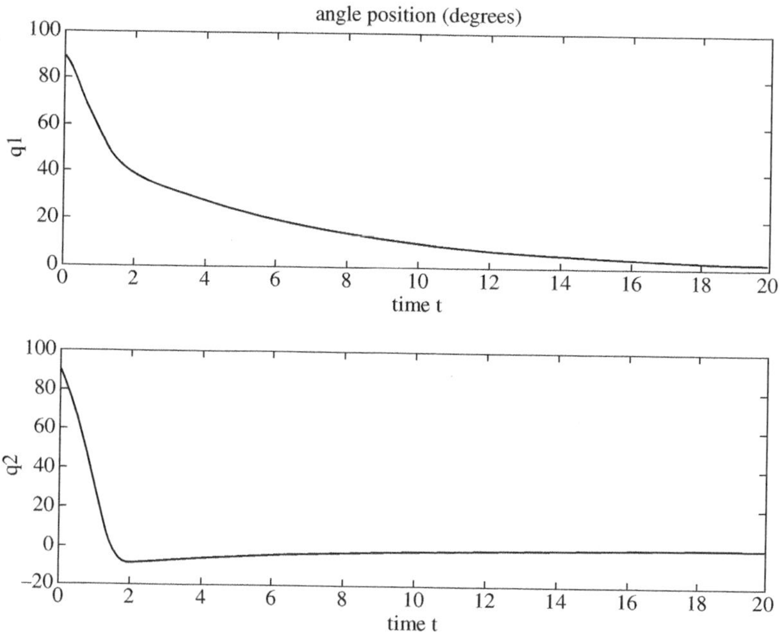

Figure 10.9 Simulation of angle positions for $m_L = 20$.

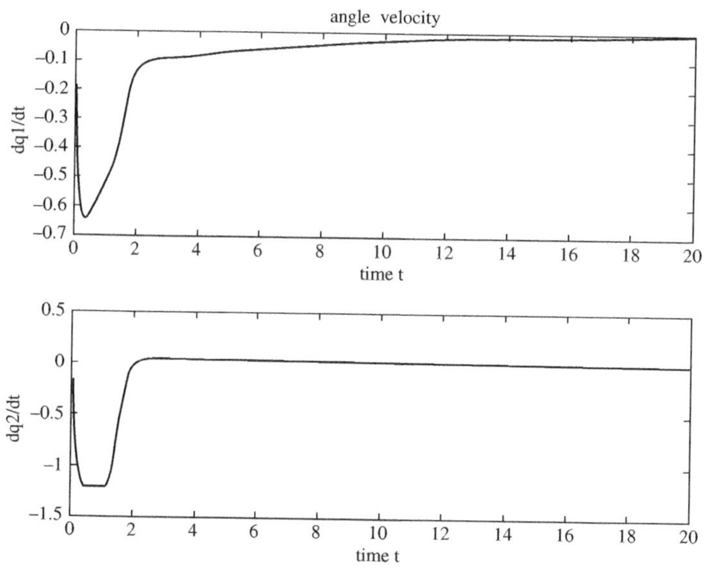

Figure 10.10 Simulation of angle velocities for $m_L = 20$.

10.5 NOTES AND REFERENCES

In this chapter, we have presented our second application of robust control design. We have considered the robust control of robot manipulators. To make this book self-contained, we derived the dynamics of robot manipulators from the basic physical laws. We then formulated the robust control problem in the framework discussed in Chapter 5. The resulting robust control problem satisfies the matching condition. However, there is uncertainty in the input matrix. We can use the method in Chapter 5 to solve the robust control problem. To test the performance of the proposed approach, we considered a two-joint SCARA-type robot, which has been used in many other approaches as well. We simulated our control law on the two-joint SCARA-type robot and the results are very nice. Our initial work on robot manipulators was published in reference [106].

Robust control of robot manipulators has been studied extensively in the literature. A survey on the subject can be found in reference [1]. Various approaches [9, 95, 152, 154] can be classified into five categories: (1) linear-multivariable approach; (2) passivity-based approach; (3) variable-structure controllers; (4) robust saturation approach; and (5) robust adaptive approach. More recently, parametric uncertainties have been dealt with [156] and the results are extended to include also nonparametric uncertainties [111]. Obviously our approach is fundamentally different from all the above approaches.

11

Aircraft Hovering Control

A vertical/short take-off and landing (V/STOL) aircraft, such as the Harrier (YAV-8B) produced by McDonnell Douglas, is a highly manoeuvrable jet aircraft. It can hover in close proximity to the ground to make lateral motion. Control of such motion is highly complex and has been the subject of many research papers.

We use the optimal control approach to design a robust control law for the lateral motion control. The resulting control law has excellent performance, as demonstrated by simulations.

11.1 MODELLING AND PROBLEM FORMULATION

The Harrier is powered by a single turbo-fan engine with four exhaust nozzles which provide the gross thrust for the aircraft. These nozzles (two on each side of the fuselage) are mechanically slaved and have to rotate together. They can move from the aft position forward approximately 100° to allow jet-borne flight and nozzle braking. Therefore, the Harrier has the following two modes of operation, in addition to the transition between the two modes.

1. Wing-borne forward flight as a fixed-wing jet aircraft. In this mode of flight, the four exhaust nozzles are in the aft position. The control is

Robust Control Design: An Optimal Control Approach F. Lin

executed by the conventional aerodynamic control surfaces: aileron, stabilator (*stabil*izer–elev*ator*), and rudder for roll, pitch, and yaw moments, respectively.

2. Jet-borne maneuvering (hovering). In this mode, the four exhaust nozzles are in the forward position, allowing the thrust to be directed vertically. In addition to the throttle and nozzle controls, the Harrier also utilizes another set of controls, using reaction control valves to provide moment generation. Reaction control valves (called puffers) in the nose, tail, and wingtips use bleed air from the high-pressure compressor of the engine to produce thrust at these points and therefore moments (and forces) at the aircraft centre of mass. Lateral motion control is accomplished through roll attitude control (rolling moment). It is this mode of flight that we concentrate on in this chapter.

Since we are interested in hovering control, we consider a prototype planar vertical take off and landing (PVTOL) aircraft. This system is the natural restriction of a V/STOL aircraft to jet-borne manoeuvre in a vertical–lateral plane.

This prototype PVTOL aircraft, as shown in Figure 11.1, has a minimum number of states and inputs, but retains many of the features that must be considered when designing control laws for a real aircraft such as the Harrier. The aircraft state is simply the positions, $\tilde{x}$, $\tilde{y}$, of the aircraft centre of mass, the roll angle, θ, of the aircraft, and the corresponding velocities, $\dot{\tilde{x}}$, $\dot{\tilde{y}}$, $\dot{\theta}$. The control inputs, U_t, U_m, are, respectively, the thrust (directed out the bottom of the aircraft) and the rolling moment about the aircraft centre of mass. Note that, we have not followed the standard variable naming conventions in aircraft dynamics. If desired, one could relabel the system by changing $\tilde{x}$, $\tilde{y}$ and θ to $-Y$, $-Z$ and ϕ, respectively.

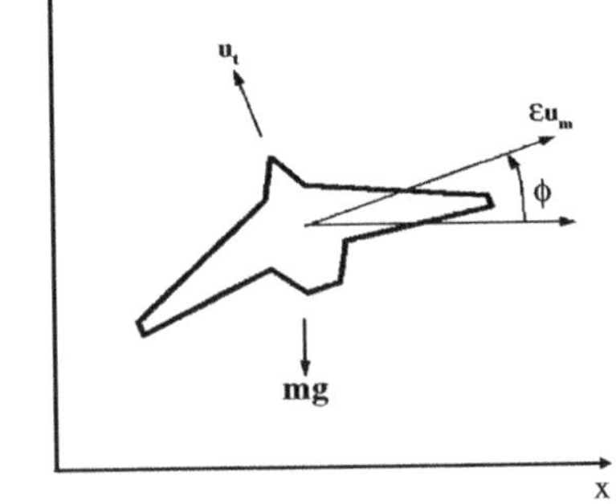

Figure 11.1 Prototype PVTOL aircraft.

In the Harrier, the roll moment reaction jets in the wingtips create a force that is not perpendicular to the $\tilde{x}$-body axis. Thus, the production of a positive rolling moment (to the pilot's left) will also produce a slight acceleration of the aircraft to the right. As we will see, this phenomenon makes the aircraft non-minimum phase. Let $\varepsilon_o > 0$ be the small coefficient describing the coupling between the rolling moment and the lateral force on the aircraft, that is, the lateral force can be written as $\varepsilon_o U_m$. Note that $\varepsilon_o > 0$ means that applying a (positive) moment to roll to the pilot's left produces an acceleration, $\varepsilon_o U_m$, to the right.

In the model of the PVTOL aircraft, we neglect any flexure effect in the aircraft wings or fuselage and consider the aircraft as a rigid body. From Figure 11.1, we can derive the following dynamic equations of the PVTOL aircraft

$$
\begin{aligned}
m\ddot{\tilde{x}} &= -U_t \sin\theta + \varepsilon_o U_m \cos\theta \\
m\ddot{\tilde{y}} &= U_t \cos\theta + \varepsilon_o U_m \sin\theta - mg \\
J\ddot{\theta} &= U_m
\end{aligned}
$$

where mg stands for the gravitational force exerted on the aircraft centre of mass and J is the mass moment of inertia about the axis through the aircraft centre of mass and along the fuselage.

For simplicity, we scale this model by dividing the first two equations by mg, and the third equation by J, to obtain

$$
\begin{aligned}
\frac{\mathrm{d}^2}{\mathrm{d}t^2}\begin{bmatrix} \dfrac{\tilde{x}}{g} \\ \dfrac{\tilde{y}}{g} \end{bmatrix} &= \begin{bmatrix} -\sin\theta & \cos\theta \\ \cos\theta & \sin\theta \end{bmatrix}\begin{bmatrix} \dfrac{U_t}{mg} \\ \dfrac{\varepsilon_o J}{mg}\dfrac{U_m}{J} \end{bmatrix} + \begin{bmatrix} 0 \\ -1 \end{bmatrix} \\
\frac{\mathrm{d}^2\theta}{\mathrm{d}t^2} &= \frac{U_m}{J}
\end{aligned}
$$

Next, let us define new variables

$$
\begin{aligned}
\begin{bmatrix} x \\ y \end{bmatrix} &= \begin{bmatrix} \dfrac{\tilde{x}}{g} \\ \dfrac{\tilde{y}}{g} \end{bmatrix} \\
\begin{bmatrix} u_t \\ u_m \end{bmatrix} &= \begin{bmatrix} \dfrac{U_t}{mg} \\ \dfrac{U_m}{J} \end{bmatrix}
\end{aligned}
$$

In addition, from now on, we replace $(\varepsilon_o J/mg)$ by ε. Then the rescaled dynamics becomes

$$\begin{bmatrix}\ddot{x}\\ \ddot{y}\end{bmatrix} = \begin{bmatrix}-\sin\theta & \varepsilon\cos\theta\\ \cos\theta & \varepsilon\sin\theta\end{bmatrix}\begin{bmatrix}u_t\\ u_m\end{bmatrix} + \begin{bmatrix}0\\ -1\end{bmatrix} \tag{11.1}$$

$$\ddot{\theta} = u_m$$

Obviously, at steady state, $\theta = 0$, $u_t = 1$, i.e., the thrust should support the aircraft weight to keep it steady.

Next, we analyse the internal stability of system (11.1) by looking at its zero dynamics. The zero dynamics of a nonlinear system are the internal dynamics of the system subject to the constraint that the outputs (and, therefore, all derivatives of the outputs) are set to zero for all time. For our PVTOL system, the outputs are the position of the aircraft centre of mass, x and y. and the internal state is the rolling angle θ and its derivative $\dot{\theta}$.

In system (11.1), the matrix operating on the controls is nonsingular (its determinant is ε). Therefore, for $\varepsilon > 0$, constraining the output x, y and their derivatives to zero results in

$$\begin{bmatrix}-\sin\theta & \varepsilon\cos\theta\\ \cos\theta & \varepsilon\sin\theta\end{bmatrix}\begin{bmatrix}u_t\\ u_m\end{bmatrix} + \begin{bmatrix}0\\ -1\end{bmatrix} = \begin{bmatrix}0\\ 0\end{bmatrix}$$

$$\Rightarrow \begin{bmatrix}u_t\\ u_m\end{bmatrix} = -\begin{bmatrix}-\sin\theta & \varepsilon\cos\theta\\ \cos\theta & \varepsilon\sin\theta\end{bmatrix}^{-1}\begin{bmatrix}0\\ -1\end{bmatrix} = \begin{bmatrix}\cos\theta\\ \sin\theta/\varepsilon\end{bmatrix}$$

Therefore, the zero dynamics of system (11.1) is given by

$$\ddot{\theta} = u_m = \frac{\sin\theta}{\varepsilon}$$

This equation describes the dynamics of an undamped pendulum. It has two sequences of equilibria. One sequence is unstable and the other is stable, but not asymptotically stable. Nonlinear systems, such as that of (11.1), with zero dynamics that are not asymptotically stable are called non-minimum phase.

Based upon this fact, it can be shown that the tracking control designed through exact input–output linearization of the PVTOL system (11.1) can produce undesirable results (periodic rolling back and forth and unacceptable control law). The source of the problem lies in trying to control modes of the system using inputs that are weakly (ε) coupled rather than controlling the system in the way it was designed to be controlled.

For the PVTOL aircraft, we should control the linear acceleration by vectoring the thrust vector (using the rolling moment, U_m, to control this vectoring) and adjusting thrust magnitude using the throttle (U_t).

Hence, we formulate the robust control problem as follows:

Robust Hovering Control Problem 11.1

For the system

$$\begin{bmatrix} \ddot{x} \\ \ddot{y} \end{bmatrix} = \begin{bmatrix} -\sin\theta & \varepsilon\cos\theta \\ \cos\theta & \varepsilon\sin\theta \end{bmatrix} \begin{bmatrix} u_t \\ u_m \end{bmatrix} + \begin{bmatrix} 0 \\ -1 \end{bmatrix}$$

$$\ddot{\theta} = u_m$$

find a feedback control law u_t and u_{m} which can accomplish the jet-borne lateral motion (hovering), say from $x = 1$ or -1 to $x = 0$. This control law has to be robust with respect to the variation of the coupling parameter ε.

From the practical point of view, any acceptable control design should satisfy the following requirements.

Requirement 11.1

The PVTOL aircraft altitude $y(t)$, in the hovering mode, should have very small deviation from the prespecified altitude, say $y = 0$. Vertical take-off and landing aircraft are designed to be maneuvered in close proximity to the ground. Therefore it is desirable to find a control law that provides exact tracking of altitude, if possible.

Requirement 11.2

$u_t > 0$, because $U_t = mgu_t$ is the thrust directed out to the bottom of the aircraft. Vectoring of the thrust is accomplished through the rolling moment U_m.

Requirement 11.3

$|\theta| << \pi/2$ or $90°$, because most V/STOL aircraft do not have a large enough 'thrust-to-weight ratio' to maintain level flight with a large roll angle θ.

Requirement 11.4

Large control inputs are not acceptable because of the limitations on the maximum thrust and rolling moment generated by bleed air from the high-pressure compressor of the engine.

Any control law which violates one of the above four requirements should be rejected. In the next section, we will seek a robust control law which satisfies the above requirements using the optimal control approach.

11.2 CONTROL DESIGN FOR JET-BORNE HOVERING

As the first step towards the robust control design for jet-borne hovering of the PVTOL aircraft, we make the following control substitution, which is obviously one-to-one

$$\begin{bmatrix} u_t \\ u_m \end{bmatrix} = \begin{bmatrix} 1 & -\varepsilon \tan\theta \\ 0 & 1 \end{bmatrix} \begin{bmatrix} 1+u_1 \\ u_2 \end{bmatrix}$$

Substituting the above into Equation (11.1), we have

$$\begin{aligned} \begin{bmatrix} \ddot{x} \\ \ddot{y} \end{bmatrix} &= \begin{bmatrix} -\sin\theta & \varepsilon\cos\theta \\ \cos\theta & \varepsilon\sin\theta \end{bmatrix} \begin{bmatrix} 1 & -\varepsilon\tan\theta \\ 0 & 1 \end{bmatrix} \begin{bmatrix} 1+u_1 \\ u_2 \end{bmatrix} + \begin{bmatrix} 0 \\ -1 \end{bmatrix} \\ &= \begin{bmatrix} -\sin\theta & \varepsilon/\cos\theta \\ \cos\theta & 0 \end{bmatrix} \begin{bmatrix} 1+u_1 \\ u_2 \end{bmatrix} + \begin{bmatrix} 0 \\ -1 \end{bmatrix} \\ &= \begin{bmatrix} -\sin\theta \\ \cos\theta - 1 \end{bmatrix} + \begin{bmatrix} -\sin\theta & 0 \\ \cos\theta & 0 \end{bmatrix} \begin{bmatrix} u_1 \\ u_2 \end{bmatrix} + \begin{bmatrix} \varepsilon/\cos\theta \\ 0 \end{bmatrix} u_2 \end{aligned} \tag{11.2}$$

and

$$\ddot{\theta} = u_2 \tag{11.3}$$

The objective of making the above control substitution is of two-fold: (1) to make the aircraft altitude $y(t)$ independent of ε and hence independent of the lateral force generated by the rolling moment u_2 (it is required that the aircraft altitude y has very small deviation from the desired altitude – through this substitution, y is no longer directly perturbed by ε); (2) to make the velocity vector $\dot{x}$, $\dot{y}$, $\dot{\theta}$, acceleration vector $\ddot{x}$, $\ddot{y}$, $\ddot{\theta}$, and the new control u_1, u_2 go to zero at steady state.

For convenience, we introduce the six-dimensional state vector

$$z = \begin{bmatrix} z_1 \\ z_2 \end{bmatrix}$$

where

$$z_1 = \begin{bmatrix} x \\ y \\ \theta \end{bmatrix} \qquad z_2 = \begin{bmatrix} \dot{x} \\ \dot{y} \\ \dot{\theta} \end{bmatrix}$$

Furthermore, we define the following matrices

$$A(\theta) = \begin{bmatrix} -\sin\theta \\ \cos\theta - 1 \\ 0 \end{bmatrix}$$

$$B(\theta) = \begin{bmatrix} -\sin\theta & 0 \\ \cos\theta & 0 \\ 0 & 1 \end{bmatrix}$$

$$C(\theta) = \begin{bmatrix} 1/\cos\theta \\ 0 \\ 0 \end{bmatrix}$$

Then Equations (11.2) and (11.3) can be written as

$$\ddot{z}_1 = A(\theta) + B(\theta)\begin{bmatrix} u_1 \\ u_2 \end{bmatrix} + C(\theta)\varepsilon u_2$$

or, equivalently

$$\begin{bmatrix} \dot{z}_1 \\ \dot{z}_2 \end{bmatrix} = \begin{bmatrix} z_2 \\ A(\theta) \end{bmatrix} + \begin{bmatrix} 0 \\ B(\theta) \end{bmatrix}\begin{bmatrix} u_1 \\ u_2 \end{bmatrix} + \begin{bmatrix} 0 \\ C(\theta) \end{bmatrix}\varepsilon u_2 \tag{11.4}$$

We can now use the results of Chapter 6 to solve Robust Control Problem 11.1. We view $f(z) = \varepsilon u_2(z)$ as uncertainty and guess $\|f(z)\| \leq k\|z\|$ for some $k > 0$ to be determined later. That is, we assume $\|f(z)\| \leq k\|z\| = g_{\max}(z)$ for some $|g_{\max}(z)$ (and we will check if this assumption is satisfied). To obtain the corresponding optimal control problem, we define

$$\widetilde{A}(z) = \begin{bmatrix} z_2 \\ A(\theta) \end{bmatrix}$$

$$\widetilde{B}(z) = \begin{bmatrix} 0 \\ B(\theta) \end{bmatrix}$$

$$\widetilde{C}(z) = \begin{bmatrix} 0 \\ C(\theta) \end{bmatrix}$$

Then, Equation (11.4) becomes

$$\dot{z} = \widetilde{A}(z) + \widetilde{B}(z)u + \widetilde{C}(z)f(z) \tag{11.5}$$

where

$$u = \begin{bmatrix} u_1 \\ u_2 \end{bmatrix}$$

Since

$$\widetilde{B}(z)^{\mathrm{T}}\widetilde{B}(z) = B(\theta)^{\mathrm{T}}B(\theta) = I_{2\times 2}$$

we immediately learn that

$$\widetilde{B}(z)^{+} = (\widetilde{B}(z)^{\mathrm{T}}\widetilde{B}(z))^{-1}\widetilde{B}(z)^{\mathrm{T}} = \widetilde{B}(z)^{\mathrm{T}}$$
$$B(\theta)^{+} = (B(\theta)^{\mathrm{T}}B(\theta))^{-1}B(\theta)^{\mathrm{T}} = B(\theta)^{\mathrm{T}}$$

Therefore,

$$\begin{aligned}(I_{6\times 6} - \widetilde{B}(z)\widetilde{B}(z)^{+})\widetilde{C}(z) &= (I_{6\times 6} - \widetilde{B}(z)\widetilde{B}(z)^{\mathrm{T}})\widetilde{C}(z)\\ &= (I_{6\times 6} - \begin{bmatrix} 0 \\ B(\theta) \end{bmatrix}[0 \quad B(\theta)^{\mathrm{T}}])\begin{bmatrix} 0 \\ C(\theta) \end{bmatrix}\\ &= \begin{bmatrix} I_{3\times 3} & 0 \\ 0 & I_{3\times 3} - B(\theta)B(\theta)^{\mathrm{T}} \end{bmatrix}\begin{bmatrix} 0 \\ C(\theta) \end{bmatrix}\\ &= \begin{bmatrix} 0 \\ (I_{3\times 3} \quad - B(\theta)B(\theta)^{\mathrm{T}})C(\theta) \end{bmatrix}\end{aligned}$$

with

$$(I_{3\times 3} - B(\theta)B(\theta)^{\mathrm{T}})C(\theta) = \begin{bmatrix} \cos\theta \\ \sin\theta \\ 0 \end{bmatrix}$$

Hence, the dynamics of the optimal control problem

$$\dot{z} = \widetilde{A}(z) + \widetilde{B}(z)u + (I_{6\times 6} - \widetilde{B}(z)\widetilde{B}(z)^{+})\widetilde{C}(z)v$$

becomes

$$\begin{aligned}\begin{bmatrix} \dot{z}_1 \\ \dot{z}_2 \end{bmatrix} &= \begin{bmatrix} z_2 \\ A(\theta) \end{bmatrix} + \begin{bmatrix} 0 \\ B(\theta) \end{bmatrix}\begin{bmatrix} u_1 \\ u_2 \end{bmatrix} + \begin{bmatrix} 0 \\ (I_{3\times 3} - B(\theta)B(\theta)^{\mathrm{T}})C(\theta) \end{bmatrix}v\\ &= \begin{bmatrix} \dot{x} \\ \dot{y} \\ \dot{\theta} \\ -\sin\theta \\ \cos\theta - 1 \\ 0 \end{bmatrix} + \begin{bmatrix} 0 & 0 & 0 \\ 0 & 0 & 0 \\ 0 & 0 & 0 \\ -\sin\theta & \cos\theta & 0 \\ \cos\theta & \sin\theta & 0 \\ 0 & 0 & 1 \end{bmatrix}\begin{bmatrix} u_1 \\ v \\ u_2 \end{bmatrix}\end{aligned}$$

Define

$$T(\theta) = \begin{bmatrix} -\sin\theta & \cos\theta & 0 \\ \cos\theta & \sin\theta & 0 \\ 0 & 0 & 1 \end{bmatrix}$$

and rewrite the dynamics of the optimal control problem as

$$\begin{bmatrix} \dot{z}_1 \\ \dot{z}_2 \end{bmatrix} = \begin{bmatrix} z_2 \\ A(\theta) \end{bmatrix} + \begin{bmatrix} 0 \\ T(\theta) \end{bmatrix} \begin{bmatrix} u_1 \\ v \\ u_2 \end{bmatrix}.$$

To derive the cost function of the optimal control problem, we estimate the following bound

$$\begin{aligned} \|\widetilde{B}(z)^+\widetilde{C}(z)f(z)\| &\le \|\widetilde{B}(z)^T\widetilde{C}(z)\| \times \|f(z)\| \\ &= \|B(\theta)^T C(\theta)\| \times \|f(z)\| \\ &= \left\| \begin{bmatrix} -\sin\theta & \cos\theta & 0 \\ 0 & 0 & 1 \end{bmatrix} \begin{bmatrix} 1/\cos\theta \\ 0 \\ 0 \end{bmatrix} \right\| \times \|f(z)\| \\ &= \left\| \begin{bmatrix} -\tan\theta \\ 0 \end{bmatrix} \right\| \times \|f(z)\| \\ &\le k\|\tan\theta\| \times \|z\| \end{aligned}$$

By Requirement 11.3: $|\theta| << \pi/2$, we can find a θ_o, $0 < \theta_o < \pi/2$, such that $|\theta| < \theta_o$. Therefore

$$\|\widetilde{B}(z)^+\widetilde{C}(z)f(z)\| \le k\|\tan\theta_o\| \times \|z\| = f_{\max}(z)$$

So, let us solve the following optimal control problem.

Optimal Control Problem 11.2

For the following system

$$\dot{z} = \widetilde{A}(z) + \widetilde{B}(z)u + (I_{6\times 6} - \widetilde{B}(z)\widetilde{B}(z)^+)\widetilde{C}(z)v \tag{11.6}$$

or

$$\begin{bmatrix} \dot{z}_1 \\ \dot{z}_2 \end{bmatrix} = \begin{bmatrix} z_2 \\ A(\theta) \end{bmatrix} + \begin{bmatrix} 0 \\ T(\theta) \end{bmatrix} \begin{bmatrix} u_1 \\ v \\ u_2 \end{bmatrix}$$

find a feedback control $u_1(z), v(z), u_2(z)$ that minimizes the cost functional

$$\int_0^\infty (f_{\max}(z)^2 + g_{\max}(z)^2 + \beta^2\|z\|^2 + \|u\|^2 + \|v\|^2)\,dt \qquad (11.7)$$

This optimal control problem is analogues to Problem 6.4 in Chapter 6 by letting $\alpha = 1$, $\rho = 1$. Again β is a design parameter whose value is to be determined.

To check if the closed system with control $u_o(z)$, $v_o(z)$, given by the solution to the above optimal control problem, is robustly stable, let $V(z_o)$ be the minimum cost of the optimal control of the system from some initial state z_o. Let us see if $V(z)$ is a Lyapunov function for system

$$\dot{z} = \widetilde{A}(z) + \widetilde{B}(z)u + \widetilde{C}(z)f(z)$$

Clearly

$$V(z) > 0, \quad z \neq 0$$
$$V(z) = 0, \quad z = 0$$

Since $u_o(z)$, $v_o(z)$ is the solution to the optimal control problem with system (11.6) and cost functional (11.7), the following Hamilton–Jacobi–Bellman equation must be satisfied.

$$\begin{aligned}\min_{u,v}(f_{\max}(z)^2 + g_{\max}(z)^2 + \beta^2\|z\|^2 + \|u\|^2 + \|v\|^2 \\ + V_z^{\mathrm{T}}(\widetilde{A}(z) + \widetilde{B}(z)u + (I - \widetilde{B}(z)\widetilde{B}(z))^+)\widetilde{C}(z)v)) = 0\end{aligned}$$

In other words, $u_o(z)$, $v_o(z)$ must satisfy:

$$\begin{aligned}&f_{\max}(z)^2 + g_{\max}(z)^2 + \beta^2\|z\|^2 + \|u_o\|^2 + \|v_o\|^2 \\ &\qquad + V_z^T(\widetilde{A}(z) + \widetilde{B}(z)u_o + (I - \widetilde{B}(z)\widetilde{B}(z))^+)\widetilde{C}(z)v_o) = 0 \\ &2u_o(z)^{\mathrm{T}} + V_z^{\mathrm{T}}\widetilde{B}(z) = 0 \\ &2v_o(z)^{\mathrm{T}} + V_z^{\mathrm{T}}(I - \widetilde{B}(z)\widetilde{B}(z))^+)\widetilde{C}(z) = 0\end{aligned}$$

We now have

$$\begin{aligned}\dot{V}(z) &= V_z^{\mathrm{T}}\dot{z} \\ &= V_z^{\mathrm{T}}(\widetilde{A}(z) + \widetilde{B}(z)u_o + \widetilde{C}(z)f(z)) \\ &= V_z^{\mathrm{T}}(\widetilde{A}(z) + \widetilde{B}(z)u_o + (I - \widetilde{B}(z)\widetilde{B}(z)^+)\widetilde{C}(z)v_o) \\ &\quad + V_z^{\mathrm{T}}\widetilde{C}(z)f(z) - V_z^T(I - \widetilde{B}(z)\widetilde{B}(z)^+)\widetilde{C}(z)v_o\end{aligned}$$

$$
\begin{aligned}
&= V_z^{\mathrm{T}}(\widetilde{A}(z)+\widetilde{B}(z)u_o+(I-\widetilde{B}(z)\widetilde{B}(z)^+)\widetilde{C}(z)v_o) \\
&\quad + V_z^{\mathrm{T}}\widetilde{B}(z)\widetilde{B}(z)^+\widetilde{C}(z)f(z)+V_z^{\mathrm{T}}(I-\widetilde{B}(z)\widetilde{B}(z)^+)\widetilde{C}(z)f(z) \\
&\quad - V_z^{\mathrm{T}}(I-\widetilde{B}(z)\widetilde{B}(z)^+)\widetilde{C}(z)v_o \\
&= -f_{\max}(z)^2-g_{\max}(z)^2-\beta^2\|z\|^2-\|u_o\|^2-\|v_o\|^2 \\
&\quad -2u_o(z)^{\mathrm{T}}\widetilde{B}(z)^+\widetilde{C}(z)f(z)-2v_o(z)^{\mathrm{T}}f(z)+2v_o(z)^T v_o
\end{aligned}
$$

Also

$$
\begin{aligned}
&-\|u_o\|^2-2u_o(z)^{\mathrm{T}}\widetilde{B}(z)^+\widetilde{C}(z)f(z)\le\|\widetilde{B}(z)^+\widetilde{C}(z)f(z)\|^2\le f_{\max}(z)^2 \\
&-2v_o(z)^{\mathrm{T}}f(z)\le\|v_o\|^2+\|f(z)\|^2\le\|v_o\|^2+g_{\max}(z)^2
\end{aligned}
$$

Substituting into the previous equation, we obtain

$$
\begin{aligned}
\dot{V}(z) &\le -f_{\max}(z)^2-g_{\max}(z)^2-\beta^2\|z\|^2-\|v_o\|^2 \\
&\quad +f_{\max}(z)^2+\|v_o\|^2+g_{\max}(z)^2+2\|v_o\|^2 \\
&= -\beta^2\|z\|^2+2\|v_o\|^2
\end{aligned}
$$

Therefore, for $V(z)$ to be a Lyapunov function for system (11.5), we need to guarantee that

$$2\|v_o\|^2\le\beta^2\|z\|^2$$

Since we have assumed $\|f(z)\|\le k\|z\|=g_{\max}(z)$, we also need to make sure that the following inequality holds

$$\|\varepsilon u_2(z)\|^2=\|f(z)\|^2\le k^2\|z\|^2$$

Because Optimal Control Problem 11.2 is for a nonlinear system, we cannot apply the standard solution to the LQR problem. However, with some approximation, we can solve this nonlinear optimal control problem. The solution is as follows.

First, there are three parameters to be determined: k, θ_o, β. The approach we take is similar to that of Chapter 6. We first pick some values for k, θ_o, β, find the solution, and then check if the sufficient conditions are satisfied.

So, let us first substitute $f_{\max}(z)$, $g_{\max}(z)$ in the cost functional (11.7)

$$
\begin{aligned}
&\int_0^\infty (f_{\max}(z)^2 \quad + g_{\max}(z)^2+\beta^2\|z\|^2+\|u_1\|^2+\|v\|^2+\|u_2\|^2)\,\mathrm{d}t \\
&\quad =\int_0^\infty (k^2\|\tan\theta_o\|^2\|z\|^2+k^2\|z\|^2+\beta^2\|z\|^2+\|u_1\|^2+\|v\|^2 \\
&\quad\quad +\|u_2\|^2)\,\mathrm{d}t
\end{aligned}
$$

$$= \int_0^\infty ((k^2\|\tan\theta_o\|^2 + k^2 + \beta^2)\|z\|^2 + \|u_1\|^2 + \|v\|^2 + \|u_2\|^2)\, dt$$

$$= \int_0^\infty (w^2\|z\|^2 + \|u_1\|^2 + \|v\|^2 + \|u_2\|^2)\, dt$$

where $w^2 = k^2\|\tan\theta_o\|^2 + k^2 + \beta^2$.

The optimal cost $V(z)$ must satisfy the Hamilton–Jacobi–Bellman equation, which is given by

$$\min_{u_1,v,u_2}\left(w^2\|z\|^2 + \|u_1\|^2 + \|v\|^2 + \|u_2\|^2 + V_z^{\mathrm{T}}\left(\begin{bmatrix} z_2 \\ A(\theta) \end{bmatrix} + \begin{bmatrix} 0 \\ T(\theta) \end{bmatrix}\begin{bmatrix} u_1 \\ v \\ u_2 \end{bmatrix}\right)\right) = 0$$

or

$$\min_{u_1,v,u_2}\left(w^2\|z\|^2 + \left\|\begin{bmatrix} u_1 \\ v \\ u_2 \end{bmatrix}\right\|^2 + V_z^{\mathrm{T}}\left(\begin{bmatrix} z_2 \\ A(\theta) \end{bmatrix} + \begin{bmatrix} 0 \\ T(\theta) \end{bmatrix}\begin{bmatrix} u_1 \\ v \\ u_2 \end{bmatrix}\right)\right) = 0$$

In particular, if u_{1o}, v_o, u_{2o} are optimal control, then

$$w^2\|z\|^2 + \left\|\begin{bmatrix} u_{1o} \\ v_o \\ u_{2o} \end{bmatrix}\right\|^2 + V_z^{\mathrm{T}}\left(\begin{bmatrix} z_2 \\ A(\theta) \end{bmatrix} + \begin{bmatrix} 0 \\ T(\theta) \end{bmatrix}\begin{bmatrix} u_{1o} \\ v_o \\ u_{2o} \end{bmatrix}\right) = 0 \tag{11.8}$$

and

$$2\begin{bmatrix} u_{1o} \\ v_o \\ u_{2o} \end{bmatrix}^{\mathrm{T}} + V_z^T\begin{bmatrix} 0 \\ T(\theta) \end{bmatrix} = 0 \tag{11.9}$$

From the definition of $T(\theta)$, it is easy to show that its transpose and inverse are equal to itself:

$$T(\theta) = T(\theta)^{\mathrm{T}} = T(\theta)^{-1} = \begin{bmatrix} -\sin\theta & \cos\theta & 0 \\ \cos\theta & \sin\theta & 0 \\ 0 & 0 & 1 \end{bmatrix}$$

Therefore, from Equation (11.9)

$$\begin{bmatrix} u_{1o} \\ v_o \\ u_{2o} \end{bmatrix} = -\frac{1}{2}\begin{bmatrix} 0 & T(\theta) \end{bmatrix} V_z$$

Substituting this into Equation (11.8), we have,

$$w^2(\|z_1\|^2 + \|z_2\|^2) - \frac{1}{4}V_z^{\mathrm{T}}\begin{bmatrix} 0 \\ T(\theta) \end{bmatrix}\begin{bmatrix} 0 & T(\theta) \end{bmatrix} V_z + V_z^{\mathrm{T}}\begin{bmatrix} z_2 \\ A(\theta) \end{bmatrix} = 0 \qquad (11.10)$$

Since we require that θ vary in a small neighbourhood of 0, $|\theta| << \pi/2$, we can linearize $A(\theta)$ around 0 as follows.

$$A(\theta) = \begin{bmatrix} -\sin\theta \\ \cos\theta - 1 \\ 0 \end{bmatrix} \approx \begin{bmatrix} -\theta \\ 0 \\ 0 \end{bmatrix} = \begin{bmatrix} 0 & 0 & -1 \\ 0 & 0 & 0 \\ 0 & 0 & 0 \end{bmatrix}\begin{bmatrix} x \\ y \\ \theta \end{bmatrix} = A_o z_1$$

where

$$A_o = \begin{bmatrix} 0 & 0 & -1 \\ 0 & 0 & 0 \\ 0 & 0 & 0 \end{bmatrix}$$

Also

$$\begin{bmatrix} 0 \\ T(\theta) \end{bmatrix}\begin{bmatrix} 0 & T(\theta) \end{bmatrix} = \begin{bmatrix} 0 & 0 \\ 0 & I_{3\times3} \end{bmatrix}$$

Hence, Equation (11.10) becomes

$$w^2\begin{bmatrix} z_1 & z_2 \end{bmatrix}\begin{bmatrix} z_1 \\ z_2 \end{bmatrix} - \frac{1}{4}V_z^{\mathrm{T}}\begin{bmatrix} 0 & 0 \\ 0 & I_{3\times3} \end{bmatrix} V_z + V_z^{\mathrm{T}}\begin{bmatrix} 0 & I_{3\times3} \\ A_o & 0 \end{bmatrix}\begin{bmatrix} z_1 \\ z_2 \end{bmatrix} = 0 \qquad (11.11)$$

In order to solve V from the above equation, we guess that V is quadratic:

$$V = z^{\mathrm{T}} S z$$

where S is some 6×6 positive definite (symmetric) matrix whose derivative is

$$V_z = 2Sz$$

Equation (11.11) can now be written as

$$w^2 z^{\mathrm{T}} z - z^{\mathrm{T}} S\begin{bmatrix} 0 & 0 \\ 0 & I_{3\times3} \end{bmatrix} Sz + z^{\mathrm{T}} S\begin{bmatrix} 0 & I_{3\times3} \\ A_o & 0 \end{bmatrix} z + z^{\mathrm{T}}\begin{bmatrix} 0 & I_{3\times3} \\ A_o & 0 \end{bmatrix}^{\mathrm{T}} Sz = 0$$

Let $Q = w^2 I_{6\times6}$, we obtain the following 'Riccati-type' equation

$$S\begin{bmatrix} 0 & I_{3\times3} \\ A_o & 0 \end{bmatrix} + \begin{bmatrix} 0 & I_{3\times3} \\ A_o & 0 \end{bmatrix}^{\mathrm{T}} S + Q - S\begin{bmatrix} 0 & 0 \\ 0 & I_{3\times3} \end{bmatrix} S = 0 \qquad (11.12)$$

A positive definite solution of the above equation exists and is unique. In terms of this S, the solution to the LQR problem is given by

$$\begin{bmatrix} u_{1o} \\ v_o \\ u_{2o} \end{bmatrix} = -\begin{bmatrix} 0 & T(\theta) \end{bmatrix} Sz = -\begin{bmatrix} 0 & 0 & 0 & -\sin\theta & \cos\theta & 0 \\ 0 & 0 & 0 & \cos\theta & \sin\theta & 0 \\ 0 & 0 & 0 & 0 & 0 & 1 \end{bmatrix} Sz \quad (11.13)$$

From the above equation it can be seen that while u_{1o}, v_o are nonlinear functions of state z, u_{2o} is indeed a linear function of z and hence is linear bounded as we guessed when we write $\|f(z)\| \leq k\|z\| = g_{\max}(z)$.

If we take $k = 3$, $\beta = 4.2426$ and $\theta_o = 45°$, then

$$w^2 = k^2 \|\tan\theta_o\|^2 + k^2 + \beta^2 = 36$$

Using MATLAB, we can solve the Riccati Equation (11.12) which leds to the following matrix

$$S = \begin{bmatrix} 41.5670 & 0 & -0.4356 & 6.0000 & 0 & -0.0004 \\ 0 & 41.5692 & 0 & 0 & 6.0000 & 0 \\ -0.4356 & 0 & 42.2129 & -0.9996 & 0 & 6.0828 \\ 6.0000 & 0 & -0.9996 & 6.9278 & 0 & -0.0721 \\ 0 & 6.0000 & 0 & 0 & 6.9282 & 0 \\ -0.0004 & 0 & 6.0828 & -0.0721 & 0 & 6.9398 \end{bmatrix}$$

The control is given by

$$\begin{aligned} u_{1o} &= (6.0000x - 0.9996\theta)\sin\theta - 6.0000y\cos\theta \\ &\quad + (6.9278\dot{x} - 0.0721\dot{\theta})\sin\theta - 6.9278\dot{y}\cos\theta \\ u_{2o} &= 0.0004x - 6.0828\theta + 0.0721\dot{x} - 6.9398\dot{\theta} \\ v_o &= -(6.0000x - 0.9996\theta)\cos\theta - 6.0000y\sin\theta \\ &\quad - (6.9278\dot{x} - 0.0721\dot{\theta})\cos\theta - 6.9278\dot{y}\sin\theta \end{aligned}$$

Since all the coefficients in the above equation are less than $\beta^2/2 = 9$, clearly

$$2\|v_o\|^2 \leq \beta^2 \|z\|^2$$

is satisfied. Similarly, $\|\varepsilon u_2(z)\|^2 \leq k^2\|z\|^2$ is satisfied for all $|\varepsilon| \leq 1$ (the typical value of ε is 0.01).

11.3 SIMULATION

To visualize the performance of the developed control law, we perform some simulations of the controlled system. To do so, let us first derive control in terms of U_t, U_m.

$$\begin{bmatrix} U_t \\ U_m \end{bmatrix} = \begin{bmatrix} mgu_t \\ Ju_m \end{bmatrix} = \begin{bmatrix} mg(1+u_{1o} - \varepsilon \tan\theta u_{2o}) \\ Ju_{2o} \end{bmatrix} = \begin{bmatrix} mg(1+u_{1o}) - \varepsilon_o J \tan\theta u_{2o}) \\ Ju_{2o} \end{bmatrix}$$

The dynamics of the PVTOL aircraft is given by

$$m\ddot{\tilde{x}} = -U_t \sin\theta + \varepsilon_o U_m \cos\theta$$

$$m\ddot{\tilde{y}} = U_t \cos\theta + \varepsilon_o U_m \sin\theta - mg$$

$$J\ddot{\theta} = U_m$$

Since $x = \tilde{x}/g, y = \tilde{y}/g$, the control in terms of $\tilde{x}, \tilde{y}$ is

$$\begin{aligned} u_{1o} =& (6.0000\tilde{x}/g - 0.9996\theta)\sin\theta - 6.0000\cos\theta\tilde{y}/g \\ &+ (6.9278\dot{\tilde{x}}/g - 0.0721\dot{\theta})\sin\theta - 6.9278\cos\theta\dot{\tilde{y}}/g \\ u_{2o} =& 0.0004\tilde{x}/g - 6.0828\theta + 0.0721\dot{\tilde{x}}/g - 6.9398\dot{\theta} \end{aligned}$$

In the simulation we set up the values of parameters to be

$$m = 50000\,\text{kg}$$

$$J = 200000\,\text{kg m}^2$$

To check the performance of the control above obtained, we simulate the actual responses of the aircraft for different ε_o. In all simulation, we use the following initial conditions.

$$\tilde{x}(0) = 1000\,\text{m}$$

$$\tilde{y}(0) = 0\,\text{m}$$

$$\theta(0) = 0^\circ$$

$$\dot{\tilde{x}}(0) = 0$$

$$\dot{\tilde{y}}(0) = 0$$

$$\dot{\theta}(0) = 0$$

For $\varepsilon_o = 0.01$, the X–Y positions of the aircraft are shown in Figure 11.2; and the angle and thrust of the aircraft are shown in Figure 11.3.

For $\varepsilon_o = 0.02$, the X–Y positions of the aircraft are shown in Figure 11.4; and the angle and thrust of the aircraft are shown in Figure 11.5.

For $\varepsilon_o = 0.05$, the X–Y positions of the aircraft are shown in Figure 11.6; and the angle and thrust of the aircraft are shown in Figure 11.7.

For $\varepsilon_o = 0.1$, the X–Y positions of the aircraft are shown in Figure 11.8; and the angle and thrust of the aircraft are shown in Figure 11.9.

From the figures, we can see that our control satisfies Requirements 11.1–11.4.

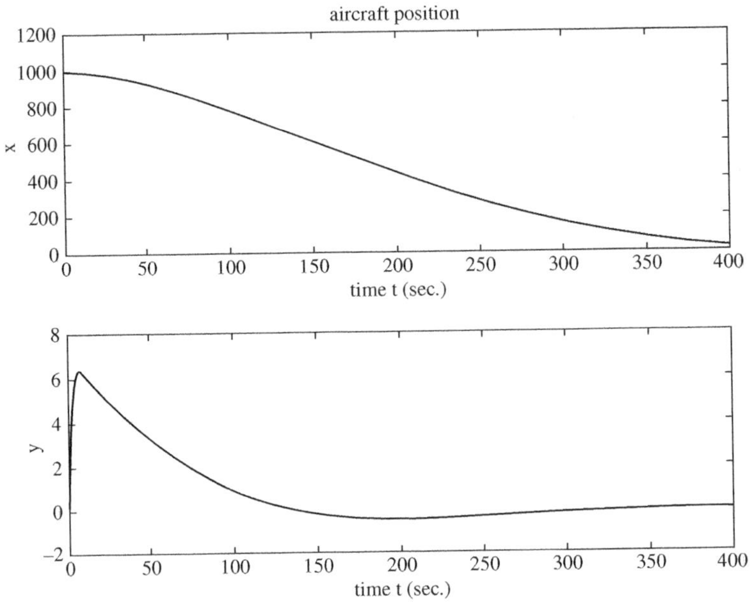

Figure 11.2 MATLAB simulation of X–Y positions for $\varepsilon_o = 0.01$.

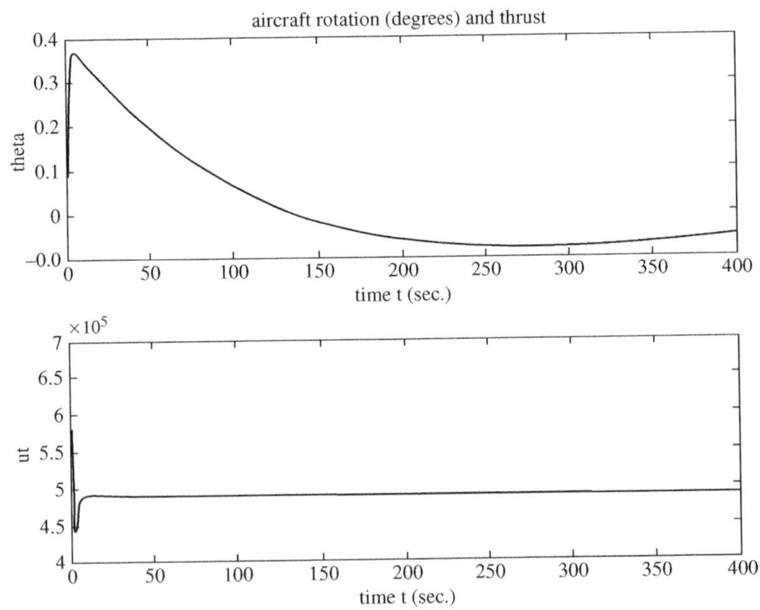

Figure 11.3 MATLAB simulation of angle and thrust for $\varepsilon_o = 0.01$.

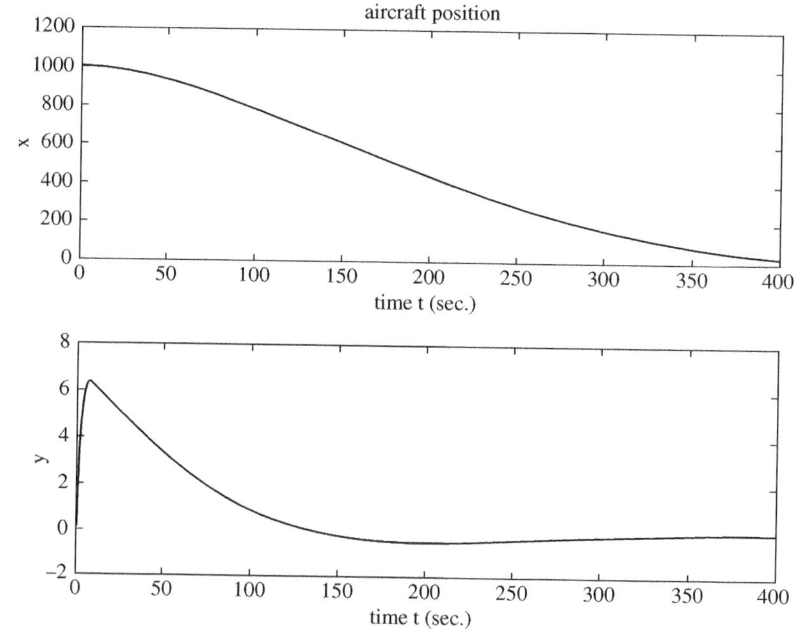

Figure 11.4 MATLAB simulation of X–Y positions for $\varepsilon_o = 0.02$.

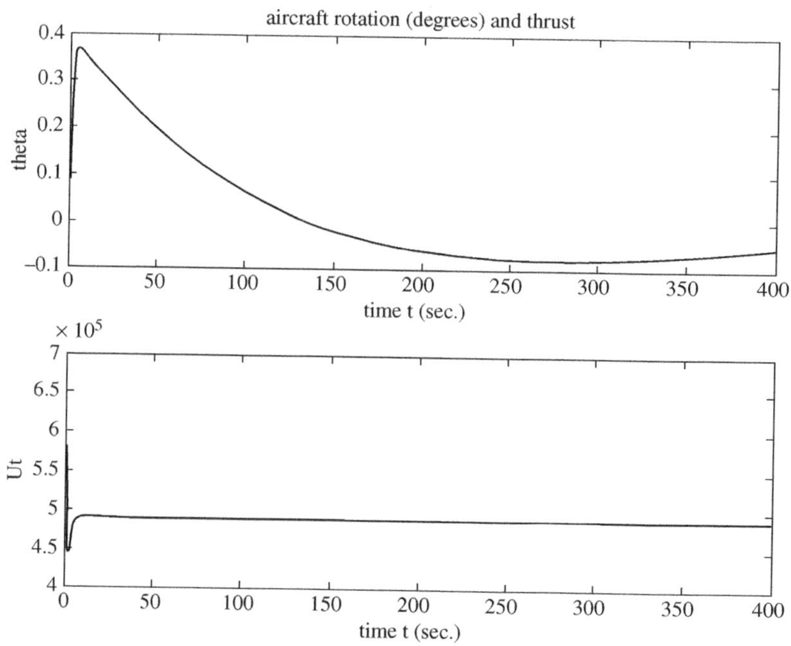

Figure 11.5 MATLAB simulation of angle and thrust for $\varepsilon_o = 0.02$.

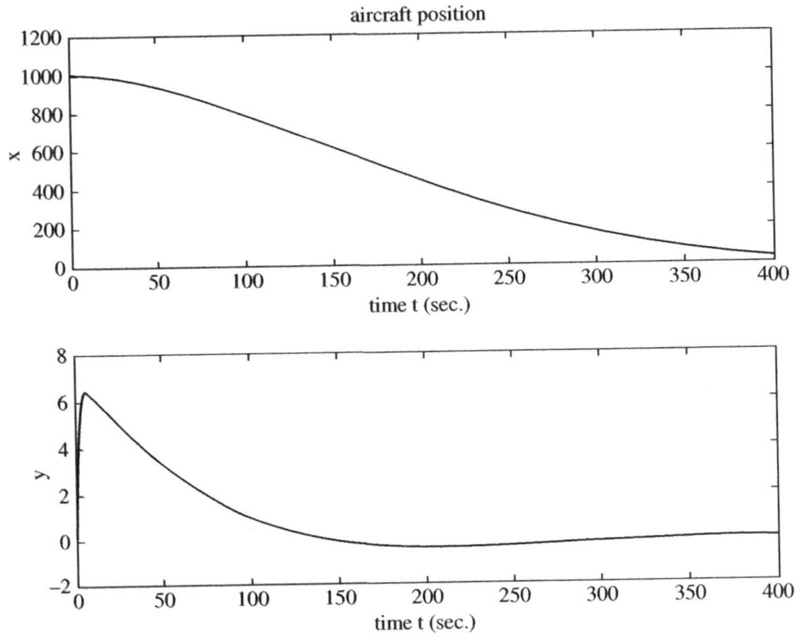

Figure 11.6 MATLAB simulation of X–Y positions for $\varepsilon_o = 0.05$.

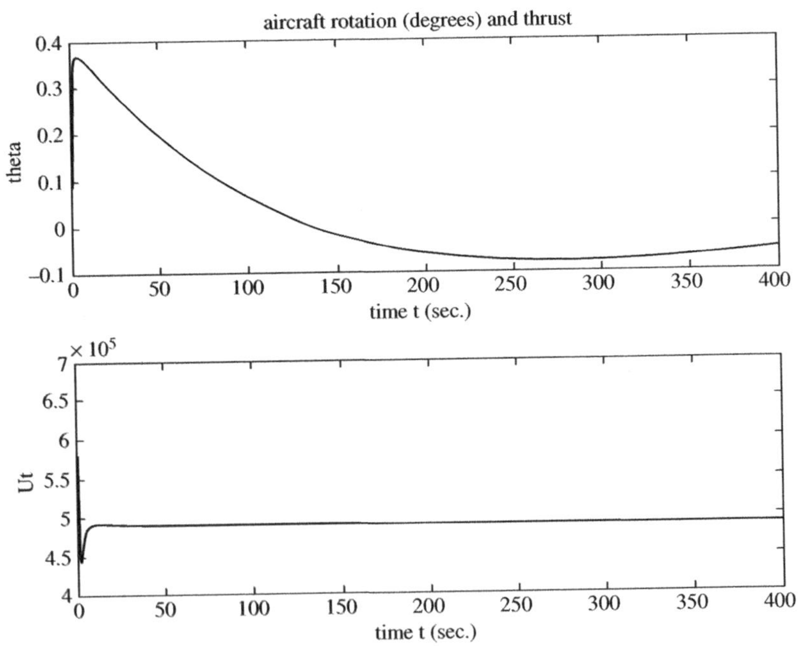

Figure 11.7 MATLAB simulation of angle and thrust for $\varepsilon_o = 0.05$.

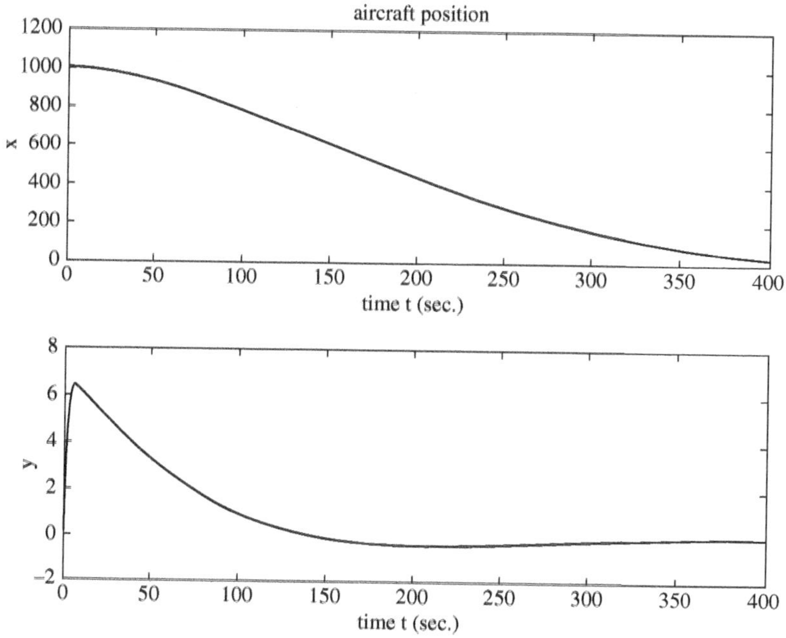

Figure 11.8 MATLAB simulation of X–Y positions for $\varepsilon_o = 0.1$.

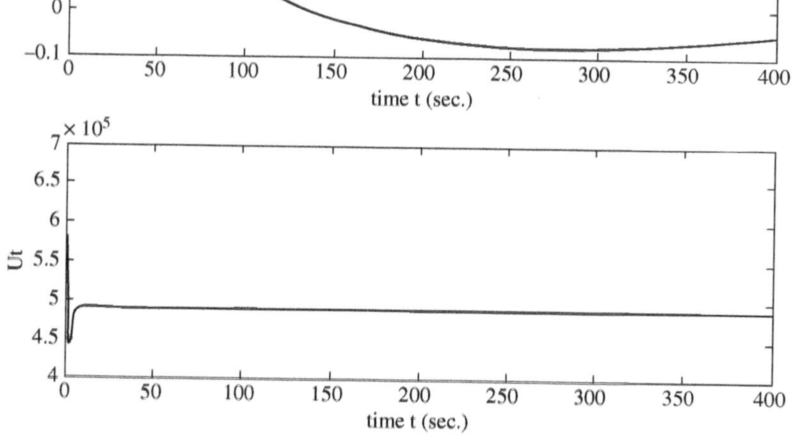

Figure 11.9 MATLAB simulation of angle and thrust for $\varepsilon_o = 0.1$.

Requirement 11.1

The PVTOL aircraft altitude $y(t)$, in the hovering mode, should have very small deviation from the pre-specified altitude, say $y = 0$. Indeed, y deviates only 6 m from 0, while x moves from 1000 m to 0.

Requirement 11.2

$U_t > 0$. Indeed, U_t is always positive.

Requirement 11.3

$|\theta| << 90°$. From the simulation θ is less than 1°.

Requirement 11.4

Large control inputs are not acceptable. From the simulation, the maximum U_t is about 135% of the thrust needed to support the aircraft. In other words, only 35% more thrust is needed to do the manoeuvre.

The trajectories of the controlled system for different $\varepsilon_o \in [0.01, 0.1]$ are very similar, demonstrating the robustness of the control.

It should be pointed out that in this work, no effort has been made to optimize the parameters involved, i.e., choosing the parameter values such that the closed-loop system has the 'best' performance. In fact, the choice $\rho = 1$ is by no means the best choice. We selected this specific numeric value solely for the purpose of being able to solve the Hamilton–Jacobi–Bellman equation explicitly. Intuitively, for better performance, a bigger ρ is preferred, because a smaller weight on v in the cost function results in an optimal control u_{1o}, u_{2o}, v_o which heavily rely on v_o, instead of on u_{1o}, u_{2o}. However, v is the augmented control which is to be discarded in forming the robust control. Therefore, a more realistic and better robust control law can be obtained by setting a much large ρ in the cost function of the corresponding optimal control problem.

11.4 NOTES AND REFERENCES

In this chapter, we have presented the last of three applications of our optimal control approach to solve real robust control problems in practical systems: hovering control of V/STOL aircraft. We have derived the dynamic equations for the V/STOL aircraft, simplified and converted the dynamic equations to state equations. We have designed a robust control to take care of the coupling between the rolling moment and the lateral force on the aircraft. We managed to solve a nonlinear optimal control problem analytically to obtain a nonlinear robust control law. Most other work on hovering control of V/STOL aircraft uses linear control which is restrictive and our solution is nonlinear and hence more general. We simulated the closed-loop system using the nonlinear robust control law. We found that the performance of the control law is excellent. Our initial work on V/STOL aircraft can be found in reference [110].

The Harrier is a non-minimum phase system. Therefore the theory for explicitly linearizing the input–output response of a nonlinear system using state feedback [30, 79] will not produce a satisfactory control law as indicated in reference [74]. In fact, one shortcoming of the exact input–output linearization theory is the inability to deal with non-minimum phase nonlinear system.

An approximate input–output linearization procedure, developed for slightly non-minimum phase nonlinear systems was used in reference [74] to design the hovering control. On the contrary, the method we proposed does not require linearization.

Another approach to aircraft hovering control was proposed in reference [138], which uses nonlinear regulator theory [80].

Appendix A: Mathematical Modelling of Physical Systems

The key to a successful control design is to have a good mathematical model of the system to be controlled. In this appendix, we will provide various examples of mathematical models of physical systems.

Example A.1

Consider the circuit given in Figure A.1. The circuit consists of two resistors R_1 and R_2; two inductors L_1 and L_2; one capacitor C and one voltage source v_{in}.

We want to derive its state equation. For circuits of this type, the state variables are usually currents in inductors and voltages on capacitors. In this example, the state variables are i_1, i_2, and v_c.

Applying Kirchhoff's voltage law to the first loop, we have

$$v_{in} = R_1 i_1 + L_1 \frac{di_1}{dt} + v_c$$

Similarly, for the second loop,

$$v_c = R_2 i_2 + L_2 \frac{di_2}{dt}$$

Robust Control Design: An Optimal Control Approach F. Lin

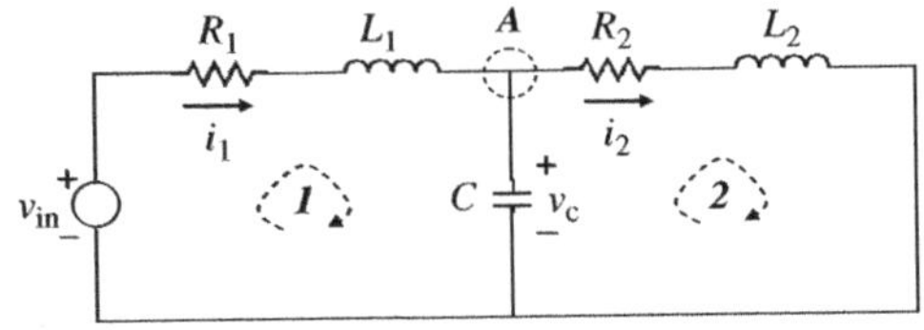

Figure A.1 Circuit diagram of Example A.1.

Applying Kirchhoff's current law to node A, we have

$$i_1 = i_2 + C\frac{dv_c}{dt}$$

From the above three equations (often called dynamic equations), we derive the following state equations.

$$\frac{di_1}{dt} = \frac{1}{L_1}(v_{in} - v_c - R_1 i_1)$$

$$\frac{di_2}{dt} = \frac{1}{L_2}(v_c - R_2 i_2)$$

$$\frac{dv_c}{dt} = \frac{1}{C}(i_1 - i_2)$$

Or, in matrix form

$$\begin{bmatrix} \dot{i}_1 \\ \dot{i}_2 \\ \dot{v}_c \end{bmatrix} = \begin{bmatrix} -\frac{R_1}{L_1} & 0 & -\frac{1}{L_1} \\ 0 & -\frac{R_2}{L_2} & \frac{1}{L_2} \\ \frac{1}{C} & -\frac{1}{C} & 0 \end{bmatrix} \begin{bmatrix} i_1 \\ i_2 \\ v_c \end{bmatrix} + \begin{bmatrix} \frac{1}{L_1} \\ 0 \\ 0 \end{bmatrix} v_{in}$$

The output equation depends on what can be measured. If we can measure the voltage v_{out} over the resistor R_2, then

$$v_{out} = [0\ R_2\ 0] \begin{bmatrix} i_1 \\ i_2 \\ v_c \end{bmatrix}$$

Let us take some realistic values for the circuit elements: $R_1 = 50\,\Omega$, $R_2 = 10\,\Omega$, $L_1 = 0.001\,\text{H}$, $L_2 = 0.002\,\text{H}$, and $C = 2\,\mu\text{F} = 2 \times 10^{-6}\,\text{F}$. Then

$$\begin{bmatrix} \dot{i}_1 \\ \dot{i}_2 \\ \dot{v}_c \end{bmatrix} = \begin{bmatrix} -50000 & 0 & -1000 \\ 0 & -5000 & 500 \\ 500000 & -500000 & 0 \end{bmatrix} \begin{bmatrix} i_1 \\ i_2 \\ v_c \end{bmatrix} + \begin{bmatrix} 1000 \\ 0 \\ 0 \end{bmatrix} v_{in}$$

$$v_{out} = \begin{bmatrix} 0 & 10 & 0 \end{bmatrix} \begin{bmatrix} i_1 \\ i_2 \\ v_c \end{bmatrix}$$

In other words

$$A = \begin{bmatrix} -50000 & 0 & -1000 \\ 0 & -5000 & 500 \\ 500000 & -500000 & 0 \end{bmatrix} \qquad B = \begin{bmatrix} 1000 \\ 0 \\ 0 \end{bmatrix}$$

$$C = \begin{bmatrix} 0 & 10 & 0 \end{bmatrix} \qquad D = 0$$

From (A, B, C, D), we can obtain the transfer function as

$$G(s) = C(sI - A)^{-1}B + D$$

This can be done using MATLAB command 'ss2tf'. For the above (A, B, C, D), we have

$$G(s) = \frac{2500\,000\,000\,000}{s^3 + 55\,000s^2 + 1000\,000\,000s + 15\,000\,000\,000\,000}$$

Example A.2

Two masses are hung from the ceiling by two strings, as shown in Figure A.2. A string can be modelled as a combination of a spring and a dashpot for friction. In the figure, y_1, y_2 are the displacements of masses M_1, M_2 from the resting position under gravity. The input to the system is the force f. K_1, K_2 are two spring constants and D_1, D_2 represent frictions. By Hook's law, the forces due to the springs are linearly proportional to the corresponding displacement; that is, they are K_2y_2, $K_1(y_1 - y_2)$ respectively.

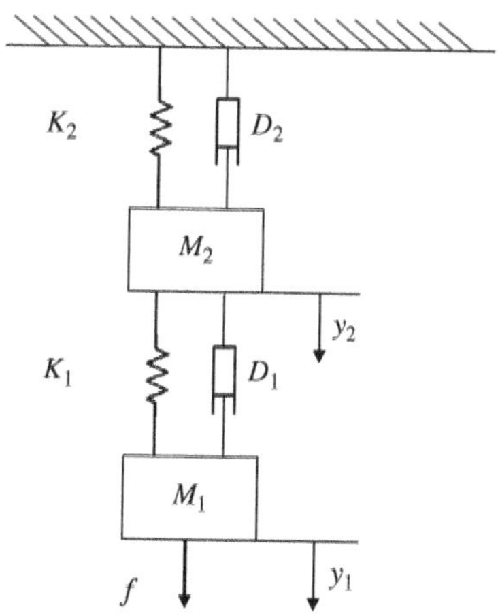

Figure A.2 The mechanical system of Example A.2.

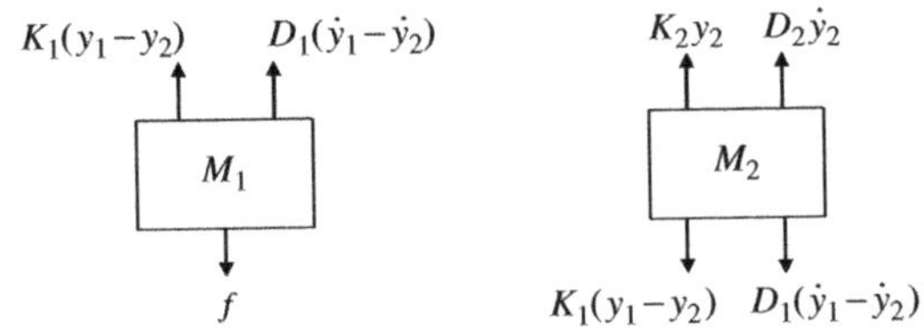

Figure A.3 Free body diagrams of the mechanical system in Figure A.2.

The forces due to friction are more complex and depend on displacements and velocities. As a first-order approximation, we assume that they are linearly proportional to the velocities; that is, they are $D_2\dot{y}_2$, $D_1(\dot{y}_1-\dot{y}_2)$ respectively.

The free body diagrams of two masses are shown in Figure A.3. We assume that gravity has been balanced by the strings as y_1, y_2 are measured from the resting position. So gravity does not show in the figure.

Applying Newton's second law to M_1, we obtain

$$M_1\ddot{y}_1 = f - K_1(y_1-y_2) - D_1(\dot{y}_1-\dot{y}_2)$$

Applying Newton's second law to M_2, we have

$$M_2\ddot{y}_2 = K_1(y_1-y_2) + D_1(\dot{y}_1-\dot{y}_2) - K_2y_2 - D_2\dot{y}_2$$

To obtain the state equations from the above dynamic equations, we first define state variables. For such mechanical systems, state variables are often displacements and velocities. In this example, we define state variables as $x_1 = y_1$, $x_2 = \dot{y}_1$, $x_3 = y_2$, $x_4 = \dot{y}_2$. This leads to the following state equations.

$$\begin{aligned}
\dot{x}_1 &= \dot{y}_1 \\
&= x_2 \\
\dot{x}_2 &= \ddot{y}_1 \\
&= \frac{1}{M_1}(f - K_1(y_1-y_2) - D_1(\dot{y}_1-\dot{y}_2)) \\
&= \frac{1}{M_1}(f - K_1(x_1-x_3) - D_1(x_2-x_4)) \\
\dot{x}_3 &= \dot{y}_2 \\
&= x_4, \\
\dot{x}_4 &= \ddot{y}_2 \\
&= \frac{1}{M_2}(K_1(y_1-y_2) + D_1(\dot{y}_1-\dot{y}_2) - K_2y_2 - D_2\dot{y}_2) \\
&= \frac{1}{M_2}(K_1(x_1-x_3) + D_1(x_2-x_4) - K_2x_3 - D_2x_4)
\end{aligned}$$

Or, in matrix form, we have

$$\begin{bmatrix} \dot{x}_1 \\ \dot{x}_2 \\ \dot{x}_3 \\ \dot{x}_4 \end{bmatrix} = \begin{bmatrix} 0 & 1 & 0 & 0 \\ -\frac{K_1}{M_1} & -\frac{D_1}{M_1} & \frac{K_1}{M_1} & \frac{D_1}{M_1} \\ 0 & 0 & 1 & 0 \\ \frac{K_1}{M_2} & \frac{D_1}{M_2} & -\frac{K_1+K_2}{M_2} & -\frac{D_1+D_2}{M_2} \end{bmatrix} \begin{bmatrix} x_1 \\ x_2 \\ x_3 \\ x_4 \end{bmatrix} + \begin{bmatrix} 0 \\ \frac{1}{M_1} \\ 0 \\ 0 \end{bmatrix} f$$

Assume that we can measure y_1, then the output equation is

$$y_1 = \begin{bmatrix} 1 & 0 & 0 & 0 \end{bmatrix} \begin{bmatrix} x_1 \\ x_2 \\ x_3 \\ x_4 \end{bmatrix}$$

Example A.3

Consider a rotational system with a motor driving two parts as shown in Figure A.4. The motor has inertia J_m and generates torque T. It drives two parts at two end via two flexible shafts, which can be modelled as two torsional springs with torsional spring constants K_1, K_2 respectively. The two parts at the two ends have inertias J_1, J_2 respectively. Denote the angular displacement and angular velocity of the motor by θ_m, ω_m respectively; the angular displacement and angular velocity of Part 1 by θ_1, ω_1 respectively; and the angular displacement and angular velocity of the Part 2 by θ_2, ω_2 respectively;

The free body diagrams of the three parts are shown in Figure A.5. Applying Newton's second law for rotational motion, we obtain

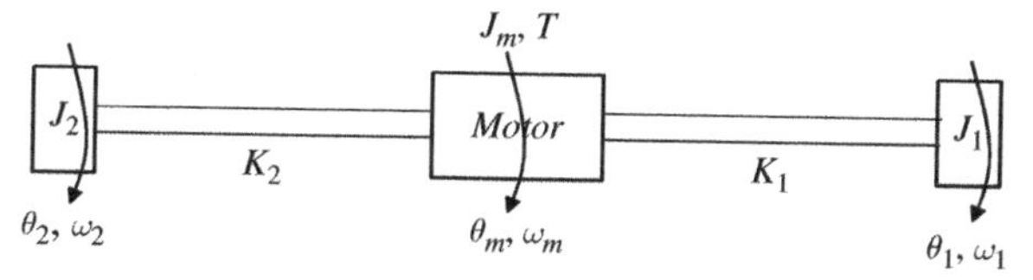

Figure A.4 The rotational system of Example A.3.

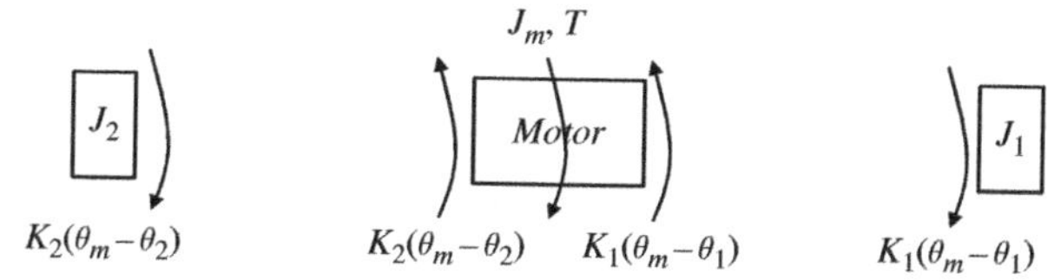

Figure A.5 Free body diagrams of the rotational system in Figure A.4.

$$J_m\ddot{\theta}_m = T - K_1(\theta_m - \theta_1) - K_2(\theta_m - \theta_2)$$
$$J_1\ddot{\theta}_1 = K_1(\theta_m - \theta_1)$$
$$J_2\ddot{\theta}_2 = K_2(\theta_m - \theta_2)$$

To obtain the state equations from the above dynamic equations, we define six state variables: θ_m, ω_m, θ_1, ω_1, θ_2, and ω_2. The state equations are derived as follows.

For such mechanical systems, state variables are often displacements and velocities. In this example, we define state variables as θ_m. This leads to the following state equations.

$$\dot{\theta}_m = \omega_m$$
$$\dot{\omega}_m = \frac{1}{J_m}(T - K_1(\theta_m - \theta_1) - K_2(\theta_m - \theta_2))$$
$$\dot{\theta}_1 = \omega_1$$
$$\dot{\omega}_1 = \frac{1}{J_1}(K_1(\theta_m - \theta_1))$$
$$\dot{\theta}_2 = \omega_2$$
$$\dot{\omega}_2 = \frac{1}{J_2}(K_2(\theta_m - \theta_2))$$

In matrix form

$$\begin{bmatrix} \dot{\theta}_m \\ \dot{\omega}_m \\ \dot{\theta}_1 \\ \dot{\omega}_1 \\ \dot{\theta}_2 \\ \dot{\omega}_2 \end{bmatrix} = \begin{bmatrix} 0 & 1 & 0 & 0 & 0 & 0 \\ -\frac{K_1+K_2}{J_m} & 0 & \frac{K_1}{J_m} & 0 & \frac{K_2}{J_m} & 0 \\ 0 & 0 & 0 & 1 & 0 & 0 \\ \frac{K_1}{J_1} & 0 & -\frac{K_1}{J_1} & 0 & 0 & 0 \\ 0 & 0 & 0 & 0 & 0 & 1 \\ \frac{K_2}{J_2} & 0 & 0 & 0 & -\frac{K_2}{J_2} & 0 \end{bmatrix} \begin{bmatrix} \theta_m \\ \omega_m \\ \theta_1 \\ \omega_1 \\ \theta_2 \\ \omega_2 \end{bmatrix} + \begin{bmatrix} 0 \\ \frac{1}{J_m} \\ 0 \\ 0 \\ 0 \\ 0 \end{bmatrix} T$$

Assume that we can measure θ_1 and θ_2, then the output equation is

$$\begin{bmatrix} \theta_1 \\ \theta_2 \end{bmatrix} = \begin{bmatrix} 0 & 0 & 1 & 0 & 0 & 0 \\ 0 & 0 & 0 & 0 & 1 & 0 \end{bmatrix} \begin{bmatrix} \theta_m \\ \omega_m \\ \theta_1 \\ \omega_1 \\ \theta_2 \\ \omega_2 \end{bmatrix}$$

Example A.4

A DC motor consists of a permanent magnet and a rotor made of wires, as shown schematically in Figure A.6.

In the figure, R and L are the resistance and inductance of the rotor respectively. θ_{m} and ω_{m} are the angular displacement and angular velocity of the motor by respectively.

There are two inputs to the system. One is the input voltage v_{in} and the other is the load torque T_{load}, representing the load to be driven by the motor.

When the input voltage v_{in} is applied to the motor, a torque T_{m} is generated by the motor. T_{m} is proportional to the current i in the motor:

$$T_{\mathrm{m}} = Ki \tag{A.1}$$

where K is some constant. As the rotor starts moving, it generates a voltage v_{back}, called back or counter electromotive force (back emf for short), v_{back} is proportional to the velocity of the motor:

$$v_{\mathrm{back}} = K\omega_{\mathrm{m}} \tag{A.2}$$

where K is the same constant as in Equation (A.1).

Hence, for the electrical part, we apply Kirchhoff's voltage law to obtain

$$v_{\mathrm{in}} = Ri + L\frac{\mathrm{d}i}{\mathrm{d}t} + v_{\mathrm{back}} \tag{A.3}$$

For the mechanical part, we apply Newton's second law to obtain

$$J_{\mathrm{m}}\ddot{\theta}_{\mathrm{m}} = T_{\mathrm{m}} - T_{\mathrm{load}} - D_{\mathrm{m}}\omega_{\mathrm{m}} \tag{A.4}$$

where J_{m} is the inertia of the rotor and D_{m} is the friction coefficient.

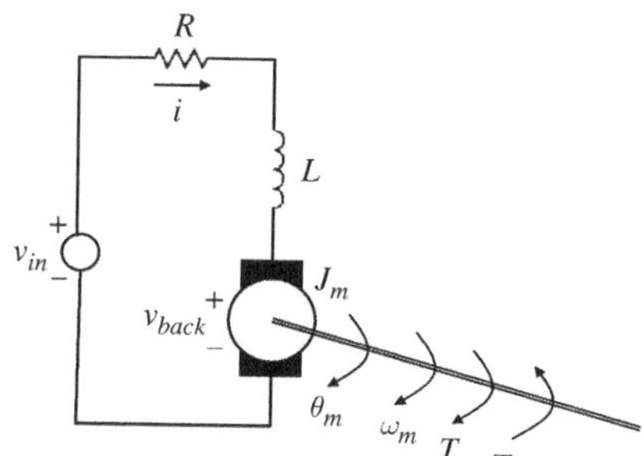

Figure A.6 The DC Motor of Example A.5.

Define state variables as i, θ_{m}, and ω_{m}. We derive the following state equations.

$$\dot{\theta}_{\mathrm{m}} = \omega_{\mathrm{m}}$$

$$\dot{\omega}_{\mathrm{m}} = \frac{1}{J_{\mathrm{m}}}(T_{\mathrm{m}} - T_{\mathrm{load}} - D_{\mathrm{m}}\omega_{\mathrm{m}}) = \frac{1}{J_{\mathrm{m}}}(Ki - T_{\mathrm{load}} - D_{\mathrm{m}}\omega_{\mathrm{m}})$$

$$\dot{i} = \frac{1}{L}(v_{\mathrm{in}} - Ri - v_{\mathrm{back}}) = \frac{1}{L}(v_{\mathrm{in}} - Ri - K\omega_{\mathrm{m}}).$$

In matrix form, we have

$$\begin{bmatrix} \dot{\theta}_{\mathrm{m}} \\ \dot{\omega}_{\mathrm{m}} \\ \dot{i} \end{bmatrix} = \begin{bmatrix} 0 & 1 & 0 \\ 0 & -D_{\mathrm{m}}/J_{\mathrm{m}} & K/J_{\mathrm{m}} \\ 0 & -K/L & -R/L \end{bmatrix} \begin{bmatrix} \theta_{\mathrm{m}} \\ \omega_{\mathrm{m}} \\ i \end{bmatrix} + \begin{bmatrix} 0 & 0 \\ 0 & -1/J_{\mathrm{m}} \\ 1/L & 0 \end{bmatrix} \begin{bmatrix} v_{\mathrm{in}} \\ T_{\mathrm{load}} \end{bmatrix}$$

With θ_m as the output, the output equation is

$$\theta_{\mathrm{m}} = \begin{bmatrix} 1 & 0 & 0 \end{bmatrix} \begin{bmatrix} \theta_{\mathrm{m}} \\ \omega_{\mathrm{m}} \\ i \end{bmatrix}$$

Example A.5

A water tank is shown in Figure A.7. There are two controls in the system: we can control the inflow F_{in} and we can control the outflow F_{out} by controlling the valve $u \in [0, 1]$. Here $u = 0$ means the valve is closed; and $u = 1$ means the valve is completely open. The valve can be partially open when u is between 0 and 1. Therefore, the inputs of the system are F_{in} and u. The water tank has a uniform area of cross-section, denoted by S. The state variable of the system is the water level x. To derive the state equation, we note that,

Rate of change in the water volume = inflow − outflow.

Obviously

$$\text{water volume} = Sx$$

The outflow F_{out} depends on the water level and the control input u as follows.

$$F_{\mathrm{out}} = K\sqrt{x}u$$

where K is some constant.

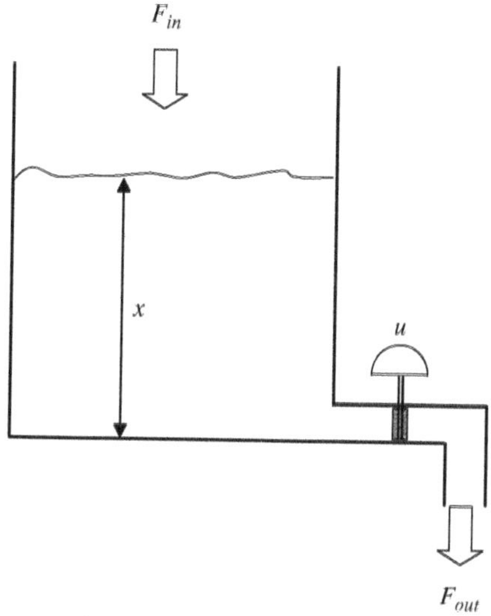

Figure A.7 The water tank of Example A.4.

Therefore

$$\frac{\mathrm{d}}{\mathrm{d}t}Sx = F_{\text{in}} - K\sqrt{x}u$$

That is

$$\dot{x} = \frac{1}{S}(F_{in} - K\sqrt{x}u)$$

This is a nonlinear system and cannot be written in matrix form.

Example A.6

Figure A.8 shows an inverted pendulum mounted on a cart. In the figure, M is the mass of the cart; m is the mass of the pendulum; L is the length of the pendulum; y is the displacement of the cart, θ is the angle of the pendulum; and u is the force acting on the cart, which is the input to the system.

We assume that the mass of the pendulum is concentrated at the end of the pendulum. We also do not consider friction. Let us derive the model of the system using Lagrange's equation, which can then be written as

$$\frac{\mathrm{d}}{\mathrm{d}t}\frac{\partial L}{\partial \dot{q}} - \frac{\partial L}{\partial q} = \tau$$

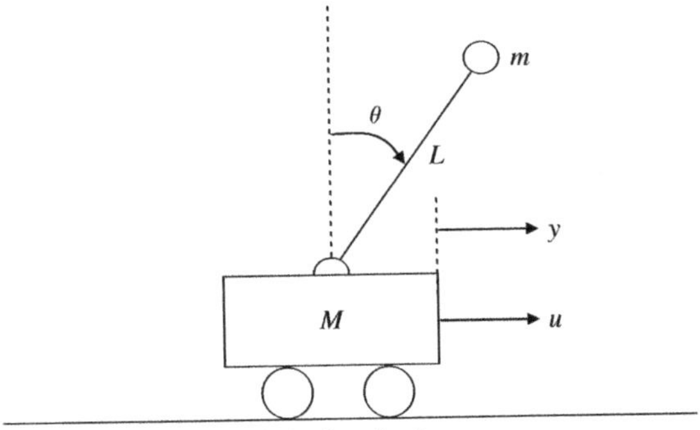

Figure A.8 The inverted pendulum of Example A.6.

where q is an n-dimensional vector of generalized coordinates q_i, τ is an n-dimensional vector of generalized forces τ_i, and $L = K - P$, the difference between the kinetic and potential energies, is the Lagrangian.

In this system, there are two generalized coordinates: the displacement of the cart y and the angle of the pendulum θ.

Let us first find the generalized forces corresponding to the generalized coordinates. The way to find the generalized force F_i corresponding to a generalized coordinate q_i is as follows. (1) Compute the work done by all nonconservative forces when q_i is changed to $q_i + dq_i$ with all other generalized coordinates held fixed. (2) Denote the work by dW_i. (3) Calculate the generalized force as $F_i = dW_i/dq_i$.

In our system, to calculate the generalized force F_θ corresponding to θ, let $\theta \to \theta + d\theta$. We have $dW_\theta = 0$. Hence $F_\theta = 0$. Similarly, to calculate the generalized force F_y corresponding to y, let $y \to y + dy$. We have $dW_y = Fdy$. Hence $F_y = F$.

The kinetic energy of the system is given by

$$K = \frac{1}{2}M\dot{y}^2 + \frac{1}{2}m\dot{z}^2$$

where $\dot{y}$ is the velocity of the cart and $\dot{z}$ is the velocity of the pendulum. The velocity $\dot{z}$ can be calculated by decomposing it in the horizontal direction

$$\dot{z}_h = \frac{d}{dt}(y + L\sin\theta) = \dot{y} + L\dot{\theta}\cos\theta$$

and the vertical direction

$$\dot{z}_v = \frac{d}{dt}L\cos\theta = -L\dot{\theta}\sin\theta$$

Since

$$\begin{aligned}\dot{z}^2 &= \dot{z}_h^2 + \dot{z}_v^2 \\ &= (\dot{y} + L\dot{\theta}\cos\theta)^2 + (-L\dot{\theta}\sin\theta)^2 \\ &= \dot{y}^2 + (L\dot{\theta}\cos\theta)^2 + 2\dot{y}L\dot{\theta}\cos\theta + (L\dot{\theta}\sin\theta)^2 \\ &= \dot{y}^2 + (L\dot{\theta})^2 + 2\dot{y}L\dot{\theta}\cos\theta\end{aligned}$$

the kinetic energy of the system is

$$K = \frac{1}{2}M\dot{y}^2 + \frac{1}{2}m(\dot{y}^2 + (L\dot{\theta})^2 + 2\dot{y}L\dot{\theta}\cos\theta)$$

The potential energy of the system is

$$P = mgL\cos\theta$$

The Lagrangian

$$L = K - P = \frac{1}{2}M\dot{y}^2 + \frac{1}{2}m(\dot{y}^2 + (L\dot{\theta})^2 + 2\dot{y}L\dot{\theta}\cos\theta) - mgL\cos\theta$$

Calculate the derivatives as follows

$$\begin{aligned}\frac{\partial L}{\partial\dot{\theta}} &= mL^2\dot{\theta} + m\dot{y}L\cos\theta \\ \frac{\mathrm{d}}{\mathrm{d}t}\frac{\partial L}{\partial\dot{\theta}} &= mL^2\ddot{\theta} + m\ddot{y}L\cos\theta - m\dot{y}\dot{\theta}L\sin\theta \\ \frac{\partial L}{\partial\theta} &= -m\dot{y}L\dot{\theta}\sin\theta + mgL\sin\theta\end{aligned}$$

The first Lagrange equation is

$$\frac{\mathrm{d}}{\mathrm{d}t}\frac{\partial L}{\partial\dot{\theta}} - \frac{\partial L}{\partial\theta} = F_\theta$$

Or

$$\begin{aligned}&mL^2\ddot{\theta} + m\ddot{y}L\cos\theta - m\dot{y}\dot{\theta}L\sin\theta + m\dot{y}L\dot{\theta}\sin\theta - mgL\sin\theta = 0 \\ &\Leftrightarrow mL^2\ddot{\theta} + m\ddot{y}L\cos\theta - mgL\sin\theta = 0 \\ &\Leftrightarrow L\ddot{\theta} + \ddot{y}\cos\theta - g\sin\theta = 0 \qquad \text{(A.5)}\end{aligned}$$

Similarly

$$\frac{\partial L}{\partial \dot{y}} = M\dot{y} + m\dot{y} + mL\dot{\theta}\cos\theta$$

$$\frac{\mathrm{d}}{\mathrm{d}t}\frac{\partial L}{\partial \dot{y}} = M\ddot{y} + m\ddot{y} + mL\ddot{\theta}\cos\theta - mL\dot{\theta}^2\sin\theta$$

$$\frac{\partial L}{\partial y} = 0$$

The second Lagrange equation is

$$\frac{\mathrm{d}}{\mathrm{d}t}\frac{\partial L}{\partial \dot{y}} - \frac{\partial L}{\partial y} = F_y$$

Or

$$M\ddot{y} + m\ddot{y} + mL\ddot{\theta}\cos\theta - mL\dot{\theta}^2\sin\theta = F \tag{A.6}$$

Solving Equations (A.5) and (A.6) for $\ddot{y}$ and $\ddot{\theta}$, we have

$$\ddot{y} = \frac{u + mL\dot{\theta}^2\sin\theta - mg\sin\theta\cos\theta}{M + m\sin^2\theta}$$

$$\ddot{\theta} = \frac{-u\cos\theta - mL\dot{\theta}^2\sin\theta\cos\theta + (M+m)g\sin\theta}{L(M + m\sin^2\theta)}$$

From the above equations, we can derive the state equations of the system. Let us define the state variables as: $x_1 = y$, $x_2 = \dot{y}$, $x_3 = \theta$, and $x_4 = \dot{\theta}$. The state equations are

$$\dot{x}_1 = x_2$$

$$\dot{x}_2 = \frac{u + mLx_4^2\sin x_3 - mg\sin x_3\cos x_3}{M + m\sin^2 x_3}$$

$$\dot{x}_3 = x_4$$

$$\dot{x}_4 = \frac{-u\cos x_3 - mLx_4^2\sin x_3\cos x_3 + (M+m)g\sin x_3}{L(M + m\sin^2 x_3)}$$

References and Bibliography

[1] C. Abdallah, D.M. Dawson, P. Dorato, and M. Jamshidi. Survey of robust control for rigid robots. *IEEE Control Systems Magazine*, **11**(2):24–30, 1991.

[2] N.U. Ahmed. Generalized solutions of HJB equations applied to stochastic control on hilbert space. *Nonlinear Analysis, Theory, Methods and Applications*, **54**(3):495–523, 2003.

[3] F. Amato, A. Pironti, and S. Scala. Necessary and sufficient conditions for quadratic stability and stabilizability of uncertain linear time-varying systems. *IEEE Transaction on Automatic Control*, **41**(1):125–128, 1996.

[4] H. Ambrose and Z. Qu. Model reference robust control for mimo systems. *International Journal of Control*, **68**(3):599–623, 1997.

[5] A.C. Antoulas. On identification for robust control: A summary. *Proceedings of the IEEE Conference on Decision and Control*, 3:3402–3409, 1998.

[6] A.C. Antoulas and B.D.O. Anderson. On the choice of inputs in identification for robust control. *Automatica*, **35**(6):1009–1031, 1999.

[7] P.J. Antsaklis and A.N. Michel. *Linear Systems*, McGraw-Hill, 1997.

[8] H.X.D. Araujo, I. Queinnec, and S. Tarbouriech. Robust control of linear uncertain systems subject to state and actuator limitations. *Journal Europeen des Systemes Automatises*, **35**(1–2):69–84, 2001.

[9] S. Arimoto and F. Miyazaki. Stability and robustness of PID feedback control for robot manipulators of sensory capability. *Proceedings of the 1st International Symposium on Robotics Research*, 783–799, 1983.

[10] E.-W. Bai and S. Dasgupta. Robust control design of sampled systems. *International Journal of Systems Science*, **21**(5):985–992, 1990.

[11] A.V. Balakrishnan. Compensator design for stability enhancement with collocated controllers, *IEEE Transactions on Automatic Control*, **36**(9):994–1007, 1991.

Robust Control Design: An Optimal Control Approach F. Lin

[12] A.V. Balakrishnan. Stability enhancement of flexible structures by nonlinear boundary feedback control, in *Boundary Control and Boundary Variations*, ed. J. P. Zolesio, 18–37. (*Lecture Notes in Control and Information Sciences*, 100), 1986.

[13] J.A. Ball, G. Groenewald, and T. Malakorn. Bounded real lemma for structured noncommutative multidimensional linear systems and robust control. *Multidimensional Systems and Signal Processing*, **17**(2–3):119–150, 2006.

[14] S.S. Banda, H.-H. Yeh, and S.A. Heise. Robust control of uncertain systems with combined H_∞ and LQG optimizations. *International Journal of Systems Science*, **22**(1):85–96, 1991.

[15] A.E. Barabanov. Operator approach to H_∞ control of linear systems. *Proceedings of the IEEE Conference on Decision and Control*, **1**:291–296, 1998.

[16] J. Baras. Frequency domain design of linear distributed systems. *Proceedings of the 19th IEEE Conference on Decision and Control*, 728–732, 1980.

[17] B.R. Barmish. Necessary and sufficient conditions for quadratic stabilizability of uncertain system. *Journal of Optimizization Theory Appl.*, **46**(4):399–408, 1985.

[18] B.R. Barmish, M. Corless and G. Leitmann. A new class of stabilizing controllers for uncertain dynamical systems. *SIAM Journal of Control Optimization*, **21**: 246–255, 1983.

[19] B.R. Barmish. A generalization of Kharitonov's four polynomial concept for robust stability problems with linearly dependent coefficient perturbations. *IEEE Transactions on Automatic Contr.*, **34**:157–165, 1989.

[20] G. Bartolini, A. Ferrara, and L. Giacomini. Robust control design for a class of uncertain non-linear systems featuring a second-order sliding mode. *International Journal of Control*, **72**(4):321–331, 1999.

[21] T. Basar. A dynamic games approach to controller design: disturbance rejection in discrete-time, *IEEE Transactions on Automatic Control*, **36**(8):936–952, 1991.

[22] S. Battilotti. Universal controllers for robust control problems. *Mathematics of Control, Signals, and Systems*, **10**(2):188–202, 1997.

[23] D.S. Bayard, Y. Yam, and E. Mettler. A criterion for joint optimization of identification and robust control. *IEEE Transactions on Automatic Control*, **37**(7):986–991, 1992.

[24] S. Beale and B. Shafai. Robust control system design with a proportional integral observer. *International Journal of Control*, **50**(1):97–111, 1989.

[25] N. Becker and W.M. Grimm. On H_2 and H_∞–stability approaches for the robust control of robot manipulators. *IEEE Transactions on Automatic Control*, **33**(1):118–122, 1988.

[26] P.R. Belanger. *Control Enginerring: A Modern Approach*, Sauders College Publishing, 1995.

[27] A. Belmiloudi. Robust control problems associated with time–varying delay nonlinear parabolic equations. *IMA Journal of Mathematical Control and Information*, **20**(3):305–334, 2003.

[28] P. Bernhard and N. Hovakimyan. Nonlinear robust control and minimax team problems. *International Journal of Robust and Nonlinear Control*, **9**(4):239–257, 1999.

[29] A.E. Bryson and Y.C. Ho. *Applied Optimal Control*, New York: Wiley, 1975.

[30] C.I. Byrnes and A. Isidori. A frequency domain philosophy for nonlinear systems, with applications to stabilization and to adaptive control. *Proceedings of the 23rd Conference on Decision and Control*, 1569–1573, 1984.

[31] T.K. Caughey and F. Ma. Complex modes and solvability of nonclassical linear systems, *Trans. ASME: J. Applied Mechanics*, **60**(1):26–28, 1993.

[32] L. Chen, J.B. Xie, and Z.Q. Su. Experimental study of robust control of a two link robot arm. *Proceeding of the Transformation of Science and Technology into Productive Power*, 459–466, 1991.

[33] M.-S. Chen and G.-H. Tarng. Robust control for siso systems with mismatched uncertainties. *Journal of Control Systems and Technology*, **5**(3):171–180, 1997.

[34] S.B. Chen. The robust optimal control of uncertain systems – state space method, *IEEE Transactions on Automatic Control*, **38**(6), 951–957, 1993.

[35] S. Chen and Z. Zhu. Robust control via state feedback for a class of uncertain bilinear systems. *Applied Mathematics and Mechanics (English Edition)*, **17**(2):181–185, 1996.

[36] X. Chen and K. Zhou. Multiobjective H_2/H_∞ control design. *SIAM Journal on Control and Optimization*, **40**(2):628–660, 2002.

[37] X. Chen and T. Fukuda. Model reference robust control for non-minimum phase systems with disturbances. *International Journal of Control*, **72**(2):123–132, 1999.

[38] Y.H. Chen. Decentralized robust control system design for large-scale uncertain systems. *International Journal of Control*, **47**(5):1195–1205, 1988.

[39] Y.H. Chen. Robust control system design. non-adaptive versus adaptive. *International Journal of Control*, **51**(6):1457–1477, 1990.

[40] Y.H. Chen. Adaptive robust control system design. Using information related to the bound of uncertainty. *Control – Theory and Advanced Technology*, **7**(1):31–53, 1991.

[41] Y.H. Chen. Complicated version of a robust control scheme. *Journal of Optimization Theory and Applications*, **84**(1):235–245, 1995.

[42] Y.H. Chen. Robust control design for a class of mismatched uncertain nonlinear systems. *Applied Mathematics and Computation*, **70**(2–3):155–165, 1995.

[43] C.-F. Cheng, W.-J. Wang, and Y.-P. Lin. Decentralized robust control of decomposed uncertain interconnected systems. *Journal of Dynamic Systems, Measurement and Control, Transactions of the ASME*, **115**(4):592–599, 1993.

[44] C.K. Chui and G. Chen. *Linear Systems and Optimal Control*, Springer–Verlag, 1989.

[45] T.-S. Chung. An inherent stability problem in Cartesian compliance and an alternative structure of compliance control. *IEEE Transactions on Robotics and Automation*, 7(1):21–30, 1991.

[46] B.L. Cooley and J.H. Lee. Integrated identification and robust control. *Journal of Process Control*, 8(5–6):431–440, 1998.

[47] R.F. Curtain. A synthesis of time and frequency domain methods for the control of infinite dimensional systems: A system theoretic approach, in *SIAM Frontiers in Applied Mathematics*, 1989.

[48] R.F. Curtain. Robust stabilizability of normalized coprime factors. The infinite–dimensional case. *International Journal of Control*, **51**(6):1173–1190, 1990.

[49] M.A. Dahleh and I.J. Diaz-Bobillo. *Control of Uncertain Systems: A linear Program Approach*, Prentice Hall, 1995.

[50] J.C. Doyle, B. Francis, and A. Tannenbaum. *Feedback Control Theory*, Macmillan Publishing Company, 1992.

[51] J.C. Doyle, K. Glover, P.P. Khargonekar, and B. Francis. State–space solutions to standard H_2 and H_∞ control problems, *IEEE Transactions on Automatic Control*, **34**(8):831–847, 1989.

[52] D.M. Dawson, Z. Qu, F.L. Lewis, and J.F. Dorsey. Robust control for the tracking of robot motion. *International Journal of Control*, **52**(3):581–595, 1990.

[53] M.C. Delfour. Linear–quadratic optimal control problem with delays in state and control variables: A state space approach. *SIAM Journal on Control and Optimization*, **24**(5):835–883, 1986.

[54] M. Dinh, G. Scorletti, V. Fromion, and E. Magarotto. Parameter dependent H_∞ control by finite dimensional lmi optimization: Application to trade-off dependent control. *International Journal of Robust and Nonlinear Control*, **15**(9):383–406, 2005.

[55] P. Dorato, R. Tempo, and G. Muscato. Bibliography on robust control. *Automatica*, **29**(1):201–213, 1993.

[56] P. Dorato. Historical review of robust control. *IEEE Control Systems Magazine*, 7(2):44–47, 1987.

[57] G.E. Dullerud, F. Paganini. *A Course in Robust Control Theory*, Springer Verlag, 2000.

[58] B. Francis, and J.C. Doyle. Linear control theory with an H_∞ optimality criterion. *SIAM J. Contr. Optimiz*, **25**:815–844, 1987.

[59] B.A. Francis. *A Course in H_∞ Control Theory*, New York: Springer–Verlag, 1987.

[60] Y.K. Foo and I. Postlethwaite. H_∞–minimax approach to the design of robust control systems. *Systems & Control Letters*, 5(2):81–88, 1984.

[61] Y.K. Foo and I. Postlethwaite. H_∞-minimax approach to the design of robust control systems, part II: All solutions, all-pass form solutions and the 'best' solution. *Systems & Control Letters*, 7(4):261–268, 1986.

[62] R.A. Freeman, P.V. Kokotovic. *Robust Nonlinear Control Design: State-Space and Lyapunov Techniques*, Springer Verlag, 1996.

[63] M. Fu, A.W. Olbrot and M.P. Polis. Robust stability for time–delay systems: the edge theorem and graphical tests. *IEEE Transaction on Automatic Control*, **34**, 813–819, 1989.

[64] M. Fujita, K. Uchida, and F. Matsumura. Gain perturbation tolerance in H_∞ state feedback control. *International Journal of Control*, **51**(2):315–328, 1990.

[65] A.R. Gaiduk. Robust control systems with delay: a synthesis procedure. *Automation and Remote Control (English translation of Avtomatika i Telemekhanika)*, **58**(2):73–83, 1997.

[66] K. Glover and J.C. Doyle. State-space formulae for all stabilizing controllers that satisfy an H_∞ norm bound and relations to risk sensitivity. *Systes and Control Letters*, **11**:167–172, 1988.

[67] L.E. Ghaoui and J.P. Folcher. Multiobjective robust control of LTI systems subject to unstructured perturbations. *Systems and Control Letters*, **28**(1):23–30, 1996.

[68] M. Green and D. J. N. Limebeer. *Linear Robust Control*, Prentice Hall, 1995.

[69] L. Guvenc and J. Ackermann. Links between the parameter space and frequency domain methods of robust control. *International Journal of Robust and Nonlinear Control*, **11**(15):1435–1453, 2001.

[70] I.J. Ha and E. Gilbert. Robust tracking in nonlinear systems. *IEEE Transactions on Automatic Control*, **32**:763–771, 1987.

[71] W.M. Haddad, V. Chellaboina, J.L. Fausz, and A. Leonessa. Optimal non-linear robust control for non-linear uncertain systems. *International Journal of Control*, **73**(4):329–342, 2000.

[72] M.C. Han and Y.H. Chen. Decentralized robust control of nonlinear systems with bounded time-varying uncertainties. *Control – Theory and Advanced Technology*, 7(4):609–628, 1991.

[73] M.C. Han and Y.H. Chen. Polynomial robust control design for uncertain systems. *Automatica*, **28**(4):809–814, 1992.

[74] J. Hauser, S. Sastry and G. Meyer. Nonlinear control design for slightly non-minimum phase systems: Application to V/STOL aircraft. *Automatica*, **28**: 665–679, 1992.

[75] X. Huang, S. Wang, and J. Dai. Robust control oriented L1 central estimation recusive algorithm. *Kongzhi yu Juece/Control and Decision*, **10**(4):364–374, 1995.

[76] A. Ichikawa. Product of nonnegative operators and infinite-dimensional H_∞ Riccati equations. *Systems and Control Letters*, **41**(3):183–188, 2000.

[77] M. Ikeda and D.D. Siljak. Robust stabilization of nonlinear systems via linear state feedback, In C.T. Leondes (Ed) *Control and Dynamic Systems, 51*, Academic Press, 1992.

[78] V. Ionescu, C. Oarabreve, M.D. Weiss, C. Oara, M. Weiss. *Generalized Riccati Theory and Robust Control. A Popov Function Approach*, Wiley, 1999.

[79] A. Isidori and C.H. Moog. On the nonlinear equivalent of the notion of transmission zeros. In C. I. Byrnes and A. Kurzhanski (eds), *Modelling and Adaptive Control*, Lecture Notes in Control and Information Sciences. Springer, Berlin, 1989.

[80] A. Isidori and C.I. Byrnes. Output regulation of nonlinear systems, *IEEE Transactions on Automatic Control*, **35**(2):131–140, 1990.

[81] I. Jaworska and S. Tzafestas. Improvement of systems reliability using robust control theory. *International Journal of Systems Science*, **22**(3):587–593, 1991.

[82] Z.P. Jiang, I. Mareels, and D. Hill. Robust control of uncertain nonlinear systems via measurement feedback. *IEEE Transactions on Automatic Control*, **44**(4):807–812, 1999.

[83] X. Jiao, H. Qin, and T. Shen. Robust control for nonlinear systems with bounded perturbation. *Kongzhi Lilun Yu Yinyong/Control Theory and Applications*, **18**(6):902–912, 2001.

[84] H.R. Karimi and M.J. Yazdanpanah. Robust control for a class of uncertain state-delayed singularly perturbed systems. *Asian Journal of Control*, 7(2):202–208, 2005.

[85] L.H. Keel, S.P. Bhattacharyya, and J.W. Howze. Robust control with structure perturbations. *IEEE Transactions on Automatic Control*, **33**(1):68–78, 1988.

[86] L.H. Keel, H. Chapellat, L.H. Keel, S.P. Bhattacharyya. *Robust Control: The Parametric Approach*. Prentice Hall, 1995.

[87] P.P. Khargonekar, I.R. Petersen and K. Zhou. Robust stabilization of uncertain linear systems: quadratic stabilization and H^∞ control theory. *IEEE Transactions on Automatic Control*, **35**(3):356–361, 1990.

[88] P.P. Khargonekar and M.A. Rotea. Mixed H_1/H_∞ control: a convex optimization approach, *IEEE Transactions on Automatic Control*, **36**(7), 824–837, 1991.

[89] V.L. Kharitonov. Asymptotic stability of an equilibrium position of a family of systems of linear differential equations. *Differential, Uravnen.*, **14**, 2086–2088, 1978.

[90] A. Kojima and S. Ishijima. H_∞ preview disturbance attenuation. *Proceedings of the IEEE Conference on Decision and Control*, **3**:3201–3203, 1998.

[91] F.I. Kraus, M. Mansour and B.D.O. Anderson. Robust Schur polynomial stability and Kharitonov's theorem. *Proceedings of the 26th Conference on Decision Control*, 2088–2095, 1987.

[92] B.C. Kuo and F.Golnaraghi. *Automatic Control Systems*, Wiley, 2003.

[93] C.-Y. Kuo and S.-P.T. Wang. Robust position control of robotics manipulator in Cartesian coordinates. *IEEE Transactions on Robotics and Automation*, 7(5):653–659, 1991.

[94] H. Kwakernaak. Robust control and H_∞-optimization – tutorial paper. *Automatica*, **29**(2):255–273, 1993.

[95] I.D. Landau, J. Langer, D. Rey, and J. Barnier. Robust control of a 360° flexible arm using the combined pole placement/sensitivity function shaping method. *IEEE Transactions on Control Systems Technology*, **4**(4):369–383, 1996.

[96] S.H. Lane and R.F. Stengel. Flight control design using nonlinear inverse dynamics. *Automatica*, **24**:471–483, 1988.

[97] M.-N. Lee and M.J. Chung. Robust control for uncertain linear systems with sinusoidal disturbance of uncertain frequencies. *IEE Proceedings: Control Theory and Applications*, **145**(6):531–536, 1998.

[98] Y.S. Lee, S.H. Han, and W.H. Kwon. Receding horizon H_∞ control for systems with a state-delay. *Asian Journal of Control*, **8**(1):63–71, 2006.

[99] G. Leitmann. Guaranteed asymptotic stability for some linear systems with bounded uncertainties. *J. Dyn. Syst., Meas. and Contr.*, **101**:212–216, 1979.

[100] F.L. Lewis, C.T. Abdallah, and D.M. Dawson. *Control of Robot Manipulators*, Macmillan, 1993.

[101] F.L. Lewis and V.L. Syrmos. *Optimal Control*, Wiley, 1995.

[102] J.-S. Liang, Y.-S. Chen, and A.Y.T. Leung. Robust control of periodic bifurcation solutions. *Applied Mathematics and Mechanics (English Edition)*, **25**(3):263–271, 2004.

[103] D.J. Limebeer. Advances in robust control. *Elektron*, **12**(5):7–17, 1995.

[104] F. Lin. An optimal control approach to robust control design. *International Journal of Control*, **73**(3):177–186, 2000.

[105] F. Lin, R. Brandt and J. Sun. Robust control of nonlinear systems: compensating for uncertainty. *International Journal of Control*, **56**:1453–1459, 1992.

[106] F. Lin and R.D. Brandt. An optimal control approach to robust control of robot manipulators. *IEEE Transactions on Robotics and Automation*, **14**(1):69–77, 1998.

[107] F. Lin, and A.W. Olbrot. An LQR approach to robust control of linear systems with uncertain parameters, *Proceedings of the 35th IEEE Conference on Decision and Control*, 4158–4163, 1996.

[108] F. Lin and W. Zhang. Robust control of nonlinear systems without matching condition, *Proceedings of the 32nd IEEE Conference on Decision and Control*, 2572–2577, 1993.

[109] F. Lin and W. Zhang. Robust active damping of vibration systems with uncertainties, *Proceedings of the 1994 American Control Conference*, 2424–2428, 1994.

[110] F. Lin and W. Zhang. Robust hovering control of a PVTOL aircraft, *IEEE Transactions on Control Systems Technology*, 7(3):43–351, 1999.

[111] G. Liu and A.A. Goldenberg. Uncertainty decomposition–based robust control of robot manipulators. *IEEE Transactions on Control Systems Technology*, **4**(4):384–393, 1996.

[112] G.-J. Liu and A.A. Goldenberg. Asymptotically stable robust control of robot manipulators. *Mechanism and Machine Theory*, **31**(5):607–618, 1996.

[113] H. Lu, H. Fang, and X. Huang. On the structure of H_∞ optimal controllers for infinite dimensional systems. *Huazhong Ligong Daxue Xuebao/Journal Huazhong (Central China) University of Science and Technology*, **25**(Suppl 2):8–11, 1997.

[114] M.S. Mahmoud. *Robust Control and Filtering for Time–Delay Systems*, Marcel Dekker, 2000.

[115] M.S. Mahmoud, P. Shi, and A. Ismail. Control of interconnected jumping systems: An approach. *Asian Journal of Control*, **6**(1):1–8, 2004.

[116] P.M. Makila. On three puzzles in robust control. *IEEE Transactions on Automatic Control*, **45**(3):552–556, 2000.

[117] K. Mizukami, H. Wu, and F. Suzumura. Procedure for designing robust control of systems with bounded uncertain parameters. *Electrical Engineering in Japan (English translation of Denki Gakkai Ronbunshi)*, **110**(4):44–50, 1990.

[118] P. Myszkorowsky. Practical stabilization of a class of uncertain nonlinear systems, *Systems & Control Letters*, **18**:233–236, 1992.

[119] U. Nurges and R. Luus. Discrete kharitonov's theorem and robust control. *Control and Intelligent Systems*, **30**(3):110–118, 2002.

[120] O. Nwokah, S. Jayasuriya, and Y. Chait. Parametric robust control by quantitative feedback theory. *Journal of Guidance, Control, and Dynamics*, **15**(1):207–214, 1992.

[121] M. Obayashi, K. Hirasawa, J. Murata, and A. Kajiwara. Robust control with pole assignment in a specified region for systems under parameter perturbations and its application to a nonlinear system. *Denki Gakkai Ronbunshi. C, Erekutoronikusu Joho Kogaku, Shisutemu*, **115**(8):984–995, 1995.

[122] T. Okada, M. Kihara, and H. Furihata. Robust control system with observer. *International Journal of Control*, **41**(5):1207–1219, 1985.

[123] A. Olbrot and J. Cieslik. A qualitative bound for robustness of stabilization by state feedback. *IEEE Trans. Automat. Contr.*, **33**(12):1165–1166, 1988.

[124] A. Olbrot. Nonrobustness of the pole placement design under small perturbations in the delays. *Proceedings of the 1992 American Control Conference*, 76–77, 1992.

[125] A. Olbrot. On robust spectral assignability: there are limits to the pole placement theorem for systems with infinitesimal perturbations. In T.E. Duncan and B. Pasik-Duncan (eds.) *Lecture Notes in Control and Information Sciences, 184, Stochastic Theory and Adaptive Control*, Springer, 1992.

[126] A. Olbrot. Arbitrary robust eigenvalue placement by a static state feedback. *Proceedings of the 33rd IEEE Conference on Decision and Control*, 561–562, 1994.

[127] Z. Pan and T. Basar. H_∞ control of large-scale jump linear systems via averaging and aggregation. *International Journal of Control*, **72**(10):866–881, 1999.

[128] I.R. Petersen and C.V. Hollot. A Riccati equation approach to the stabilization of uncertain linear systems. *Automatica*, **22**(4):397–411, 1986.

[129] I.R. Petersen. A new extension to Kharitonov's theorem. *Proceedings of the 26th Conference on Decision Control*, 440–444, 1987.

[130] I.R. Petersen and D.C. McFarlane. Optimal guaranteed cost control and filtering for uncertain linear systems, *IEEE Transactions on Automatic Control*, **39**(9):1971–1977, 1994.

[131] B.T. Polyak and M.E. Halpern. Robust stability and design of linear discrete–time siso systems under L1 uncertainties. *IEEE Transactions on Automatic Control*, **44**(11):2076–2080, 1999.

[132] E.M. Potapenko. Robust control systems with second-order observers. *Automation and Remote Control (English translation of Avtomatika i Telemekhanika)*, **57**(2):233–243, 1996.

[133] Z. Qu. Global stabilization of nonlinear systems with a class of uncertain nonlinear systems, *Systems & Control Letters*, **18**:301–307, 1992.

[134] Z. Qu. *Robust Control of Nonlinear Uncertain Systems*, Wiley, 1998.

[135] B. Qu and X. Huang. H_∞ robust control for multi-time delay system. *Kongzhi Lilun Yu Yinyong/Control Theory and Applications*, **16**(6):836–841, 1999.

[136] Z. Qu, J.F. Dorsey, X. Zhang, and D.M. Dawson. Robust control of robots by the computed torque law. *Systems & Control Letters*, **16**(1):25–32, 1991.

[137] I. Rhee and J.L. Speyer. A game theoretic approach to a finite-time disturbance attenuation problem. *IEEE Transactions on Automatic Control*, **36**(9):1021–1032, 1991.

[138] J. Roltgen and C.I. Byrnes. Hover control of a PVTOL using nonlinear regulator theory. *Proceedings of the 1993 American Control Conference*, 1126–1130, 1993.

[139] H. Rotstein, J. Bandoni, A. Desages, and J. Romagnoli. Mathematical programming with linear uncertain constraints. Application to robust control. *Computers and Chemical Engineering*, **14**(4–5):373–379, 1990.

[140] W.J. Rugh. *Linear System Theory*, Prentice Hall, 1996.

[141] A.P. Sage and C.C. White. *Optimum Systems Control*, Prentice Hall, 1977.

[142] R.S. Sanchez-Pena and M. Sznaier. *Robust Systems Theory and Applications*, Wiley, 1998.

[143] H. Sano. Residual mode filters and H_∞ control for a class of infinite-dimensional discrete-time systems. *International Journal of Systems Science*, **34**(3):205–214, 2003.

[144] J.M.A. Scherpen and M.H.G. Verhaegen. H_∞ output feedback control for linear discrete time-varying systems via the bounded real lemma. *International Journal of Control*, **41**:177–188, 1996.

[145] G. Schmitt-Braess. Feedback of passive systems: Synthesis and analysis of linear robust control systems. *IEE Proceedings: Control Theory and Applications*, **150**(1):83–91, 2003.

[146] V.N. Shashikhin. Robust control using interval analysis. *Reliable Computing*, **7**(3):219–230, 2001.

[147] P. Shi, L. Xie, and Y. Wang. Robust control design for interconnected uncertain dynamical systems under sampled measurements. *Computers and Electrical Engineering*, **23**(2):81–94, 1997.

[148] R. Shoureshi, M.E. Momot, and M.D. Roesler. Robust control for manipulators with uncertain dynamics. *Automatica*, **26**(2):353–359, 1990.

[149] M.J. Sidi. *Design of Robust Control Systems: From Classical to Modern Practical Approaches*, Krieger, 2001.

[150] D.D. Siljak. Parameter space methods for robust control design: a guided tour. *IEEE Transactions on Automatic Control*, **34**(7):674–688, 1989.

[151] M. Slemrod. Feedback stabilization of a linear control system in hilbert space with an a priori bounded control. *Mathematics of Control, Signals, and Systems*, **2**(3):265–285, 1989.

[152] J.-J.E. Slotine and W. Li. On the adaptive control of robot manipulators. *International Journal of Robotics Research*, **6**(3):49–59, 1987.

[153] A.G. Sparks, H.-H. Yeh, and S.S. Banda. Mixed H_2 and H_∞ optimal robust control design. *Optimal Control Applications & Methods*, **11**(4):307–325, 1990.

[154] M.W. Spong, J.S. Thorp, and J.M. Kleinwaks. Robust microprocessor control of robot manipulators. *Automatica*, **23**:373–379, 1987.

[155] M.W. Spong and M. Vidyasagar. *Robot Dynamics and Control*, Wiley, 1989.

[156] M.W. Spong. On the robust control of robot manipulators. *IEEE Transactions on Automatic Control*, **37**(11):1782–1786, 1992.

[157] H.L. Stalford and C.H. Chao. Necessary and sufficient condition in lyapunov robust control. *Journal of Optimization Theory and Applications*, **63**(2):191–204, 1989.

[158] A. Stoorvogel. *The H_∞ Control Problem: A State Space Approach*, Prentice Hall, 1992.

[159] H. Su, X. Ji, and J. Chu. Delay-dependent robust control for uncertain singular time-delay systems. *Asian Journal of Control*, **8**(2):180–189, 2006.

[160] S.M. Swei and M. Corless. On the necessity of the matching condition in robust stabilization. *Proc. of the 30th IEEE Conference on Decision and Control*, Brighton, England, 2611–2614, 1991.

[161] G. Tadmor. Worst–case design in the time domain. the maximum principle and the standard H_∞ problem. *Mathematics of Control, Signals, and Systems*, **3**(4):301–324, 1990.

[162] G. Tadmor. Uncertain feedback loops and robustness in general linear systems. *Automatica*, **27**(6):1039–1042, 1991.

[163] O. Taiwo. Design of robust control systems for plants with recycle. *International Journal of Control*, **43**(2):671–678, 1986.

[164] K.C Tan and Y. Li. Evolutionary L_∞ identification and model reduction for robust control. *Proceedings of the Institution of Mechanical Engineers. Part I: Journal of Systems and Control Engineering*, **214**(3):231–237, 2000.

[165] P.M. Thompson. Classical/H2 solution for a robust control design benchmark problem. *Journal of Guidance, Control, and Dynamics*, **18**(1):160–169, 1995.

[166] A.V. Timofeyev. Properties of invertible dynamical models and synthesis of a high-quality robust control. *Soviet Journal of Computer and Systems Sciences (English translation of Tekhnicheskaya Kibernetika)*, **29**(6):125–136, 1991.

[167] M.-C. Tsai and Y.-P. Shi. H_∞ optimal robust control of feedback systems. *Journal of Dynamic Systems, Measurement and Control, Transactions ASME*, **111**(2):336–339, 1989.

[168] C.-C. Tsui. *Robust Control System Design: Advanced State Space Techniques*, Marcel Dekker, 2003.

[169] K. Tsumura and M. Kitamura. Topological structure of robust control systems. *Proceedings of the IEEE Conference on Decision and Control*, **2**:1201–1206, 1998.

[170] K. Uchida, A. Kojima, and M. Fujita. H inf control attenuating initial-state uncertainties. *International Journal of Control*, **66**(2):245–252, 1997.

[171] V.A. Ugrinovskii, I.R. Petersen, A.V. Savkin, and E.Y. Ugrinovskaya. Decentralized state-feedback stabilization and robust control of uncertain large-scale systems with integrally constrained interconnections. *Systems and Control Letters*, **40**(2):107–119, 2000.

[172] J.G. Vanantwerp, R.D. Braatz, and N.V. Sahinidis. Globally optimal robust control for systems with time-varying nonlinear perturbations. *Computers and Chemical Engineering*, **21**(Suppl. 1):125–130, 1997.

[173] S.R. Venkatesh, A. Megretski, and M.A. Dahleh. On robust control synthesis and analysis in a hilbert space. *Systems and Control Letters*, **39**(1):1–12, 2000.

[174] M. Vidyasagar. *Nonlinear Systems Analysis*, Prentice Hall, 1978.

[175] H. Wang and B.K. Ghosh. New robust control for a class of uncertain discrete-time systems. *IEEE Transactions on Automatic Control*, **42**(9):1252–1254, 1997.

[176] W.-J. Wang and Y.H. Chen. Decentralized robust control design with insufficient number of controllers. *International Journal of Control*, **65**(6):1015–1030, 1996.

[177] Y. Wang, L. Xie and C. Souza. Robust control of a class of uncertain nonlinear systems, *Systems & Control Letters*, **19**:139–149, 1992.

[178] Y.-H. Wang and S.-Y. Zhang. Robust control for nonlinear similar composite systems with uncertain parameters. *IEE Proceedings: Control Theory and Applications*, **147**(1):80–86, 2000.

[179] Y. Wang, L. Xie, and C.E. de Souza. Robust control of a class of uncertain nonlinear systems. *Systems & Control Letters*, **19**(2):139–149, 1992.

[180] Z. Wang and L. Zhang. Robust control of linear systems with part uncertain parameters. *Applicable Algebra in Engineering, Communications and Computing*, **5**(5):92–102, 1994.

[181] C. Wanyi. On the H_∞ control for linear time-delay systems. *Journal of Systems Engineering and Electronics*, **9**(3):23–28, 1998.

[182] H. Wu. Decentralized robust control for a class of large-scale interconnected systems with uncertainties. *International Journal of Systems Science*, **20**(12):2597–2608, 1989.

[183] M. Wu, Z. Peng, C. Tang, and W. Gui. mu analysis and mu synthesis for nonlinear robust control systems. *Journal of Central South University of Technology (English Edition)*, **5**(1):68–73, 1998.

[184] M. Wu, L. Zhang, and G. Liu. Robust control of a class of nonlinear systems with L_∞-bounded disturbance. *Kongzhi Lilun Yu Yinyong/Control Theory and Applications*, **19**(3):369–379, 2002.

[185] M. Wu, L. Zhang, Z. Peng, and W. Gui. μ synthesis method for robust control of uncertain nonlinear systems. *High Technology Letters*, **6**(4):1–6, 2000.

[186] Q.-H. Wu and M. Mansour. Decentralized robust control using H_∞-optimization technique. *Information and Decision Technologies*, **15**(1):59–76, 1989.

[187] Y. Xia and Y. Jia. Robust control of state delayed systems with polytopic type uncertainties via parameter–dependent lyapunov functionals. *Systems and Control Letters*, **50**(3):183–193, 2003.

[188] J. Xu and Q. Zhou. Robust control of robot systems with unconstracted uncertainties. *Huanan Ligong Daxue Xuebao/Journal of South China University of Technology*, **24**(2):53–63, 1996.

[189] J. Xu, Q. Zhou, and X. Fan. H_∞/L_2 robust control problem of linear systems and its simplification. *Huanan Ligong Daxue Xuebao/Journal of South China University of Technology*, **23**(6):70–80, 1995.

[190] J. Xu, Q. Zhou, and S. Xu. Robust control of a kind of nonlinear systems. *Huanan Ligong Daxue Xuebao/Journal of South China University of Technology*, **23**(6):22–32, 1995.

[191] X.-G. Yan and G.-Z. Dai. Decentralized output feedback robust control for nonlinear large-scale systems. *Automatica*, **34**(11):1469–1472, 1998.

[192] X.-G. Yan, J. Lam, and G.-Z. Dai. Decentralized robust control for nonlinear large-scale systems with similarity. *Computers and Electrical Engineering*, **25**(3):169–179, 1999.

[193] G.-H. Yang and S.-Y. Zhang. Decentralized robust control for interconnected systems with time-varying uncertainties. *Automatica*, **32**(11):1603–1608, 1996.

[194] P.M. Young and M.A. Dahleh. Infinite-dimensional convex optimization in optimal and robust control theory. *IEEE Transactions on Automatic Control*, **42**(10):1370–1381, 1997.

[195] S.H. Zak. On the stabilization and observation of nonlinear/uncertain dynamic systems, *IEEE Transactions on Automatic Control*, **35**:604–607, 1990.

[196] G. Zames. Feedback and optimal sensitivity: Model reference transformations, multiplicative seminorms, and approximate inverses. *IEEE Transactions on Automatic Control*, **26**:301–320, 1981.

[197] Q. Zhang. Decentralized robust control for interconnected descriptor system. *Kongzhi yu Juece/Control and Decision*, **10**(1):74–84, 1995.

[198] W. Zhang. Active damping of flexible structures via saturating actuators, *International Journal of Systems Science*, **24**(9):1621–1652, 1993.

[199] R. Zhong and Z. Yang. Delay–dependent robust control of descriptor systems with time delay. *Asian Journal of Control*, **8**(1):36–44, 2006.

[200] K. Zhou. *Robust and Optimal Control*, Prentice Hall, 1996.

[201] K. Zhou, J.C. Doyle. *Essentials of Robust Control*, Prentice Hall, 1997.

Index

Robust Control Design: An Optimal Control Approach F. Lin

Printed and bound by CPI Group (UK) Ltd, Croydon, CR0 4YY
16/09/2026
14963379-0006